Fundamentals of
Virology

Fundamentals of Virology

Today's Persistence and Response

Hao Nguyen

Library of Congress Control Number: 2025936383

Fundamentals of Virology: Today's Persistence and Response

1 2 3 4 5 6 7 8 9   DSS   30 29 28 27 26 25

ISBN 978-1-265-42976-8
MHID 1-265-42976-6

Sponsoring Editor
Hilary Maybaum

Copy Editors
KnowledgeWorks Global Ltd.

Production Supervisor
Richard C. Ruzycka

Proofreader
KnowledgeWorks Global Ltd.

Acquisitions Coordinator
Olivia Higgins

Indexer
KnowledgeWorks Global Ltd.

Project Manager
Rishi Arora,
KnowledgeWorks Global Ltd.

Composition
KnowledgeWorks Global Ltd.

I dedicate this book to my children, Le Mai Eme and Hai Son Griffin, who are the loves of my life. Life isn't life without them.

I am forever grateful to my parents who gave me life and the chance to fulfill many dreams, and thank my siblings (Huyen, Huong, Huy, Hoan, and Hiep) and my extended family for giving me the wonderful love and support through the years.

Table of Contents

Preface xv
Acknowledgments xvii

1 Introduction **1**
Historical Perspective 1
So, What Is a Virus? And, What Is Its
   Relationship to All Living Things? 2
Viral Structure 4
Strategies for How to Replicate Each Type of Viral Genome 5
A Quick Tour of the Movement of a Virus From
   One Host to the Next Host 7
A Quick Discussion of the Different Portals of Exit
   From the Host Organism 7
A Quick Discussion of the Different Modes of Transmission 7
A Quick Discussion of the Different Portals of Entry
   Into the Next Host Organism 8
A Quick Discussion of the Movement of a Virus From Entry
   Into a Host Cell to Exit From That Same Host Cell 8
Dissemination Strategy 9
How Do Viruses Cause Diseases? 10
Transmission of Virus From One Host to the Next Host and So On . . . 10
Factors Affecting an Infection and Ability to Cause Diseases 11
Further Readings 12

2 Host Immune Responses to Viral Infection **15**
Historical Perspective 15
What Is an Immune Response and When Is It Activated? 16
What Does Your Body Do Once It Perceives the Invasion of a
   Pathogen Like a Virus? 16
     Mucosal Immune System 16
     The Epithelium 17
     The Lamina Propria 19
     The Muscularis Mucosa 19
Innate Immune System 19
     Phagocytosis 21
     Signal Transduction Pathways That Lead to the
       Regulation of Gene Expression 23
Inflammatory Response Pros and Cons 25
     What Is an Inflammatory Response or Inflammation? 25
     How Is the Inflammatory Response Activated? 25
     Additional Effect of the Innate Immune System 29

Protection of Host Cell Against Damage From
the Complement Cascade 33
Adaptive Immune System 33
B- and T-Lymphocyte Maturation 34
Naïve, Mature T-Lymphocyte Activation 37
Naïve, Mature B-Lymphocyte Priming 39
Affinity Maturation and Isotype Switching 41
Summary of the Three Levels of Human Immune
Systems Against a Viral Infection 42
Additional Virus-Specific Defense Mechanisms 43
Interferons Are Specific Against a Viral Infection 43
RNA Interference 44
Bacterial Defense Against Bacteriophage 45
Restriction Digestion 45
CRISPR 47
Subversion Mechanisms of Pathogens 49
Further Readings 49

3 Plus-Strand Single-Stranded RNA 51
Poliovirus... 51
Historical Perspective 51
Classification and Structure of Poliovirus 52
Host Range, Transmission, Tropism, and Susceptibility 53
Infecting the Host Susceptible Cell, Viral Particle
Replication, and Tissue Damage 54
Dissemination in the Host Body 56
Pathogenesis and Clinical Manifestation 57
Diagnosis, Treatment, and Prevention 58
Current Status 60
Scientific Significance and Discoveries 60
SARS-CoV-2 61
Historical Perspective 61
Classification and Structure of SARS-CoV-2 63
Host Range, Transmission, Tropism, and Susceptibility 64
Infecting the Host Susceptible Cell, Viral Particle
Replication, and Tissue Damage 66
Dissemination in the Host Body 69
Pathogenesis and Clinical Manifestation 70
Diagnosis, Treatment, and Prevention 71
Current Status 72
Scientific Significance and Discoveries 73
Further Readings 73

4 Minus-Strand Single-Stranded RNA 75
Influenza A Virus 75
Historical Perspective 75
Classification and Structure of Influenza A Virus 76

Host Range, Transmission, Tropism, and
Susceptible Host Cell   80
Infecting the Host Susceptible Cell, Viral Particle
Replication, and Tissue Damage   81
Dissemination in the Host Body   85
Pathogenesis and Clinical Manifestation   85
Diagnosis, Treatment, and Prevention   86
Current Status   90
Scientific Significance and Discoveries   91
Ebola Virus   91
Historical Perspective   91
Classification and Structure of Ebola Virus   92
Host Range, Transmission, Tropism, and
Susceptible Host Cell   93
Infecting the Host Susceptible Cell, Viral
Particle Replication, and Tissue Damage   94
Dissemination in the Host Body   97
Pathogenesis and Clinical Manifestation   98
Diagnosis, Treatment, and Prevention   99
Current Status   100
Scientific Significance and Discoveries   101
Further Readings   101

5   Double-Stranded RNA (dsRNA) **103**
Rotavirus   103
Historical Perspective   103
Classification and Structure of Rotavirus   104
Host Range, Transmission, Tropism, and Susceptible Host Cell . . .   106
Infecting the Host Susceptible Cell, Viral
Particle Replication, and Tissue Damage   106
Dissemination in the Host Body   112
Pathogenesis and Clinical Manifestation   113
Diagnosis, Treatment, and Prevention   114
Current Status   115
Scientific Significance and Discoveries   115
Picobirnavirus   116
Historical Perspective   116
Classification and Structure of Picobirnavirus   116
Infection Cycle of Picobirnavirus   117
Pathogenesis and Clinical Manifestation   118
Diagnosis, Treatment, and Prevention   119
Scientific Significance and Discoveries   119
Further Readings   120

6   Single-Stranded DNA (ssDNA) **121**
Erythrovirus B19   121
Historical Perspective   121

Classification and Structure of Erythrovirus B19 122

Host Range, Transmission, Tropism, and
Susceptible Host Cell 124

Infecting the Host Susceptible Cell, Viral
Particle Replication, and Tissue Damage 124

Dissemination in the Host Body 128

Pathogenesis and Clinical Manifestation 128

Diagnosis, Treatment, and Prevention 129

Current Status 130

Scientific Significance and Discoveries 131

Porcine Circovirus 2 131

Historical Perspective 131

Classification and Structure of Porcine Circovirus 132

Infecting the Host Susceptible Cell, Viral Particle
Replication, and Tissue Damage 132

Pathogenesis and Clinical Manifestation 134

Diagnosis, Treatment, and Prevention 135

Current Status 136

Scientific Significance and Discoveries 136

Further Readings 136

7   Double-Stranded DNA (dsDNA) **139**

Herpes Simplex Virus 139

Historical Perspective 139

Classification and Structure of Herpes Simplex Virus 140

Host Range, Transmission, Tropism, and
Susceptible Host Cell 142

Infecting the Host Susceptible Cell, Viral
Particle Replication, and Tissue Damage 148

Dissemination in the Host Body 157

Pathogenesis and Clinical Manifestation 158

Diagnosis, Treatment, and Prevention 160

Current Status 161

Scientific Significance and Discoveries 162

Human Polyomavirus 162

Historical Perspective 162

Classification and Structure of Human Polyomavirus 163

Host Range, Tropism, and Susceptible Host Cell 165

Infecting the Host Susceptible Cell, Viral Particle
Replication, and Tissue Damage 165

Dissemination in the Host Body 169

Pathogenesis and Clinical Manifestation 169

Diagnosis, Treatment, and Prevention 171

Current Status 172

Scientific Significance and Discoveries 173

Further Readings 173

8   **RNA With Reverse Transcriptase** **177**
    Human Immunodeficiency Virus 177
        Historical Perspective 177
        Classification and Structure of Human
            Immunodeficiency Virus 179
        Host Range, Tropism, and Susceptible
            Host Cell 181
        Infecting the Host Susceptible Cell, Viral
            Particle Replication, and Tissue Damage 182
        Dissemination in the Host Body 188
        Pathogenesis and Clinical Manifestation 190
        Diagnosis, Treatment, and Prevention 190
        Current Status 192
        Scientific Significance and Discoveries 192
    Rous Sarcoma Virus 193
        Historical Perspective 193
        Classification and Structure of Rous Sarcoma Virus 193
        Infection Cycle of Rous Sarcoma Virus 194
        Scientific Significance and Discoveries 195
    Further Readings 197

9   **DNA With Reverse Transcriptase** **199**
    Hepatitis B Virus 199
        Historical Perspective 199
        Classification and Structure of Hepatitis B Virus 200
        Host Range, Transmission, Tropism, and Susceptible
            Host Cell 202
        Infecting the Host Susceptible Cell,
            Viral Particle Replication, and Tissue Damage 203
        Dissemination in the Host Body 208
        Pathogenesis and Clinical Manifestation 209
        Diagnosis, Treatment, and Prevention 209
        Current Status 210
        Scientific Significance and Discoveries 213
    Hepatitis Delta Virus 216
        Historical Perspective 216
        Classification and Structure of Hepatitis D Virus 216
        Host Range, Transmission, Tropism, and Susceptible
            Host Cell 217
        Infecting the Host Susceptible Cell, Viral Particle
            Replication, and Tissue Damage 218
        Dissemination in the Host Body 220
        Pathogenesis and Clinical Manifestation 220
        Diagnosis, Treatment, and Prevention 221
        Current Status 221
    Further Readings 222

10   Subviral Pathogens: Prion, Viroid, and Satellite **223**
Prions 223
    Historical Perspective 223
    Prion Disease 224
    Structure and Function of Nonpathogenic Prion Protein 224
    Host Range, Transmission, Tropism, Susceptibility,
        Replication, and Dissemination 226
    Pathogenesis and Clinical Manifestation 228
    Diagnosis, Treatment, and Prevention 228
    Current Status 229
Viroid 229
    Historical Perspective 229
    Viroid Disease 230
    Structure and Function of Viroid 231
    Host Range, Transmission, Tropism, Susceptibility,
        Replication, and Dissemination 232
    Pathogenesis and Clinical Manifestation 235
    Intracellular Host Defense Against Viroids 236
    Diagnosis, Treatment, and Prevention 237
    Current Status 237
Satellite 238
    Historical Perspective 238
    Satellite Nucleic Acid Disease 238
Structure and Function of Satellite Nucleic Acid 238
    Structure and Function of Satellite Virus 239
    Pathogenesis and Clinical Manifestation 240
Further Readings 241

11   Giant Virus and Virophage: Discoveries in the 21st Century. **243**
Acanthamoeba polyphaga Mimivirus 243
    Historical Perspective 243
    Giant Virus Overview 244
    Classification and Structure of Mimivirus 245
    Infection Cycle of APMV in Amoeba 247
    Infection Cycle of APMV in Humans 248
    Current Status 249
Virophage 250
    Historical Perspective 250
    Classification of Virophage 250
    Structure of Virophage 251
    Infection Cycle of Virophage 252
    Current Status 253
Further Readings 254

12   Bacteria, Archaea, and Plant Viruses **255**
       Viruses That Infect the Bacterial Host 255
           Lambda (λ) Phage 257
           Emesvirus zinderi 2 (MS2) Phage 261
       Viruses That Infect the Archaeal Host 263
           Acidianus Filamentous Virus 1 265
       Viruses That Infect the Plant Host 266
           Transmission of Virus to Plant Host 266
           Tobacco Mosaic Virus 268
       Further Readings 270

13   Viruses in Biotechnology and Vaccine Development **271**
       Application of Viruses in Biotechnology 271
           Historical Perspective 271
           Viruses in Bacteriophage Biotechnology 272
           Viruses in Eukaryotic Cell Biotechnology 273
           Baculovirus Expression Vector System (BEVS) in
               Biotechnology 280
       Vaccine Development 282
           Historical Perspective 282
           What Is the Function of Vaccination? 283
           Considerations When Designing a Vaccine 284
           Types of Vaccines 286
       Further Readings 290

14   Classification of Viruses Based on Epidemiological Criteria **291**
       Arbovirus 291
           Historical Perspective 291
           Common Features of Arboviruses 292
       Blood-Borne Virus 292
           Historical Perspective 292
           Common Features of Blood-Borne Viruses 295
       Emerging Viruses 299
           Historical Perspective 299
           Common Features of Emerging Viruses 300
       Gastroenteritis Virus 303
           Historical Perspective 303
           Common Features of Gastroenteritis Viruses 303
       Respiratory Virus 305
           Historical Perspective 305
           Common Features of Respiratory Viruses 305
       Tumor Virus 307
           Historical Perspective 307
           Common Features of Tumor Viruses 307
       Further Readings 310

Preface

We live in a world where there is a wide range of socioeconomic, racial and ethnic status, and geographic disparities. These disparities very often lead to access disparities to vaccines. For instance, developing countries face threats from "poverty diseases," such as hookworm and leprosy. These nations are too poor to afford the resources to develop vaccines against these diseases. On the other hand, affluent countries do not experience the same type of infections and therefore do not have high or urgent impetus to make vaccines against these infections. Other problems leading to disparity include the difficulty in delivery to certain regions, the lack of proper storage, and vaccination resistance due to personal and religious beliefs.

There are ways to approach these challenges and solve these problems. The first obvious approach is to gain a better understanding of the life cycle of viruses. Technological advances can increase the efficiency of research and development of vaccines while reducing the cost. Methods used for the production of different vaccines against specific viral infections can be streamlined by using common knowledge and methodology. A case in point is when the messenger RNA vaccine against SARS-CoV-2 was developed quickly and efficiently.

This textbook is aimed at the first approach. The information presented here is intended to be used as a guide for teaching and learning, rather than an encyclopedia of detailed descriptions of all viruses. It is intended to be used to teach an entire 15-week semester course, one chapter per week, with time reserved for assessment.

Chapter 1 is an overview of virology and descriptions of terms and processes that apply to the life cycle of a virus and how it might cause a disease in the host organism. The virulence and contagiousness of a virus depends on a number of factors that are described in this chapter. Chapter 2 is an overview of the immune systems of the host organism, specifically humans since this textbook is focused mainly on human viruses. It is important to understand that most symptoms that are experienced by the host organism are due to one's immune responses to the pathogenic entity. Many viral diseases are also caused by the activities of one's immune system.

Chapters 3 to 9 contain descriptions of examples of viruses belonging to the seven Baltimore classifications. Examples that are chosen are mainly human viruses. The description of each virus begins with a historical perspective of how and when this virus was discovered. Structure-and-function relationships are an important aspect of cell and molecular biology (CMB). The structure of a virus, both physical and genomic, is directly correlated to its life cycle. A simple, small virus tends to replicate more quickly and produce a higher volume of clones, whereas a larger virus possessing numerous open reading frames tends to replicate more slowly and produce fewer viral particles. The reader will be guided through the cellular and molecular basis of symptoms and disease formation. This is where having at least a limited knowledge of immunology is useful. Last, the reader will be informed of the current status of each viral infection in terms of morbidity, mortality, and/or up-to-date technology for diagnosis and treatment.

Chapter 10 contains descriptions of nonvirus, pathogenic agents. These pathogenic agents behave like viruses where they are obligate intracellular parasites that can be transmitted from the original host organism to the next host. They are not classified as viruses, however, because they lack one or more criteria that define a virus. Chapter 11 describes the giant virus and virophage, which is essentially a virus of a virus. These are the most recent pathogens identified and are discoveries of the 21st century. Chapter 12 contains general descriptions of bacteriophages, archaeal viruses, and plant viruses. This textbook is rather human-centric; however, I want to show similarities to and differences from human viruses. Chapter 13 identifies biological knowledge gathered through studying viruses, development of modern biotechnology for research into viruses, and development of technology using viruses for research and disease treatment. Finally, Chapter 14 contains descriptions of different groups of viruses that are classified according to epidemiological criteria instead of the Baltimore classification. These groups contain viruses that have similar features. This type of classification is important because similar general diagnostic and treatment methods can be used to fight against such viral infections.

This textbook tells multiple stories, each with a beginning and an end, and each chapter is intertwined with other chapters. The information flows from the CMB aspect, to the symptomatic manifestation, then links to the actual diseases caused by these viruses. This is not a comprehensive textbook of all things virology; instead, it's one that can be used as a guide for teaching as well as learning. I think of this textbook as being comparable to a PowerPoint slide presentation but in the storytelling format. The story must flow from beginning to end, and all events in between must be connected and explained. All the what, why, when, where, how, how much, how frequent, how fast, and more are discussed to link the cause to the effect, the structure to the function. Each chapter begins with a special historical perspective. The mood of this textbook resembles that of a detective novel. The discovery of a historical incident pulls the reader into the scene of the crime; in this case, the crime is the viral infection. The reader will be introduced to a plethora of important information to connect the virus to the disease.

This work is intended to be used as an introductory virology textbook where students have a firm foundation of advanced CMB and microbiology. For graduate research and clinical students, more detailed descriptions, analyses, and additional information can be obtained through the referenced scientific articles listed in the Further Readings section of each chapter.

Acknowledgments

I've devoted my professional life striving to gain knowledge and understand the etiology of diseases, disorders, and adverse human conditions. My guiding lights are loved ones and those who are near and dear to my heart.

I'd also like to acknowledge the many great scientists, mentors, and friends who shone the light to guide me through my professional journey. I'd like to thank Dr. Claude Rupert for introducing me to my very first research experience at The University of Texas at Dallas (UTD); Dr. John Burr for mentoring me through my graduate work, also at UTD; Dr. Gordon Luk for mentoring me during my postdoctoral fellowship at the Veterans Affair Medical Center in Dallas; my friend and colleague Dr. Thomas Landerholm for his support and help during my tenure at California State University, Sacramento; and Dr. Victor Depta at the University of Tennessee at Martin for teaching me the art of telling a story.

Finally, I'd like to thank Dr. Huong Nguyen Corbett and Dr. Ron Corbett for their guidance and support while writing this textbook. I am forever grateful.

CHAPTER 1

Introduction

Historical Perspective

Tobacco mosaic virus (TMV) was the first virus that was discovered and studied in 1892, even though the smallpox vaccine was developed in 1798 by Dr. Edward Jenner and the rabies vaccine was developed in 1885 by Dr. Louis Pasteur. Interestingly, these vaccines were developed without either of these great researchers understanding the nature of these disease-causing agents at the time. TMV is the agent that causes the tobacco mosaic disease.

Why is this the case? Why is it that scientists managed to study and understand the causative agent TMV that affected a plant, yet did not know anything about viruses that caused great harm to the human population? Our understanding of biology and other great discoveries often coincides with economic demands. If we lose money, we want to get to the root of this loss and eradicate it. This is an unfortunate reality.

Let's discuss the circumstance that prompted researchers to discover TMV. Tobacco crop production originated 6000 to 8000 years ago in the Andes region of South America. It was widely used not only for its medicinal narcotic effect but also for special rituals and even for pleasure. This crop was eventually introduced to North America and was so highly priced in colonial America that it was often used as legal tender. Of course, due to rigorous cultivation, often in soil that is depleted of essential nutrients, these plants became more susceptible to disease. Being such an economically valuable commodity, finding a solution to treat this disease was of utmost importance, especially to those who financially benefitted from it.

In the mid-1800s, tobacco growers in the Netherlands identified a new disease that did not appear to be caused by a bacterial or fungal pathogenic agent. Adolph Mayer named this disease "the mosaic disease of tobacco" in 1879 and reported that this disease can be transmitted to healthy tobacco plants merely through direct contact. He also found that the causative pathogenic agent was neither bacteria nor fungi. In 1892, Dmitrii Ivanowski showed that this infectious agent was able to pass through porcelain filters designed to retain pathogens as small as bacteria (Figure 1.1). In 1898, Martinus Beijerinck confirmed the presence of this extraordinarily small infectious soluble agent that can migrate through an agar gel, which he referred to as *contagium virum fluidum* (meaning "slimy liquid poison") or "virus," whereas bacteria are referred to as *contagium vivum fixum* because of their inability to migrate through an agar gel. This agent appeared to be able to pass through filters that would normally retain pathogens like bacteria. However, it wasn't until 1935 that TMV was purified and crystallized by Wendell M. Stanley, who won the 1946 Nobel Prize in Chemistry.

Relative Size of Viruses

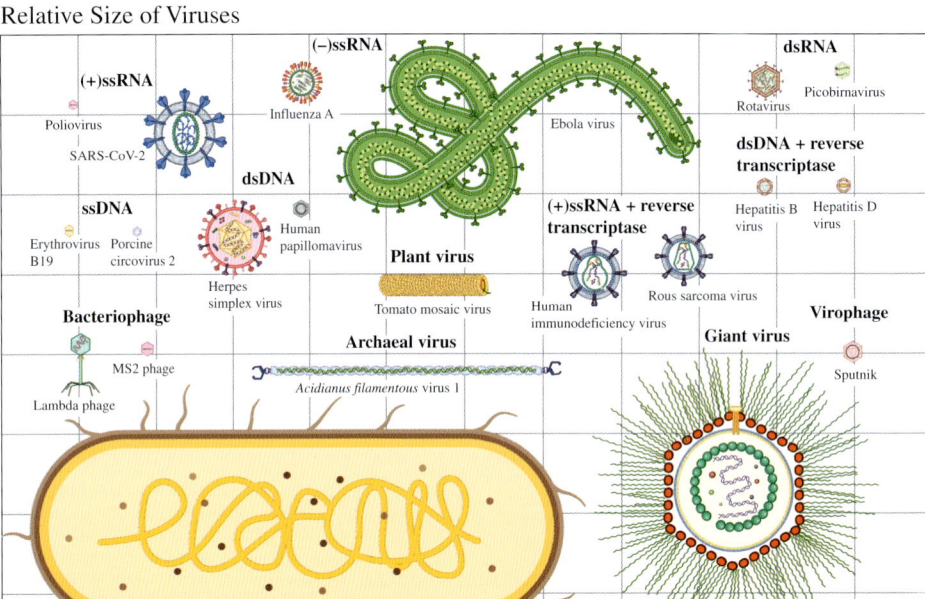

FIGURE 1.1 *Contagium virum fluidum.* Schematic of relative sizes of viruses discussed in this textbook. The background represents the 0.22-μm pore size filter mesh. The schematic of a bacterial cell illustrates the relative size of a viral particle compared to a host cell. Most viral particles cannot be removed by the typical filter mesh with a 0.22-μm pore size that is used to preclude bacteria. The major exceptions are the giant viruses, which are 3 to 10 times the pore size. It is very difficult for Ebola viral particles to fit through this pore size as well. *Credit:* Figure partially created in BioRender.

So, What Is a Virus? And, What Is Its Relationship to All Living Things?

The original term for a virus is ultrafilterable "virus," where virus is Latin for "poison." Viruses are identified as obligate intracellular parasites, which means that they are absolutely and totally dependent on a living host cell for reproduction. This does not mean that they cannot survive without a host cell or even a host organism. Many can survive for a long period of time in the natural environment. However, they can only replicate themselves and reproduce inside a host cell. Why? The host cell possesses important machineries, for example enzymes and substrates for the synthesis of the viral genome and proteins and the proper microenvironment for virion assembly. Host cells that are infected by cognate virus can be any cell type across all 3 domains: bacteria, archaea, and eukarya (Figure 1.2).

The International Committee on Taxonomy of Viruses set guidelines for naming conventions and classification of viruses. As of 2021, this organization had identified 10,434 viral species spanning 6 realms, 10 kingdoms, 17 phyla, 2 subphyla, 39 classes, 65 orders, 8 suborders, 233 families, 168 subfamilies, 2606 genera, and 84 subgenera. Taxonomic classification categorizes viruses based on physical properties, typically delineated by 16S ribosomal RNA gene sequence data at the species level. In 1971, David Baltimore developed a more modern approach for the classification of viruses,

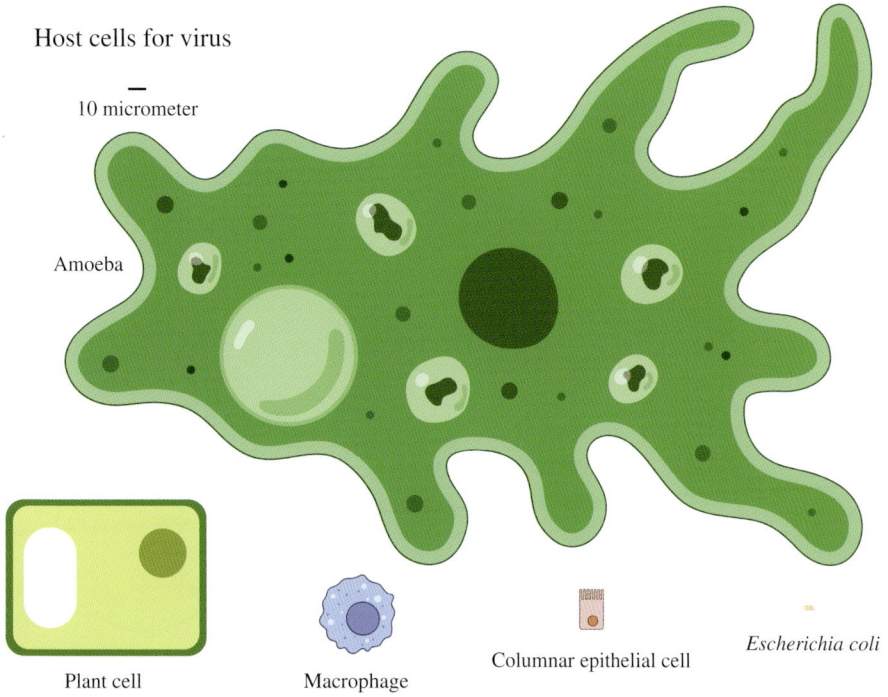

Host cells for virus

—
10 micrometer

Amoeba

Plant cell   Macrophage   Columnar epithelial cell   *Escherichia coli*

Figure 1.2   Different host cells targeted by virus. The diameter of an amoeba is 4× the length of a typical plant cell, which is 10× the diameter of a human macrophage, which is 20× the average height of a human columnar epithelial cell, which is 200× the length of the *Escherichia coli* bacterium. *Credit:* Figure partially created in BioRender.

which groups viruses into families according to their genome structure. A viral genome may be DNA or RNA, which may be double stranded or single stranded. There are 7 Baltimore classifications, and viral genome structure can be (+) single-stranded RNA (ssRNA), (−) single-stranded RNA, double-stranded RNA (dsRNA), single-stranded DNA (ssDNA), double-stranded DNA (dsDNA), RNA (with reverse transcriptase), or gapped dsDNA (with reverse transcriptase) (Figure 1.3). Furthermore, these viral genomes can be linear, segmented, or circular.

Additional characterizations used to classify viruses include symmetry of the protein coat/shell called the capsid, presence or absence of lipid membrane called the envelope, and dimension of the capsid and virion. Strategies for viral genome replication, viral messenger RNA (mRNA) synthesis, and viral protein synthesis are dependent on the type of viral genome itself. These strategies ultimately depend on two major factors. For the first, does the permissible host cell possess the appropriate machinery to carry out these processes for the virus? If the host cell does not possess all the appropriate machinery, then the virus has to provide the missing components encoded in its genome. Second, if the host cell does indeed possess the machinery, the sublocalization of each mechanism is correlated with the presence of substrates, enzymes, and other molecular components required to perform the particular mechanism. For example, a dsDNA virus can use host cell DNA polymerases to replicate its genome. However, this

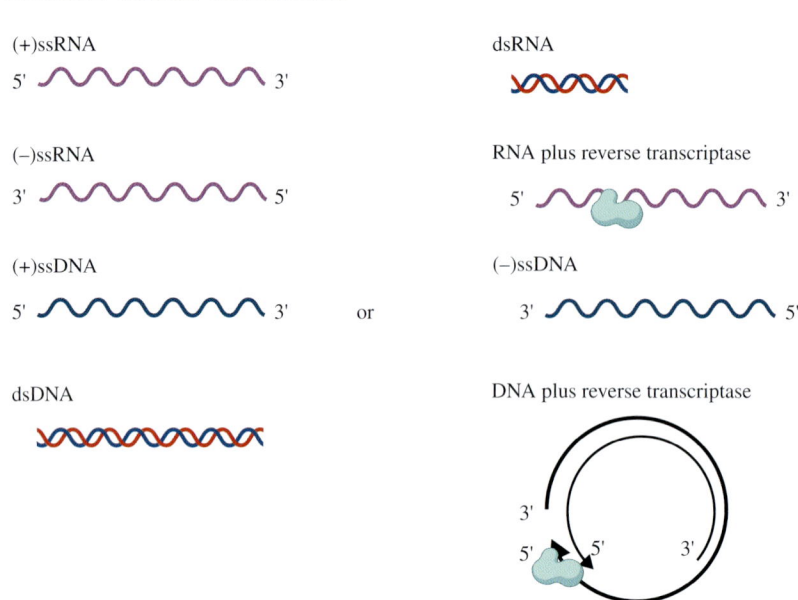

Baltimore Genome Classification

FIGURE 1.3 Seven types of viral genome. Schematics showing different types of viral genome in the Baltimore classification. *Credit:* Figure partially created in BioRender.

activity must occur in the nucleus since all eukaryotic DNA polymerases and deoxyribonucleotides are located there. As another example, since eukaryotes do not use an RNA template for RNA synthesis, the virus has to import its own RNA-directed RNA polymerase. This activity must occur in a location where ribonucleotides are available.

Viral Structure

Virus is the singular term. Virus is also the plural term when quantifying viral particles of the same genus, whereas viruses is used as the plural term when counting viral particles belonging to more than 1 genus (plural is genera). A viral particle can be non-enveloped or enveloped. A non-enveloped viral particle consists of the viral genome encased by a capsid. In turn, a capsid is identified as a protein shell that encases the viral genome (Figure 1.4). The capsid is composed of protein subunits known as capsomeres or capsid proteins. Structurally, a fully formed capsid containing the viral genome inside is referred to as a nucleocapsid.

A virion is the assembled nucleocapsid formed inside the host cell. The terms virus and viral particle are often used interchangeably with virion. Viral particles are extremely small, and their size ranges from 20 to 150 nm compared to a ribosome (20 nm) and mitochondrion (1–10 μm). Viral genome size ranges from 2 kilobases (kb) to 1 megabase (Mb), containing between 3 and 200 open reading frames (ORFs) or genes. Most viral particles possess 10 to 15 ORFs encoding special viral proteins, predominantly capsomeres that are used to construct the viral capsid. For most identified viral particles, the capsid structure may form helical symmetry, polyhedral

Virion structures

a) Non-enveloped icosahedral capsid

b) Enveloped icosahedral capsid

c) Enveloped helical/conical capsid

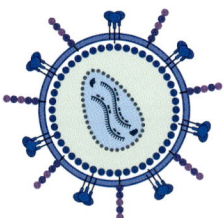

FIGURE 1.4 Types of virus structures. Schematics showing capsid structures. (a) Non-enveloped icosahedral capsid, (b) enveloped icosahedral capsid, (c) enveloped helical/conical capsid. *Credit:* Figure partially created in BioRender.

(more specifically, icosahedral) symmetry, or complex symmetry. The helical/conical symmetry is always enclosed by an envelope. The lipid component of the viral envelope is derived from the previous host cellular membranes, most often the plasma membrane. However, matrix proteins that are embedded into the envelope are glycoproteins that are encoded by viral genes.

Two viruses are considered to be distinct species when their genomic sequences differ significantly. When the genetic code of a viral particle varies slightly from the natural reference virus due to mutations, it is called a variant, while a strain refers to physical or behavioral variations from the original, ancestral virus causing altered phenotype of infection leading to a different kind of disease.

Strategies for How to Replicate Each Type of Viral Genome

There are a number of important fundamental cell and molecular considerations that must be followed during viral DNA and RNA synthesis. Strategies for viral genome replication, viral mRNA synthesis, and viral protein synthesis are dependent on the type of viral genome itself. Above all, polymerases play a central role in viral genome replication and transcription, leading to protein synthesis. A polymerase can be encoded by either host cell genes or viral genes. The four types of polymerases are (Figure 1.5):

1. DNA-dependent/-directed RNA polymerase (host RNA polymerase)
2. DNA-dependent/-directed DNA polymerases (host DNA polymerases)
3. RNA-dependent/-directed RNA polymerase (viral RdRP)
4. RNA-dependent/-directed DNA polymerase (viral RdDP)

DNA and RNA polymerases

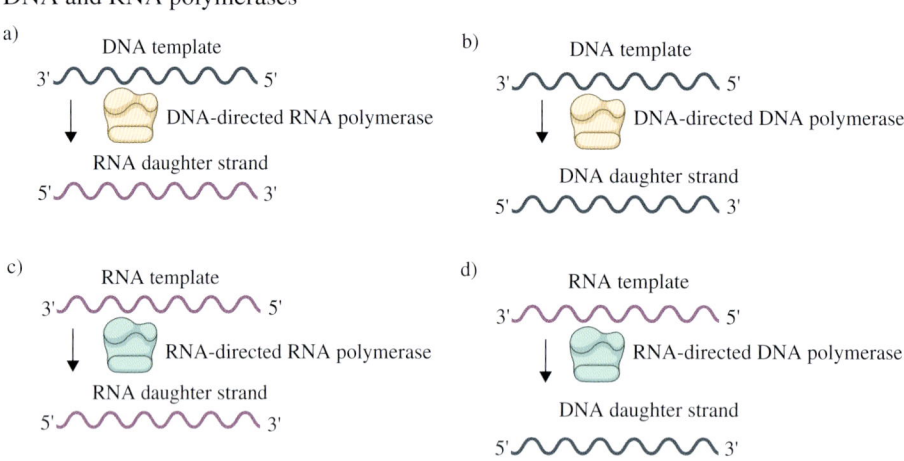

FIGURE 1.5 Four types of polymerases. Each polymerase reads a particular nucleic acid template to synthesize either DNA or RNA. *Credit:* Figure partially created in BioRender.

RdRP initiates RNA synthesis via two possible mechanisms: de novo synthesis or primer-dependent synthesis. A single nucleotide is required to initiate de novo synthesis, whereas an oligonucleotide or a protein-linked oligonucleotide is required for primer-dependent synthesis. One of the most important considerations is the fact that a (+) strand RNA viral genome may (but not always) perform the same function as an mRNA, meaning that it can be bound by a ribosome and used directly as a template for translation in the cytosol. However, the complementary (−) ssRNA must still be synthesized to be used as a template for (+) ssRNA viral genome replication, whereas a (−) ssRNA viral genome must first be used as the template to synthesize the complementary (+) ssRNA, forming a dsRNA, which is recognized and bound by RdRP for mRNA synthesis before translation. Another important consideration is the fact that ribosomes cannot recognize and bind to dsRNA; therefore, a dsRNA viral genome cannot be used immediately for translation, which requires the synthesis of the actual viral mRNA first. Single-stranded DNA viruses are also interesting. A mixed population of virions that belongs to a particular species can contain both (+) strand and (−) strand viral genomes, roughly 50% containing 1 type and 50% containing the other. This is because, once in the host cell, the complementary strand must be synthesized to form a dsDNA before mRNA synthesis can occur. Host cell transcription factors can only recognize response elements found within dsDNA.

Always remember that subcellular localization of activities is important. The localization of each mechanism is correlated with the presence of specific substrates, enzymes, and other molecular components required to perform this particular activity. For example, DNA replication only occurs in the nucleus or mitochondrial matrix because these are locations where DNA polymerases and deoxyribonucleotides are found; translation only occurs in the cytosol or mitochondrial matrix because these are locations where ribosomes and amino acids are found. If a virus requires unique mechanisms and/or components, it must bring them into the host cell, of course, because the host cell will not have these components or perform these specialized mechanisms.

A Quick Tour of the Movement of a Virus From One Host to the Next Host

Once reproduced, viral particles are transmitted from the original host to a new host organism to start the cycle again. The mode of transmission varies depending on where the virus resides in the environment (eg, in food or drink or in the air). Viral entry into a mammalian host organism, like a human, may be through the respiratory tract (through the mouth and nose), digestive tract (through both ends), urogenital tract, eyes, open wounds, or even directly into the bloodstream (ie, through injection with a needle or a mosquito bite). Direct entry into the bloodstream is referred to as passive viremia. A specific virus targets specific host organisms for infection. A host range refers to the ability of the virus to infect a particular organism or group of organisms. A zoonotic virus is one that can be transmitted from a nonhuman animal (eg, a mosquito, ape, or rat) to a human; therefore, zoonoses are viral diseases that are shared by humans and nonhumans. The host organism does not always acquire or exhibit a disease or symptoms. A natural reservoir or reservoir of infection is a long-term carrier where a particular viral species naturally lives and reproduces and then ultimately is transmitted to the next host. The reservoir host does not experience symptoms of the disease when infected by the virus, whereas a nonreservoir host does exhibit symptoms of the disease. A reservoir is often thought of as a host organism; however, some virologists also consider a reservoir to be an environment external to the host organism (eg, air, food, water, etc.). A vector is a short-term or small host organism (like mosquitos) that carries pathogens from organism to organism and place to place. Vectors generally do not exhibit symptoms of the disease.

A Quick Discussion of the Different Portals of Exit From the Host Organism

The various exit portals of the virus from its host organism correlate with tropism, in other words, with the infected tissue where the virus reproduces. This route is in turn linked to the external environment, which determines the mode of transmission. The most common portals of exit are the skin and mucous membranes, which include the respiratory, urogenital, and gastrointestinal tracts. Virus can escape through an injury or blood drawn through the skin. Coughing, sneezing, and even talking can expel viral particles from the respiratory tract. Waste products containing viral particles are expelled through the urinary or digestive tract. Exit of the virus through the genital tract can also occur during sexual intercourse. Another possible route of exit can be the transmission from the mother's blood into her fetus via the blood-placenta barrier.

A Quick Discussion of the Different Modes of Transmission

Transmission is the process by which a virus is transported from 1 host to the next host. The various modes of transmission depend on where the virus resides in the external environment and is directly linked to the mode of entry into the next host organism. Transmission can be direct or indirect. Direct transmission occurs when the virus is transferred from an infected host to a susceptible host by physical contact, such as through skin-to-skin contact, kissing, and sexual intercourse. Direct transmission can also occur when the susceptible host contacts the soil or vegetation that harbors the virus. Contact via droplets carrying viral particles from sneezing, coughing, or talking is also classified as direct transmission (or direct spread or droplet spread) only if the droplets have not fallen to the ground. Otherwise, coming into contact with viral particles on the ground is classified as indirect transmission. Indirect transmission can be airborne and carried by dust, vehicle-borne and carried by fomites, or vector-borne and carried by a host vector such as a mosquito or tick. A fomite is defined as an inanimate object that serves as a vehicle or carrier of infection, such as the wall of a hospital, contaminated eating utensils, or contaminated bedsheets. Other types of vehicles include food, water, and biologic products like blood.

A Quick Discussion of the Different Portals of Entry Into the Next Host Organism

Portal of entry refers to the manner in which a virus enters a susceptible host. The various portals of entry correlate with the mode of transmission. For example, the poliovirus is transmitted through the fecal-oral route, where feces from the source host is the vehicle. The portal of entry is via the oral route when the susceptible host eats using hands that were not properly washed. Similar to the idea that the portal of exit correlates with the infected tissue where the virus reproduced, the portal of entry must also be linked to tropism. That is, the portal of entry must provide the virus relatively easy access to the target tissue. The same portal of entry into a new host is very often the same as the portal of exit from the source host.

A Quick Discussion of the Movement of a Virus From Entry Into a Host Cell to Exit From That Same Host Cell

Once inside the host organism, the virus also targets very specific tissue and cell types, referred to as tropism. The infection of the host cell, viral particle reproduction, and subsequent release of viral particles follows seven steps: attachment, entry into the host cell, uncoating, viral protein synthesis, viral genome synthesis, assembly or packaging, and release or exit from the host cell.

1. Attachment occurs when the virus comes into physical contact with its target host cell. Only susceptible cells possess the appropriate receptor or receptors to recognize and bind to specific proteins on the viral particle surface.

2. Entry of a non-enveloped virus into a host cell almost always occurs through clathrin-dependent, receptor-mediated endocytosis. In this case the entire enveloped virion is placed in the lumen of the endosome. The lower pH environment (around pH 6) of the endosomal lumen activates fusion proteins found in both the viral envelope and endosomal membranes, which in turn induce membrane fusion. The nucleocapsid is then placed into the host cell cytosol, and uncoating may occur. Viruses may also enter the host cell through caveolin-dependent endocytosis. The virus is then placed in the lumen of the caveosome. Different from the endosome, the pH in the lumen of a caveosome is physiologically neutral (7.2–7.4); therefore, a different mechanism is used to place the virus into the cytosol. Some viruses can also enter the host cell through clathrin- and caveolin-independent endocytosis. Even enveloped virus can also

enter the host cell through this mechanism; however, they usually enter via the fusion of the viral envelope to the host cell plasma membrane. This fusion is catalyzed or aided by fusion proteins found on the surface of both the virus and the host cell. Once fusion has occurred, the nucleocapsid is released into the host cell cytosol. After the entry of the virus, only cells that are permissive or permissible have the capability to carry out viral replication (ie, produce more virions). Permissibility refers to the fact that these cells must possess the appropriate intracellular environment and molecular machinery to carry out the synthesis of new virions.

3. Uncoating is the process by which the viral capsid is either partially or completely removed, allowing the viral genome to be released and transported to the appropriate environment for replication and mRNA synthesis.

4. The translation event always occurs in the host cell cytosol, where ribosomes can be found. Some viral polypeptides also undergo translational modification like glycosylation.

5. Viral genome replication can occur in a number of different locations, depending on the type of polymerase and substrates required.

6. Assembly or packaging of the virions occurs once both the viral genome and viral proteins are synthesized. This process usually occurs, but not always, in the cytosol near the plasma membrane.

7. Strategy for the release of viral particles from an animal host cell includes exocytosis, membrane budding, cell lysis, and activation of the intrinsic apoptotic pathway.

The virus must be able to use or take over certain host cell's machineries. This in turn means that the host cell machineries must be able to recognize and use viral components to carry out transcription, translation, and viral genome replication. Of course, the synthesis of viral components must occur at very specific locations within the host cell. Various viral components are then transported to a particular subcellular location for packaging. Most molecules and/or structures that need to be transported from 1 location to another inside the cell require the use of cytoskeletons, either microtubules and/or microfilaments.

Dissemination Strategy

Once the virus manages to cross the epithelial barrier of the host organism, it needs to migrate toward the target cell. At this point, the virus has infected this host. The dispersion of the viral particles throughout the host is referred to as dissemination. From the original, primary site of infection, the viral particles migrate through loose connective tissues and would eventually reach blood vessels. The major mode of viral dissemination is through the circulatory system, most likely into capillaries, and is called hematogenous dissemination or lymphatic dissemination. The presence of viral particles in the bloodstream as the result of hematogenous dissemination is referred to as active viremia, whereas passive viremia is the presence of viral particles in the bloodstream as the result of direct introduction via either a vector like mosquitoes or a fomite like a needle. These viral particles can extravasate (or leave the bloodstream) to establish a secondary site of infection in another location in the host. Neurogenous or neuronal dissemination can also occur, where the virus can migrate into either the central or peripheral nervous system, most often via nerve endings.

How Do Viruses Cause Diseases?

All known viruses cause diseases or disorders. Diseases and problems arise mainly through damage of specific tissues that are targeted by the virus; therefore, diseases caused by a particular virus are linked to tropism. Symptoms of a viral infection, however, are often the result of the activation of our own immune system, especially the inflammatory response. Symptoms are not necessarily linked to tropism.

Let's start by exploring the many factors that affect how a virus can lead to characteristic symptoms and cause its cognate disease. First, we need to discuss how researchers measure infectious units (or viral particles). The plaque-forming unit (PFU) is used mainly to determine the quantity of bacteriophage, measured as PFU per milliliter. This technique involves growing bacteria on a solid culture medium, like an agar plate, and adding a specific volume of liquid medium containing phages. When the lawn of bacteria is grown, plaques would be formed where colonies cannot grow because the initial host cell lysed as the result of phage infection. These plaques are counted as PFUs, which is a close approximation of the phage concentration in the liquid medium. The titer of virus is expressed as PFU per milliliter. Multiplicity of infection (MOI), on the other hand, is determined to find the average number of virus particles required to infect a target host cell. MOI is used mainly to identify animal virus infecting animal cells in culture. An MOI of 1 indicates that the virus can infect the host cell on a 1-to-1 basis, whereas an MOI of 10 indicates that the population of virions must be 10-fold that of the host cell population for efficient infection. This means that an MOI of 1 is more *infectious* than an MOI of 10.

Having a technique to quantify viral particles is important to determine, or at least estimate, the size of the virus population in the host during infection. An exact assessment is not possible since we don't yet have an advanced enough tool for this measurement; however, the amount of virus in the infected host's blood can be quantified as the viral load. The viral load is expressed as the number of viral particles per milliliter of blood drawn from the host, as determined by the amount of viral genetic material. Different viral species have different abilities to cause tissue damage leading to a disease at different viral loads. The viral load is also affected by how accessible the target tissue is as well as how quickly the particular virus can reproduce in the target host cell. Viruses are assessed by their likelihood and ability to cause a disease. How quickly they can activate an immune response is also significant.

Viral infection always eventually causes host cell and tissue damage. The severity, speed, and efficiency of tissue damage usually determine the severity of the disease produced. In other words, the disease caused by a particular virus is directly linked to the tissue and organ that is damaged. However, symptoms that arise from a viral infection, at least the initial reaction, are usually the result of the activation of one's own immune response; most of the time, it's the inflammatory response.

Transmission of Virus From One Host to the Next Host and So On

After exiting from a host organism, the virus now resides in the external space and is at the mercy of this environment. In order to make it to the next host, the virus must survive in this environment, be it gentle or harsh. This is the transmission stage of a viral cycle. How viruses are transmitted depends on the most suitable environment where they can survive outside of a host. How viruses are disseminated depends on the most suitable environment where they can survive inside the host. How viruses can continue to reproduce depends on their adaptability in different environments as well as their mode of reproduction. We've already discussed the many factors that enable the survival and propagation of a viral lineage: route of transmission, mode of entry into host, dissemination and accessibility, infection into target host cell (attachment; entry into host cell; uncoating; [virion reproduction: viral genome synthesis, viral protein synthesis, assembly/packaging]; and release/exit), and mode of exit from host.

The term pathogenicity refers to the ability of a virus to infect and establish itself within a host and cause disease, where pathogenesis is the process by which a viral infection can lead to a disease. This includes the ability to gain access to the target host cell, the severity of tissue damage, and the level of harm incurred to the host. The more pathogenic virus causes greater harm and more severe pathology more easily. On the other hand, the term virulence (also known as infectiousness or infectivity) refers to the ability of a virus to survive and be transmitted from one host to the next. This passage includes the mode of transmission, route of entry into the host, ability to survive inside this host, rate of reproduction inside the host cell, route of exit from the host organism, and mode of transmission to the next host. In other words, pathogenicity is the process by which an infection leads to a specific disease, whereas virulence is the speed at which the virus can reproduce inside an infected host, and disease-causing is only a by-product of this process. Last, the term contagiousness is the ease by which an infectious disease can be transmitted from one host to the next. A contagious virus can spread throughout a population quickly and with ease.

Factors Affecting an Infection and Ability to Cause Diseases

Factors that affect the likelihood of infection include mode of transmission, dose of exposure, route of entry, accessibility and dissemination, tropism, infectious dose, viral dose, R nought (R_0) (defined at the end of this section), mode of exit from the host cell, and route of exit from the host organism.

The mode of transmission determines how the next host organism might be exposed to the virus. The amount of viral particles to which an individual is exposed is referred to as the dose of exposure. This contact can affect how likely this individual is to become infected and the severity of the symptoms; however, an infection is never absolute, and the severity of symptoms is variable depending on a number of other factors. The amount of viral particles to which an individual is exposed is directly correlated with the amount entering the body, which correlates with the likelihood that the individual is infected. In other words, if the pathogen is in *low levels*_in the area where you are exposed, you are less likely to be infected; however, if the pathogen is available in large quantities, you are more likely to be infected. The next factor to be considered is the route of infection or route of entry into the host organism. How a virus enters the host affects the range of tropism as well as the route of dissemination. In the host body, the virus must make its way to the target host cell. Accessibility is the next hurdle for the viral journey.

The amount of a pathogen that is required to induce an immune response in the host is referred to as the infectious dose. Infectious dose is measured as ID_{50} or 50% infectious dose, where it represents the number of viral particles needed to infect half of the population of the host organism, whereas LD_{50} or 50% lethal dose represents the number of viral particles needed to kill half of the population of the host organism. These measurements infer that more virulent pathogens correlate with a lower ID_{50} or LD_{50}; therefore, those with a higher ID_{50} or LD_{50} are less virulent. Another way of looking at this is that more virulent viruses have a lower infectious dose and can infect the host cell and start reproduction at a lower concentration.

However, take note that a higher infectious dose, even though it correlates to a less-virulent virus, means that the host immune system would respond at a higher viral load, which can impact the severity of symptoms. The infectious dose is affected by the initial dose of infection. The amount of particles a person is exposed to can affect how likely this individual is to become infected. Once an infection has occurred, the total amount of viral particles in the host organism defines the viral load. The viral load is actually the measurement of viral particles in the host blood, which includes both the amount of viral particles that entered the host and the amount that was produced through replication inside infected cells. For example, when a person is infected with SARS-CoV-2 (severe acute respiratory syndrome coronavirus 2), this individual's viral load becomes very high for the first few days, more than likely due to a very high rate of reproduction. Remember that a high infectious dose means that a higher viral load is in the individual, which can impact the severity of symptoms, although pathogens with a high infectious dose tend to be less virulent. Therefore, their population inside a host will start to rise earlier (maybe quicker, but not always). The larger the infectious dose is, the higher the individual's viral load is likely to become before the host's immune system is capable of controlling the infection. The connection between the infectious dose and the viral load is the rate of viral reproduction, expressed as R nought (R_0). R_0 is defined as the average number of secondary infections produced by the infectious person. The designation of an epidemic by the Centers for Disease Control and Prevention is somewhat based on R_0. If R_0 is greater than 1, the number of infected cases would have increased exponentially and cause an epidemic or even a pandemic.

An analogy that has been used by many scientists to compare the viral load to the infectious dose is as follows: The viral load is a measure of how bright the fire is burning in an individual, whereas the infectious dose is the spark that gets that fire going.

Further Readings

Baltimore D. Expression of animal virus genomes. *Bacteriol Rev*. 1971;35:235–241.

Beijerinck MJ. Concerning a *contagium vivum fluidum* as cause of the spot disease of tobacco leaves. *Verbandlelingen der Koninkyke akademie Weltenschappen te Amsterdam*. 1898;65:3–21.

Goodspeed TH. *The Genus* Nicotiana: *Origins, Relationships, and Evolution of Its Species in the Light of Their Distribution, Morphology, and Cytogenetics*. Chronica Botanica; 1954.

Iwanowski D. Concerning the mosaic disease of the tobacco plant. *St. Petersb. Acad. Imp. Sci. Bul*. 1892;35:67–70. Translation published in English in *Phytopathological Classics*, No. 7. American Phytopathological Society Press; 1942.

Lecoq H. Decouverte du premier virus, le virus de la mosaique du tabac: 1892 ou 1898? Discovery of the first virus, the tobacco mosaic virus: 1892 or 1898?. *C R Acad Sci Paris III*. 2001;324(10):929–933.

Mayer A. Concerning the mosaic disease of tobacco. Translation published in *Phytopathological Classics*, No.7; 1968. American Phytopathological Society Press; 1886.

Ranjith-Kumar CT, Gutshall L, Kim MJ, Sarisky RT, Kao CC. Requirements for de novo initiation of RNA synthesis by recombinant flaviviral RNA-dependent RNA polymerases. *J Virol*. 2002;76(24):12526–12536.

Stanley WM. Soviet studies on viruses. *Science*. 1944;99:136–138.

Tushingham S, Snyder CM, Brownstein KJ, Gang DR. Biomolecular archaeology reveals ancient origins of indigenous tobacco smoking in North American plateau. *Proc Natl Acad Sci U S A*. 2018;11(46):11742–11747.

CHAPTER 2

Host Immune Responses to Viral Infection

Historical Perspective

The phenomenon of "immunity" (from the Latin word *immunis*, which means "exempt from public service") can be traced back to the plague in 430 BC in Athens, Ancient Greece, during the Peloponnesian War. Most of the city's inhabitants were infected by pathogens (most likely smallpox, but not confirmed), and 25% of the population perished. Thucydides, a historian of this war, noted that those who recovered from the plague could not contract the illness a second time and remained healthy through the duration of this period of hardship. These are the only individuals who were allowed to come into contact with and nurse the sick.

However, Dr. Louis Pasteur is generally considered the father of immunology even though it was Dr. Edward Jenner who developed the very first vaccine against the smallpox virus, almost a century before Dr. Pasteur developed the rabies vaccine. The reason is mainly because Dr. Pasteur's work confirmed and completed the germ theory of infectious diseases. He also introduced the concept of prophylactic vaccine versus therapeutic vaccine. The germ theory opposed the view of Dr. Rudolf Virchow, the founder of cellular pathology, who theorized that all diseases are the consequence of defective host cellular functions. Even though Dr. Pasteur's research actually explained how vaccination might be used against microbial diseases and showed how both prophylactic and therapeutic vaccinations could be applied to prevention and treatment via immunology, he and Dr. Robert Koch favored the idea that the host was defenseless against such illness.

The concept of how the host uses the immune system to fight against microbial invaders was supported by the work of Elias Metchnikoff, Emil Behring, and Paul Ehrlich. Dr. Metchnikoff discovered the role of phagocytosis in the intracellular destruction of pathogens, leading to the concept of innate immunity. Drs. Behring and Ehrlich discovered that antibodies can neutralize the toxic activities of bacteria, leading to the concept of acquired immunity.

What Is an Immune Response and When Is It Activated?

Immunology is the study of the host organism's defense reactions to substances identified as "nonself"; most often, these nonself substances are foreign entities, but not always. That is, substances produced by one's own body are sometimes identified as nonself as well. There are also entities that cannot be identified as self or nonself by one's immune system. These substances are referred to as "missing self" and can also activate an immune response.

The identification of substances from one's own body as nonself is the fundamental basis for autoimmunity. Therefore, it is essential that all substances are identified as either self or nonself, while nonself substances can be viewed as either innocuous or harmful. Typically, an immune response is activated only against harmful substances, usually molecules from a pathogen.

What is a pathogen? A pathogen is a microorganism that is capable of causing a disease or disorder in the host organism, which includes viruses, bacteria, fungi, and parasites.

What Does Your Body Do Once It Perceives the Invasion of a Pathogen Like a Virus?

The answer to the question of what the body does on perceiving invasion of a pathogen depends on where the virus settles. That is, what is the route of entry? The layer of sebum outside of the skin epithelium and mucous layer of the mucous membrane is the outermost protective barrier of the body. Before breaching the epithelial lining, pathogens are trapped in these viscous fluids and would be met by a mucosal immune response, which is the first line of defense (Figure 2.1).

The major functions of the mucosal immune system are (1) to protect the mucous membrane by trapping the pathogens and preventing them from invading the internal tissues; (2) to act as a surveillance system marking harmful antigens for destruction while allowing beneficial substances, such as digested food particles, to be absorbed; and (3) to alert the innate and adaptive immune systems of the presence of pathogens and prepare for a possible infection. In fact, this immune system works synergistically with both the innate and adaptive immune systems.

The other two major immune systems, innate and adaptive, are discussed further in this chapter. The innate immune system responds quickly to an infection that usually lasts a day or 2. Innate immunity does not generate long-term protective immunological memory. This system includes the inflammatory response, complement cascades, and acute-phase response. The adaptive (or acquired) immune system has a slower response but lasts until the infection is cleared. In most cases, adaptive immunity generates long-term protective immunological memory. This system includes T-cell–mediated immune response and humoral (or B-cell–mediated) immune response.

The gastrointestinal (GI), respiratory, reproductive, and urinary tracts, as well as the surface of the eye, are lined with a mucous membrane, which is exposed to the exterior environment and is easily accessible for pathogens. The respiratory and GI tracts are the most common routes of entry into a human host.

Mucosal Immune System

The mucous membrane that lines these internal tracts consists of three layers: epithelium, lamina propria, and muscularis mucosa.

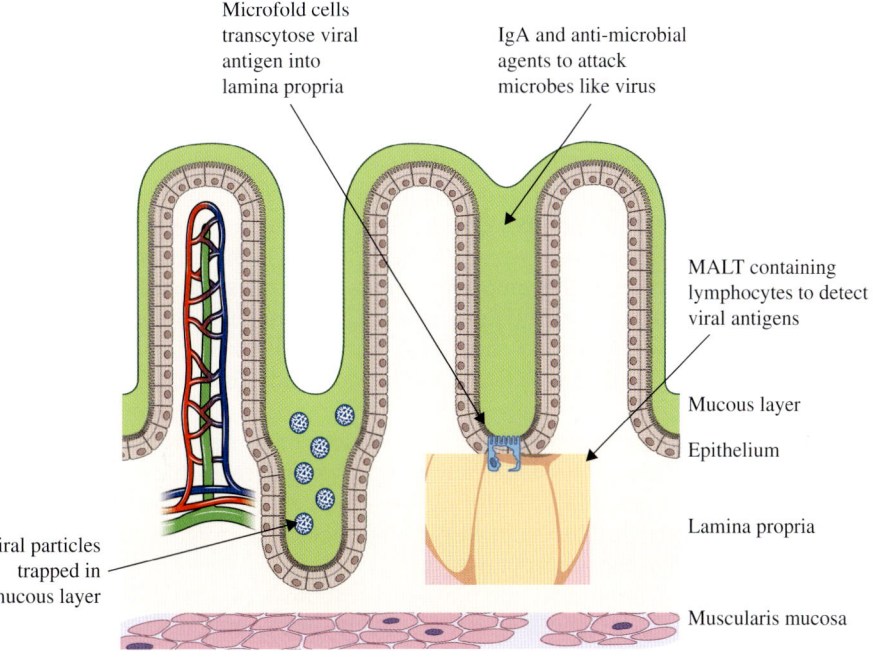

Microfold cells transcytose viral antigen into lamina propria

IgA and anti-microbial agents to attack microbes like virus

MALT containing lymphocytes to detect viral antigens

Mucous layer

Epithelium

Lamina propria

Viral particles trapped in mucous layer

Muscularis mucosa

Figure 2.1 Mucosal immune system. Pathogens are trapped by the mucous layer and attacked by IgA and other antimicrobial agents. Viral antigens are transcytosed from the mucous layer to the MALT in the lamina propria by microfold cells. *Credit:* Figure partially created in BioRender.

The Epithelium

The epithelial lining consists of a number of cells that are involved in the mucosal immune system, including goblet cells (or mucous cells in the stomach lining), microfold (M) cells, innate lymphoid cells (ILCs), Paneth cells, and intraepithelial lymphocytes (IELs). Goblet cells secrete mucus, forming the protective physical layer that not only prevents pathogens from reaching and breaching the surface of the epithelium but also protects the mucous membrane from its hazardous internal environment, such as protecting the stomach from the low pH environment and digestive enzymes.

The mucus itself contains a number of defensive mechanisms, including mucin glycoprotein, which causes mucus to thicken, and the resulting increased viscosity helps trap and disrupt the mobility of cellular microbes while retarding the dispersion of viruses once they've entered the host.

Goblet cells are regulated by cytokines, the most common of which are interleukin (IL) 1β, IL-4, IL-6, IL-13, and tumor necrosis factor (TNF) (eg, TNF-α) to produce both soluble and membrane-bound forms of mucin. The membrane-bound mucin can also aid the selection of substances, like digested food substances, for endocytosis by epithelial cells.

A viral infection often induces the secretion of an abnormal amount of mucin and mucus in general. As the result of a respiratory viral infection, this activity can lead to airway obstruction and respiratory distress. However, an infection may also lead to the disruption of goblet cell function, resulting in the reduction, or even inhibition, of mucin synthesis and secretion.

In addition to trapping and somewhat immobilizing the pathogen, these fluids also contain antifungal agents, antibacterial agents, and a significant concentration of immunoglobulin A (IgA).

Immunoglobulin A is constitutively produced by the adaptive immune system even in the absence of an infection and has low affinity for pathogenic antigens. The function of IgA is mainly to clear the pathogen by opsonizing the viral particles, thereby rendering them avirulent or neutral. The neutralized pathogens are then eliminated from the body along with waste material.

Microfold cells are antigen-sampling cells that carry out a process called transcytosis: They uptake pathogenic antigens from the small intestinal tract mucous layer into the epithelial cell through endocytosis, then continue to secrete these antigens into the mucosa-associated lymphoid tissues (MALTs) located in the lamina propria and can extend into the submucosa of the mucous membrane. MALTs are secondary (or peripheral) lymphoid tissues that carry out the activation and differentiation of B and T lymphocytes. Peyer patches, nasopharynx-associated lymphoid tissues, and bronchus-associated lymphoid tissues are MALTs found in the small intestine, nasal cavity, and bronchus, respectively.

In the MALT, pathogenic antigens are recognized by dendritic cells and lymphocytes. Dendritic cells are professional antigen-presenting cells that are required to prime naïve, mature T lymphocytes during the adaptive immune response. Since the lysosomes are poorly developed in M cells, endocytosed antigens remain unmodified as they pass through the cell's cytoplasm.

On the basolateral side, the plasma membranes of M cells invaginate to create a pocket-like structure that allows dendritic cells, and sometimes lymphocytes, to occupy this space. These dendritic cells readily and efficiently endocytose the unmodified antigens, then in turn process them to activate the adaptive immune system. Some dendritic cells can actually extend projections, called dendrites, across the epithelium to take up antigens in the mucous layer in lieu of transcytosis by M cells.

Furthermore, antigens from the mucous layer that are delivered to the MALT favor IgA production by activated B lymphocytes. Isotype switching to IgA occurs mainly in MALT regulated by the IgA-associated cytokines: transforming growth factor (TGF) β, IL-2, IL-4, IL-5, IL-6, and IL-10 produced by both dendritic cells and effector T-helper type 2 (Th2) cells. The process of B-lymphocyte priming and isotype switching is described when the adaptive immune system is discussed.

Soluble IgA (S-IgA) molecules are only secreted into the mucous layer and sebum of mucous membranes and the skin, respectively. S-IgA opsonizes viral particles that have invaded the mucous membrane, thereby neutralizing their biological activity and rendering them avirulent, resulting in the inhibition of viral dissemination and proliferation.

Innate lymphoid cells are activated and regulated primarily by metabolites that are produced by the host's nonpathogenic microbiota. Once activated, ILCs synthesize and secrete pro-inflammatory cytokines to fight against pathogens. One of these cytokines is IL-22, which promotes homeostasis and healing of the mucous membrane during an infection in the GI tract.

Paneth cells synthesize and secrete a number of antimicrobial agents in the form of peptides and/or proteins. IL-22 that is secreted by ILCs can induce Paneth cells to produce RegIIIa, which is a proteinaceous agent that targets and kills gram-positive bacteria. RegIIIa, or regenerating islet-derived protein 3A, is a tissue-specific transcription factor, that is primarily involved in tissue repair in response to an injury or inflammation, particularly as the result of skin damage. During this process, regIIIa stimulates cell growth and cell migration of specific cell types such as those in the liver, pancreas, and skin.

The IELs are T lymphocytes that can be activated to synthesize and secrete cytokines, like interferon-γ (IFN-γ) and keratinocyte growth factor, to protect epithelial cells from injury. These cells exhibit the pattern recognition receptors (PRRs), toll-like receptors (TLRs), and NOD (nucleotide oligomerization domain)-like receptor (NLR), on their surface.

The Lamina Propria

Macrophages and pro-inflammatory Th17 cells populate the lamina propria of the mucous membrane. Both cell types secrete cytokines that control the innate immune responses. Nonpathogenic microbiota are important in regulating the activity of both cell types. The major roles of these microbes are to promote homeostasis and protect the host organism from pathogenic microbes. One means to protect the host internal tracts is to promote the differentiation of naïve CD4 T lymphocytes into Th17 cells, which in turn would upregulate the inflammatory response.

To do this, nonpathogenic microbiota release signaling molecules to regulate macrophages in the lamina propria to secrete IL-1β and/or IL-10. IL-1β, along with TNF-α and IL-6, is an important regulator of the innate immune response. IL-6 (or IL-21) can induce naïve CD4 T lymphocytes to differentiate into Th17 cells. However, naïve CD4 T lymphocytes are driven to differentiate into regulatory T cells (Tregs) by IL-10 and in the absence of IL-6. Tregs are leukocytes that respond to the body's internal and external signals to regulate immune responses.

Why is this regulatory step needed? Th17 and Treg cells share a common precursor cell, which is the naïve CD4 T lymphocyte. However, the two terminally differentiated cells perform opposite functions. Th17 cells produce and secrete pro-inflammatory cytokines IL-17, IL-22, and IL-23 to recruit neutrophils and promote inflammation to fight against a pathogenic infection. By contrast, Treg cells produce anti-inflammatory cytokines IL-10 and TGF-β to suppress Th17 cell development.

The idea is that there is a common population of naïve CD4 T lymphocytes that can be induced to become either Treg or Th17 cells. An increase in the Treg population results in the decrease of the Th17 population and vice versa. This regulation allows the host to maintain an appropriate population of Th17 and thereby an appropriate level of inflammation. This is important because a high level of IL-17 secreted by Th17 causes too much inflammation, which may lead to autoimmunity. Of course, the nonpathogenic microbiota only regulates the balance between Treg and Th17 cells in a localized area, therefore is not systemic.

The Muscularis Mucosa

The muscularis mucosa is composed of a thin layer of smooth muscle in the digestive tract. Its function is to contract and relax continuously to keep the mucosa in constant motion. This activity helps the digestive system perform its job. However, the muscularis mucosa has no apparent function in the mucosal immune system.

Innate Immune System

The innate immune system is the first to respond to the primary establishment of an infection. If the innate immune system is successful within 1 or 2 days, the viral infection would be completely cleared. If not, then the adaptive immune system would be activated to clear the infection. If the viral infection is not efficiently cleared by the adaptive immune response, the viral infection would become a persistent or chronic infection, where virions would continue to be produced for long periods. Chronic inflammation may also occur in which a number of problems and disorders may arise. Chronic inflammation may lead to tissue hypoxia, leading to oxidative stress, leading to damaged organs, and finally leading to disorders related to damaged organs.

The innate immune system is an early response to an infection and includes the inflammatory response, complement cascades, and acute-phase response. These responses usually last roughly q day, no more than 2. This immune system occurs in 2 steps: recognition and response.

The first step is to recognize the pathogen by distinguishing self from nonself entities. Each type of pathogens exhibits a fairly specific marker called the pathogen-associated molecular pattern (PAMP), which can be recognized by a host molecule called the pattern recognition receptor (PRR) located on the surface of specialized phagocytic immune cells, mostly leukocytes (Figure 2.2). Examples of PAMP are lipopolysaccharide (LPS) from gram-negative bacteria, lipoteichoic acid (LTA) from gram-positive bacteria, flagellin of bacterial flagella, double-stranded and unmethylated DNA from a variety of viruses, and viral surface glycoproteins. PRRs found on the surface of a small number of phagocytic immune cells (macrophage in tissue and monocyte in blood, dendritic cell, and B lymphocyte) recognize and bind PAMPs. Once bound by a particular PAMP, the activated PRR in turn regulates intracellular signal transduction pathways, leading to one or more effector functions. These effector functions lead to the second, response, step performed by these immune cells.

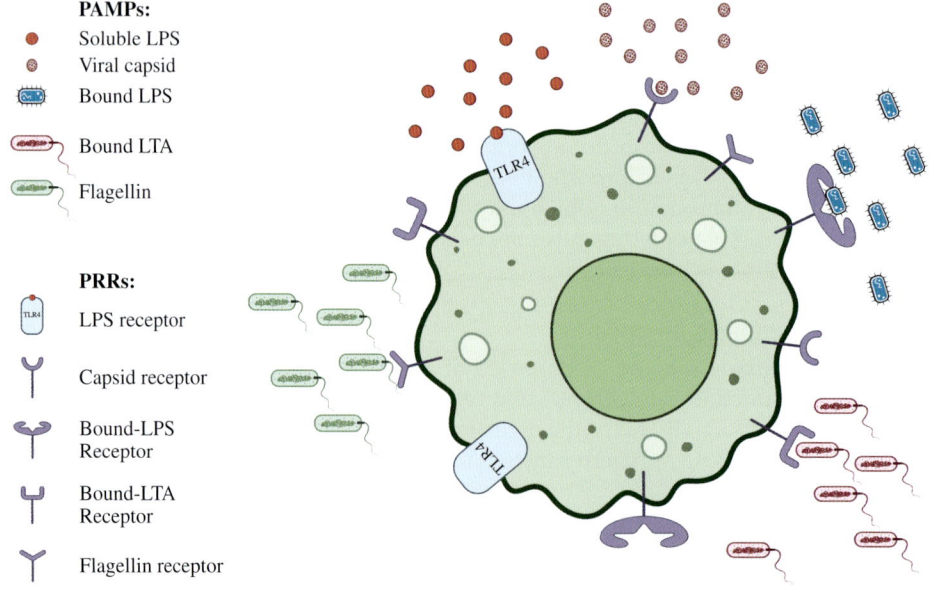

FIGURE 2.2 Phagocytes exhibit many pathogen recognition receptors on their surface. A phagocyte, like a macrophage, exhibits multiple copies of different PRRs. Each PRR recognizes and binds a specific PAMP. *Credit: Figure partially created in BioRender.*

During the response step, the nonself-entities are targeted to be cleared (or destroyed) or neutralized. This immune system responds to intracellular pathogens differently from its response to extracellular pathogens. Also, specific PRRs recognize specific PAMPs, not the actual pathogen.

This is an essential concept for the activity of M cells in the epithelium. M cells transcytose pathogenic antigens (including PAMPs), not the whole pathogen, to be used for innate immune system activation. This is also an important basis and focus for the development of vaccines.

The 5 types of PRRs are TLRs, NLRs, retinoic acid–inducible gene I (RIG-I)–like receptors (RLR), C-type lectin receptors (CLRs), and absent in melanoma 2 (AIM2)–like receptors (ALRs).

An activated PRR upregulates two major effector functions: (1) phagocytosis to clear extracellular pathogens and (2) signal transduction pathways that lead to the regulation of gene expression of a variety of molecules, including signaling molecules (like cytokines, chemokines), cell adhesion molecules (CAMs), and immunoreceptors. These molecules in turn regulate the various innate immune responses described above.

Phagocytosis

The function of phagocytosis is carried out by phagocytes to clear extracellular pathogens like viruses before they can infect a host cell. Phagocytes include macrophages, neutrophils, dendritic cells, and B lymphocytes. Realize that both phagocytosis and clathrin-dependent receptor–mediated endocytosis are processes to internalize extracellular substances, but they have different outcomes, especially with relation to virus (Figure 2.3). The function of clathrin-dependent receptor–mediated endocytosis is to internalize macromolecules to be used as food for the cell or for signaling.

Receptor-mediated (usually clathrin-dependent) endocytosis is used by most of our cells to take in substances, like food, to be utilized in the cell instead of destroyed. Also, endocytosis is the most common mode of entry for viral particles into the host cell. Since this is a carefully choreographed process somewhat directed by the virus to benefit survival, it's not at all an acceptable strategy for the phagocyte to internalize and clear pathogens.

Phagocytosis is completely regulated by the host cell's machinery and therefore is optimal for pathogen protection. Once internalized (or phagocytosed), the pathogen resides in a phagosome in the cytoplasm of the host cell. The most immediate step after internalization of the pathogen is the fusion of the phagosome with a lysosome, forming a phagolysosome. The low pH (3.5–5.0) in the phagolysosome enables the digestion of most of the content in the lumen. The optimal pH of acidic digestive enzymes inside this organelle is within this range.

However, some viruses not only can survive in this acidic environment, but also exploit the low pH to escape the phagolysosome. For instance, the low pH can induce uncoating of the viral particle as well as creating pores in the membrane of this organelle, allowing the nucleocapsid to escape into the host cell's cytosol.

In addition to creating an acidic environment, the phagolysosome also contains bacteriostatic and bactericidal agents that target bacteria. Once the contents have been digested and neutralized, the phagolysosome would eventually be discharged from the cell.

There are a number of other mechanisms to destroy pathogens inside the phagolysosome. Toxic oxygen-derived molecules (also known as reactive oxygen species, ROS), such as

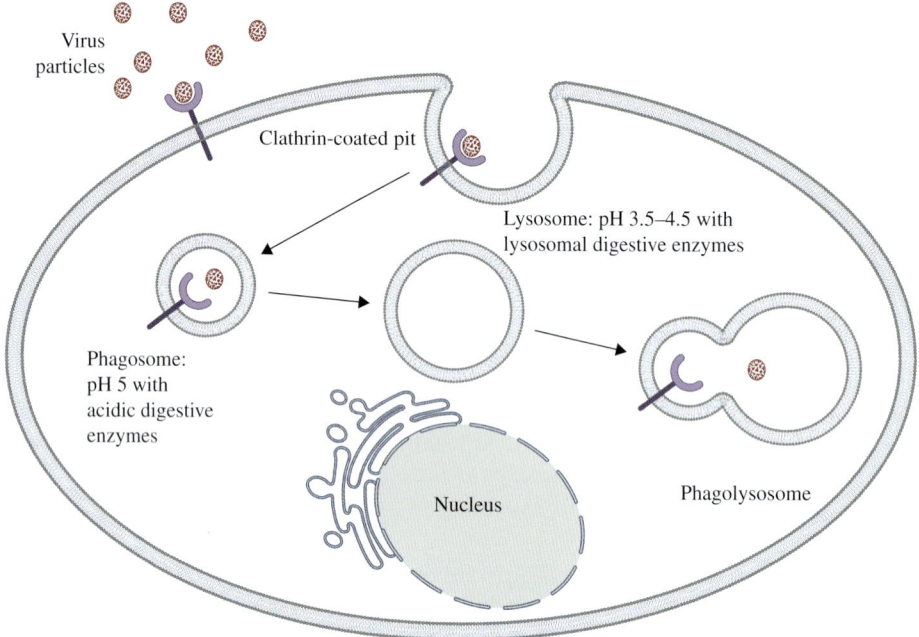

FIGURE 2.3   Phagocytosis by phagocyte. A virus is phagocytosed by a macrophage and placed into a phagosome. Partial breakdown of the virus by acidic digestive enzymes occurs in the phagosome. Fusion of the phagosome and lysosome results in the formation of a phagolysosome for further, more thorough, digestion of the virus by lysosomal digestive enzymes.
Credit: Figure partially created in BioRender.

superoxide (O_2^-), hydrogen peroxide (H_2O_2), hydroxyl radical ($-OH$), hypohalite (OCl^-), and hypobromite (OBr^-), can react with biological molecules, causing oxidative stress, which can destroy pathogenic structures. Nitric oxide (NO) is another reactive substance that has a similar function and outcome as ROSs. Chelators and binding proteins like lactoferrin (binds iron ions) and vitamin B_{12}-binding protein limit the availability of substances that are essential for microbial survival and growth.

Respiratory Burst

Respiratory (or oxidative) burst is another very effective activity that may be activated within the phagosome or phagolysosome (Figure 2.4). After phagocytosis, nicotinamide adenine dinucleotide phosphate (NADPH) oxidase (NOX2 in human) is assembled in the phagosomal or phagolysosomal membrane. NOX2 stimulates respiratory burst by producing reactive oxygen species (ROS) to destroy intracellular pathogens. The NADPH complex and superoxide dismutase (SOD) are brought into the phagosome through fusion with vesicles that exited the Golgi. During this process, a massive amount of oxygen is consumed (or oxidized) by the NADPH oxidase in a very short period of time. A 10- to 20-fold increase in oxygen consumption by NADPH oxidase activity results in the production of large amounts of free radicals or ROS, such as O_2^-, H_2O_2, $-OH$, NO, OCl^-, and OBr^-. The function of SOD restricts the biological oxidant cluster enzyme system in the body in response to cellular oxidative stress, lipid metabolism, inflammation, and oxidation. SOD converts superoxide anions into H_2O_2 in an effort to neutralize ROSs created through respiratory burst. Glycogen breakdown is vital to

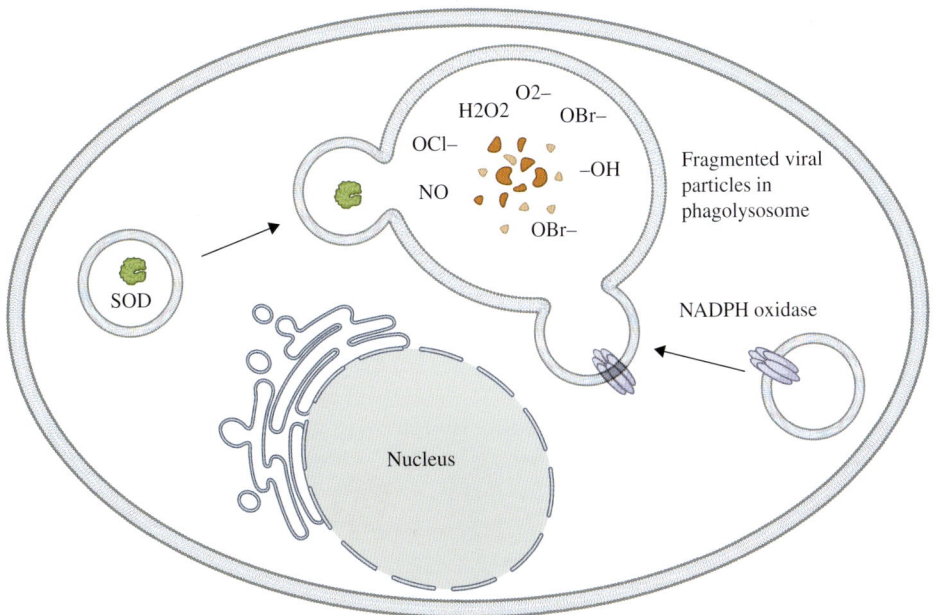

FIGURE 2.4   Respiratory burst. NADPH oxidase is assembled in the membrane of the phagosome. SOD is also delivered to the phagosome. The NADPH oxidase complex synthesizes large amounts of ROS to breakdown phagocytosed pathogens. SOD converted superoxide anions into H_2O_2. *Credit:* Figure partially created in BioRender.

produce NADPH. Individuals who have genetically defective phagocyte NADPH oxidase complexes exhibit chronic granulomatous disease due to their inability to clear pathogenic infections through respiratory burst. Pathogens, especially bacteria, can survive as intracellular bacteria despite being phagocytosed.

Signal Transduction Pathways That Lead to the Regulation of Gene Expression

Viral PAMPs are mainly recognized by TLR, RLR, and NLR and, in turn, activate major signal transduction pathways to regulate effector functions. These pathways usually induce the expression of common tissue-specific transcription factors (TS-TFs), such as tissue-specific transcription factor (IRF) 3, IRF-7, nuclear factor–κB (NFκB), and activator protein 1 (AP1), which in turn, upregulate the expression of signal molecules involved in innate immune responses. AP1 is a heterodimeric tissue-specific transcription factor that is involved in regulating cell proliferation, differentiation, and apoptosis. Macrophages are generally the first responders to a pathogenic infection, that is, once the pathogen manages to breach the epithelium. They possess PRRs that can recognize and bind viral PAMPs.

One family of PRRs specific for viral particles is the TLRs. Each TLR can recognize different virus-specific components, such as surface viral envelope proteins, capsomeres, spike proteins, and viral genome (Figure 2.5).

TLR-2 forms heterodimers with TLR-1 and TLR-6 located in the plasma membrane. TLR-2 complexes recognize a wide range of PAMPs, mostly from bacteria. However, TLR-2 complexes can also detect fungal PAMPs as well as viruses, including human cytomegalovirus, hepatitis C virus, and measles virus.

Classes of toll-like receptors

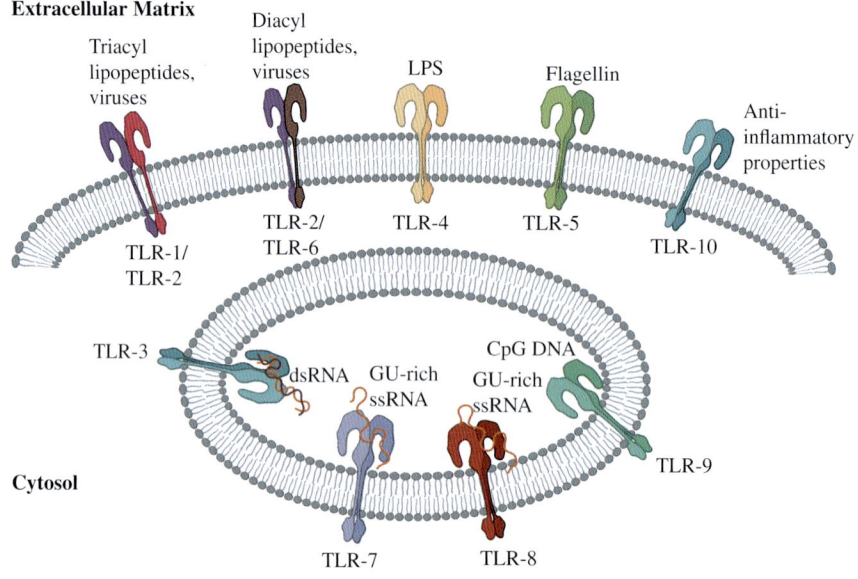

FIGURE 2.5   Toll-like receptors. TLRs are membrane-bound receptors that can detect PAMPs. TLRs 1, 2, 4, 5, and 6 detect extracellular pathogen molecules. TLRs 3, 7, 8, and 9 detect intracellular pathogen molecules and are especially important in identifying a viral infection. *Credit:* Figure partially created in BioRender.

TLR-3 may be located in either the endosomal membrane to recognize double-stranded RNA (dsRNA), a common intermediate product of RNA virus replication, or released from dying cells into the extracellular space or in the endosomal membrane to identify endocytosed virus (eg, from influenza and West Nile viruses).

TLR-4 is located in the plasma membrane and recognizes both viral and bacterial glycoproteins (eg, from HIV [human immunodeficiency virus]).

TLR-7 and -8 are located in the endosomal membrane and recognize G/U-rich single-stranded RNA from viruses that are endocytosed (eg, from HIV and hepatitis C virus.

TLR-9 is located in the endosomal membrane and recognizes unmethylated (CpG) DNA (eg, herpes simplex virus).

TLR-10 exhibits anti-inflammatory properties. It is the latest TLR discovered, and the mechanism for its function is under investigation.

The endosomal localization of TLR-3, -7, and -8 in natural killer (NK) cells and of TLR-7 and -9 in plasmacytoid dendritic cells is essential for upregulating the expression and secretion of IFNs, especially IFN-α and IFN-β to signal a viral infection in the host. Activated TLRs recruit adaptor proteins, such as myeloid differentiation primary response 88 (MyD88), TIR domain-containing adaptor inducing interferon-β (TRIF), Toll/interleukin-1 receptor domain-containing adaptor protein (TIRAP/MAL), or TRIF-related adaptor molecule (TRAM). These adaptor proteins regulate pathways that activate the transcription factor NFκB and mitogen-activated protein kinase (MAPK) cascades for the upregulation of inflammatory cytokine genes.

The nucleotide oligomerization domain NLRs can recognize PAMPs from intracellular pathogens. Activated NLR recruits a number of proteins to form an inflammasome, which is a protein complex that mediates the activation of caspase 1. Caspase 1 is known to induce both pyroptosis and necroptosis. Pyroptosis is a type of regulated (or programmed) cell death pathway that is activated in response to an infection by intracellular pathogens. Necroptosis is a type of regulated cell death that resembles necrosis but also has features of apoptosis. In either case, the infected cell is identified by NLR and programmed to be extinguished.

Last, the RLRs recognize dsRNA and use the RNA helicase RIG-I to recruit the adaptor protein MAVS (mitochondrial antiviral signaling protein), localized in the mitochondrial membrane, leading to the activation of NFκB and IRFs. MAVS performs a number of immunologic functions including activating NF-κB and interferon regulatory factors, aiding the production of antiviral mediators, and regulating NLR3 inflammasome activity.

Inflammatory Response Pros and Cons

The inflammatory response is the "Dr. Jekyll and Mr. Hyde" of the immune system. While it responds against infections and tissue damages in the host, a pro, it can cause overactive responses, a con. Also, while an inflammatory response is usually short term, long-term and inappropriate responses may occur, leading to diseases, the ugly.

What Is an Inflammatory Response or Inflammation?

The inflammatory response is one of the innate immune responses that's activated to clear microorganisms as the result of an infection or cellular debris as the result of an injury or damage to the localized tissue. The function of inflammation is to recruit leukocytes from the bloodstream to the site of infection to clear the pathogens and send out more signals to prolong the response, if necessary, as well as to recruit plasma proteins from the bloodstream to the site of infection to regulate the complement cascades and/or acute-phase reaction/responses.

An infection occurs once the pathogen has breached the epithelial barrier. Macrophages are almost always the first cell type that patrols and encounters a viral infection. Mast cells are quick responders as well. As discussed, PRRs on the surface of leukocytes bind to pathogenic PAMPs to induce the macrophage to undergo phagocytosis, as well as synthesize and secrete pro-inflammatory mediators. The mast cell undergoes degranulation, releasing a massive amount of signaling molecules that are retained in secretory vesicles. Phagocytosis occurs as an attempt to clear the infection. If successful, no further action is required. If not, both macrophages and mast cells release several pro-inflammatory mediators to gain support to control and eradicate the infection.

How Is the Inflammatory Response Activated?

Considering macrophagic activity, concurrent with phagocytosis, other PRRs on the surface of the macrophage, like TLR, can be activated to upregulate the expression and secretion of pro-inflammatory mediators. Such mediators include cytokines, chemokines, and lipid inflammatory mediators to induce an inflammatory response. Similar types of lipid inflammatory mediators are released by both macrophages and mast cells.

The inflammatory response involves the following steps: (1) vasodilation; (2) vascular permeability leading to the extravasation of leukocytes and leakage of plasma along with solubilized plasma proteins; (3) rolling adhesion, leading to the migration of leukocytes into the subendothelial space; and (4) leukocyte homing or specificity depending on the type of pathogen that had infected the host organism.

1. Vasodilation is induced by histamine (from mast cells) and vascular endothelial growth factor (VEGF), which occurs in the capillaries in the local area of infection. Signaling molecules stimulate both the synthesis and the secretion of NO from the endothelial cells to induce smooth muscle cell (or pericyte in the capillary wall) relaxation, thereby increasing capillary lumen diameter.

   Additional downstream functions of histamine include regulating: constricting of smooth muscles around veins (leading to vasoconstriction), lung passage (leading to bronchoconstriction), uterus, and stomach wall; stimulating sensory nerves in the airways; and stimulating gastric acid secretion from the stomach lining. These functions involve the cellular target of the peripheral nervous system.

   Vasoconstriction causes increased heart rate and contractility and contraction of smooth muscles in the intestine and airways. To regulate these functions, histamine works as a neurotransmitter and immunomodulator.

   Histamines can also induce the host to carry out the act of sneezing, which is a useful mechanism to rid some of the pathogens physically from the body. Histamine prompts goblet cells to produce and secrete more mucus. The increased amount of mucus in the respiratory tract, coupled with smooth muscle contraction, leads to sneezing.

   The function of vasodilation is to decrease blood flow rate while increasing the blood volume in the local vascular area. The increase in blood volume generates two of the five common characteristics (or symptoms) of inflammation, known as rubor (or redness) and calor (or heat) in the area of infection.

2. Both histamine and VEGF are also strong regulators of vascular permeability. Histamine-activated expression and secretion of VEGF occurs in a number of cells, including granular tissue and newly formed connective tissue during wound healing. Vascular permeability is the result of the loss of vascular integrity leading to the exposure of the subendothelium to allow the efflux of plasma from the intravascular lumen.

   Cytokines released by the macrophage such as TNF-α, IL-1β, and IL-6 can also regulate vascular permeability by activating the same pathways that are affected by VEGF. These signals regulate activate the phospholipase C–MAPK) pathways. Effector functions of these pathways include increased intracellular calcium, Src kinase, and protein kinase B/akt activation, and stimulation of the mitogen-activated protein kinases p42/p4MAPK and phosphoinositide 3 (PI3) kinase pathways. Akt, p42 (which is a part of the MAPK cascade), and PI3K are important in the regulation of cell survival, proliferation, and differentiation. Downstream effects of these pathways, including the disassembly of vascular endothelial cadherin (forming adherens junctions) and occludens (forming tight junctional zona occludens), thereby reduce the presence of endothelial cell junctions leading to vascular permeability.

Vasodilation and vascular permeability are important processes in inflammation by decreasing the rate of blood flow and the transport of blood cells and plasma protein through the bloodstream. This creates better conditions for cells and proteins to extravasate (or escape from the bloodstream) into the neighboring tissue where the infection occurred. However, since the endothelium is a fairly impermeable barrier that blocks the extravasation of blood material, vascular permeability is also induced to create pores for the diapedesis (migration by crawling) or leukocytes and leakage of plasma (along with platelets and plasma proteins) into the area of infection in the neighboring tissue.

The leaked plasma and plasma proteins accumulate in the tissue, causing the symptom tumor (or edema), defined as the retention of fluid in the area of infection. The accumulated fluid also puts pressure on nerve endings in the tissue, causing dolor (or pain) in the local area. The chemical activity of bradykinin, a by-product of the kinin system of tissue repair, can also contribute to dolor. Prolonged tumor and dolor can lead to a fifth symptom of inflammation, *function laesa*, defined as immobility or stiffness or loss of disturbance of function in the local inflamed region.

3. Rolling adhesion is a multistep mechanism that enables the blood vessel wall to trap and recruit circulating leukocytes and platelets for extravasation into the local site of infection (Figure 2.6). The same cytokines that regulate vascular permeability (TNF-α, IL-1β, and IL-6) also regulate changes in endothelial cells,

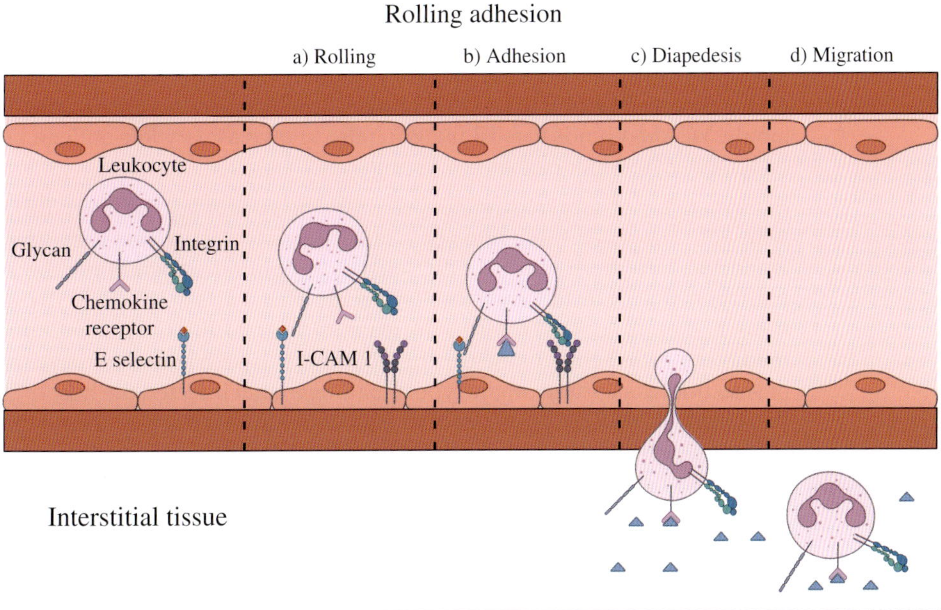

FIGURE 2.6 Extravasation of leukocyte through the process of rolling adhesion. (a) The "rolling" portion of this process occurs when the leukocytes in the bloodstream interact with the endothelium via the interaction between leukocyte surface glycan and selectin on the surface of an endothelial cell. (b) "Adhesion" occurs when the integrin on the leukocyte surface is captured by ICAM-1 of the endothelial cell. (c) The leukocyte extravasate into the interstitial tissue via diapedesis. (d) The leukocyte migrates toward the invading pathogen. *Credit:* Figure partially created in BioRender.

a process called endothelial cell (or endothelium) activation, so that they can participate in the above-mentioned multistep mechanism. The activated endothelial cell is the pro-inflammatory state of the endothelial cells, whose function is to facilitate the recruitment of specific circulating leukocytes to the capillary wall. Once activated, endothelial cells upregulate the expression and exhibition of adhesion molecules like E-selectin, P-selectin, intercellular adhesion molecule 1 (ICAM-1), and vascular cell adhesion molecule 1 (VCAM-1) on their surface to allow leukocytes to adhere to the endothelium before migrating into the tissue. This process is characterized by increased interactions between leukocytes in the bloodstream and endothelial cells.

The initial, rapid, low-affinity formation and dissociation of adhesive bonds between endothelial cells and their target cells combined with the force of blood flow result in the tethering (or capture) and rolling portion of the leukocytes along the vascular wall. This rolling portion is controlled by the binding of selectin molecules on the surface of endothelial cells to their respective ligands. E-selection binds to sialyl-Lewis^X, a carbohydrate structure that is constitutively expressed on monocytes and granulocytes like neutrophil. P-selectin, on the other hand, binds P-selectin glycoprotein ligand 1, which is a glycoprotein on the surface of most leukocytes, especially expressed by all T lymphocytes.

The adhesion portion of this mechanism is controlled by the binding of CAMs on the surface of endothelial cells to their respective ligands to cause the leukocytes to attach to the endothelium and remain stationary. ICAM-1 and VCAM-1 are endothelial adhesion proteins that belong to the immunoglobulin (Ig) superfamily. These surface proteins bind to their ligands to mediate the adhesion portion of this mechanism. ICAM-1 binds to macrophage adhesion ligand 1, fibrinogen, and leukocyte function associated antigen 1, which is a member of the integrin family. VCAM-1 binds to a4β1 integrin (also known as very late antigen 4) found on lymphocytes, monocytes, eosinophils, and basophils.

4. Extravasation of targeted leukocytes into the tissue follows after the rolling adhesion. However, there must be a mechanism to ensure that the appropriate leukocytes are recruited. Specific leukocyte homing depends on the type of pathogen that infected the host organism. Regardless, NFκB is the general transcription factor that induces the expression of additional pro-inflammatory mediators, called chemokines or chemoattractants, to trigger leukocyte homing or selection, depending on the type of pathogen that infected the host organism. IL-8 recruits neutrophils from the bloodstream. Monocyte chemoattractant protein 1 recruits monocytes from the bloodstream. Once in the connective tissue, monocytes differentiate into macrophages, which are then recruited by macrophage inflammatory protein 1a. Eotaxin 1 recruits eosinophils from the bloodstream. Stromal cell-derived factor 1 (SDF-1) binds C-X-C chemokine receptor type 4 (CXCR4) to recruit lymphocytes from the bone marrow. Leukocytes extravasate from the blood vessel through a physical movement called diapedesis, which is defined as cell migration carried out by treadmilling and actin-myosin filament interaction.

Since NFκB is a potent regulator of inflammation, therapeutic potentials are being exploited to develop anti-inflammatory drugs such as glucocorticoids and nonsteroidal anti-inflammatory drugs (NSAIDs) to inhibit or reduce its effect.

Additional Effect of the Innate Immune System

Cytokines released by the macrophage, such as TNF-α, IL-1β, and IL-6 that regulate vascular permeability and endothelial cell activation, have additional downstream functions, including targeting the hypothalamus to degrade multilocular adipose tissues (or brown fat) to increase body temperature (or fever) to decrease pathogenic replication; hepatocytes to express acute-phase proteins (APPs) to activate complement cascades; bone marrow endothelium to induce neutrophil mobilization to carry out phagocytosis; and dendritic cells to migrate into neighboring lymph nodes to regulate T-lymphocyte activation to initiate the adaptive immune response.

Pyrexia

Pyrexia, also known as a "fever," usually occurs during the latter stage of the innate immune response. This is an indication that the immune response had expanded in a systemic or whole-organism level instead of being localized in the area of infection. However, pyrexia can also be activated as the result of septicemia or viremia, which is the infection of the circulatory system by bacteria or by virus, respectively.

A systemic pyrexia response refers to an entire body now alert to the infection. Any of the innate immune responses described in this chapter against the localized infection is now active in all parts of the body and hence is systemic. A systemic response like pyrexia can result in harmful effects to the body.

Pyrexia is a beneficial response to infection by raising the body temperature from normothermia (normal temperature) to high-grade elevated temperature and beyond, referred to as hyperthermia (above normal temperature). This condition acts as an alert system to activate immune cells to fight against the infection. The febrile state (state of a fever) is regulated within a "set-point" temperature. The normothermia level is 37°C (98.6°F). Low-grade elevated temperature is 37.3°C to 38.0°C (99.1°F to 100.4°F); moderate-grade elevated temperature is 38.1°C to 39.0°C (100.6°F to 102.2°F); and high-grade elevated temperature is 39.1°C to 41°C (102.4°F to 105.8°F). Hyperthermia occurs when the body's core temperature rises beyond the set point, which is above 41°C (105.8°F). This is definitely not good because extreme temperature can cause damage to tissues.

There are a number of metabolic and physiological effects associated with pyrexia. The body's heart rate and respiratory rate increase in response to the increased oxygen demand to carry out metabolic activities. Instead of glycolysis, there's an increase in the breakdown of body proteins and fat to be used as an energy source. Pyrexia enhances the immune system by increasing the production, mobility, and activation of leukocytes. Last, the elevated body temperature may kill or inhibit the growth and activity of many bacteria and viruses since most pathogens can tolerate only a narrow temperature range at around the physiological temperature of 37°C. Again, remember that this is a very well-controlled elevation in body temperature that reduces pathogen viability and stimulates immune cell responses.

There are a number of potentially harmful and even lethal effects associated with pyrexia, especially when sustained. Hyperthermia may lead to acute neurologic and cognitive dysfunction. Abnormalities in cardiovascular activities may occur. During a pyrexic episode, a reduction in blood flow to the GI tract may also occur. An increase in liver enzyme production may lead to permanent liver damage. An increase in plasma creatinine and decrease in the glomerular filtration rate may increase the risk for acute kidney injury. Cell damage in various tissues due to the high level of inflammation may occur.

Pyrogen-induced fever is activated by the same cytokines that are released by macrophages and mast cells to regulate vascular permeability. There are endogenous pyrogens or pyrogenic cytokines (eg, TNF-α, IL-1β, IL-6, and IFN) and exogenous pyrogens (eg, LPS and LTA). However, only high amounts of pyrogens in the bloodstream can induce pyrexia. A high concentration of released pyrogenic cytokines is referred to as a "cytokine storm." Pyrogens stimulate hepatic Kupffer cells to synthesize prostanoids, especially prostaglandin E_2 (PGE$_2$), which then enters the bloodstream (Figure 2.7).

Note that even though viruses do not exhibit known pyrogens, there are a number of other ways these pathogens can cause pyrexia. These mechanisms include direct infection of macrophages; immunological reaction to viral components involving antibody formation or opsonization; and regulation by IFN-α or -β); and cell death caused by viruses.

PGE$_2$ synthesis involves the processing of membrane phospholipid regulated by the phospholipase A_2–cyclooxygenase (COX) pathway. Therefore, medications that inhibit COX activity (eg, NSAIDs) are used as treatment for a fever. PGE$_2$, in turn, diffuses across the blood-brain barrier to act on specific receptors within the median preoptic nucleus of the anterior hypothalamus, which contains thermosensitive neurons. The reduction in the firing rate of the warm-sensitive neurons results in an increase in body temperature. Essentially, the pyrogens inhibit heat-sensing neurons while exciting the cold-sensing neurons, which tricks the hypothalamus into thinking that the body is cooler than it actually is.

Pyrogen-induced fever is regulated by neuronal activity from the hypothalamic thermoregulatory center, which regulates heat production by directing both non-shivering and shivering thermogenesis. The generated heat is then conserved through vasoconstriction. Skeletal muscle, cardiac muscle, and multilocular or brown adipose

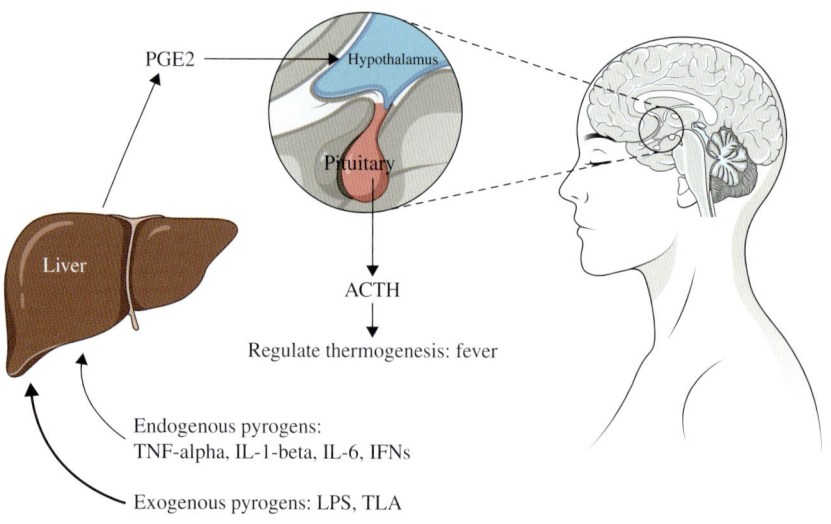

Figure 2.7   Pyrogen-induced fever. High levels of pyrogens stimulate hepatic Kupffer cells to produce and secrete PGE$_2$. PGE$_2$ in turn stimulates the hypothalamus to regulate hormone (adrenocorticotropic hormone, ACTH) release by the pituitary gland to regulate thermogenesis. *Credit:* Figure partially created in BioRender.

tissue (BAT) are the three primary tissues that target for thermogenesis by pyrogenic cytokines. To induce shivering thermogenesis, the neurotransmitter acetylcholine is released to stimulate repeated cycles of muscle contraction and relaxation to convert stored chemical energy in thermal energy. Acetylcholine also upregulates the body's overall metabolic rates.

To induce nonshivering thermogenesis, the neurotransmitter norepinephrine increases thermogenesis in multilocular adipose tissue (also known as brown fat) and induces vasoconstriction in the extremities to prevent heat loss, which in turn helps retain the core temperature. The rate of oxidation in BAT mitochondria is a significant source of heat production because heat generation is a typical by-product of mitochondrial ATP (adenosine triphosphate) production and of ATP utilization. Additionally, BAT is also a storage for high-energy substrates, lipid and glycogen, that can provide the immediate fuel for this process.

Hepatocytes Express Acute-Phase Proteins and Complement Proteins

At the acute-phase response, plasma proteins extravasated during the inflammatory response, via the leakage of plasma as a conduit, further activate a number of innate immune responses and complement cascades.

The acute-phase response is a delayed (hours or days) metabolic change that follows the onset of a fever in response to an infection. These changes are induced by the same set of pyrogenic cytokines, especially IL-6, as discussed in this chapter. This response involves the synthesis of APPs by hepatocytes, then released into circulation in large amounts. These proteins belong to two groups, APPs that are upregulated during inflammation and APPs that are downregulated. Upregulated APPs include C-reactive protein (CRP), mannan-binding lectin (MBL), LPS-binding protein (LBP), procalcitonin, serum ferritin, fibrinogen, plasminogen, hepcidin, granulocyte colony-stimulating factor (G-CSF), secreted phospholipase A2, and serum amyloid A. Downregulated APPs include albumin, prealbumin, transferrin, retinol-binding protein, and antithrombin.

The APPs have a variety of functions. CRP, MBL, and LBP are involved in pathogen recognition, which in turn activate complement cascades. G-CSF, secreted phospholipase A2, and serum amyloid A are involved in prolonging the inflammatory response. Fibrinogen and plasminogen induce the coagulation cascade.

There are a number of adverse effects of the acute-phase response, including somnolence (sleepiness or drowsiness), lethargy, anorexia; changes in plasma protein synthesis like increased CRP, ferritin, and decreased albumin; changes in hormone synthesis; inhibition of bone formation; negative nitrogen balance, changes in lipid metabolism; decreased serum iron and zinc; and elevated leukocytes and platelets, decreased synthesis of smaller, mature red blood cells.

There is a link between APPs and complement cascades. The three complement cascades are classical complement cascade, MBL complement cascade, and alternative complement cascade (Figure 2.8).

CRP (or IgM/IgG) and MBL activate the classical and MBL complement cascades, respectively. The alternative complement cascade is activated spontaneously in an aqueous environment. Instead of requiring an APP to initiate the process, the alternative complement cascade starts with the spontaneous hydrolysis, or tickover, of the soluble C3 protein in solution. The hydrolyzed C3 in turn is involved in the formation of a soluble form of C3 convertase, leading to the production of a high amount of soluble C3b fragments. The C3b fragment can opsonize the cell membrane of either pathogens

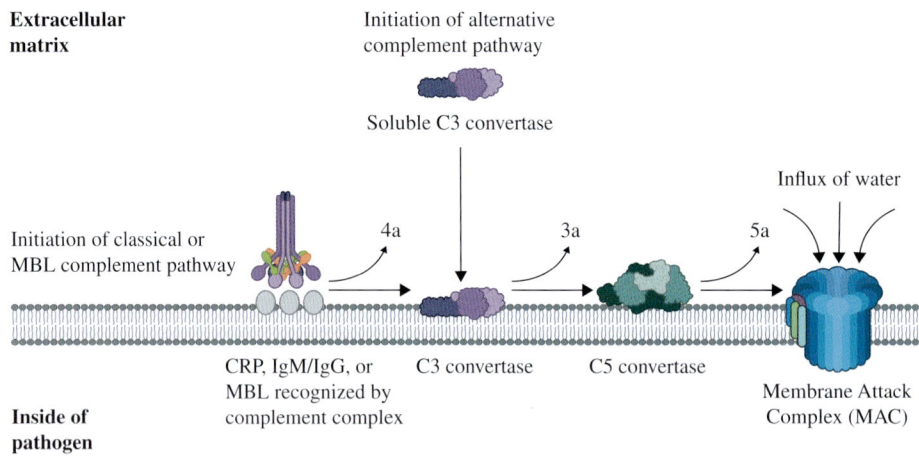

Figure 2.8   Complement cascades. The classical complement cascade is activated by CRP, IgM, or IgG. The MBL complement cascade is activated by MBL. The alternative complement cascade is activated spontaneously in the extracellular aqueous environment. All 3 cascades converge at the formation of the C3 convertase and use the same pathway to produce the MAC.
Credit: Figure partially created in BioRender.

or host cells, by forming thioester bonds with either surface proteins or carbohydrates, to initiate the assembly of the membrane-bound form of C3 convertase.

From here, all three complement cascades converge, then continue along the same pathway. The effector functions of all three complement cascades are recruitment of inflammatory and immunocompetent cells, opsonization of pathogens, and destruction of pathogens.

The classical complement cascade is initiated by the binding of CRP to phosphocholine on the surface of pathogens to mark the pathogen for attack. The binding of IgM or IgG to specific surface antigen can also activate this complement cascade.

While the MBL complement cascade is initiated by the binding of exposed mannose residues to MBL, human cells have mannose residues on their surface, but these residues are concealed and not exposed.

The alternative complement cascade is a rare, spontaneous event that does not directly involve the APPs. The ultimate major goal of a complement cascade is to assemble multiple membrane attack complexes (MACs) to destroy the pathogen through osmotic lysis.

Let's take a brief walk through these cascade processes. After binding to their specific binding partner, CRP and MBL recruit a number of complement proteins synthesized by hepatocytes. Complement proteins are named C1 through C10. Each complement protein has a specific function during the specific portion of the pathway, and their functions are well coordinated with one thing in "mind": destroy the pathogens.

The binding of CRP, IgM, IgG, or MBL leads to the recruitment of different complement proteins to the surface of the pathogen. These complement proteins are then cleaved into two pieces: one small (labeled "a") and one large fragment (labeled "b"; except for C2 where C2a is actually the large fragment and C2b is the small fragment.) Each fragment, in turn, has a specific function in the innate immune system. The assembly of the C3 convertase and C5 convertase are major amplification steps, producing high amounts of C3b and C5b fragments, respectively.

C3a and C5a fragments are peptide mediators of inflammation that bind to receptors C3a and C5a, respectively, on endothelial cells, mast cells, and phagocytes. They act as chemotactic agents to recruit newly extravasated leukocytes to the area of infection. They are also involved in phagocyte recruitment to pathogens. C3b fragment is classified as an opsonin, which is a molecule that can, via opsonization, cover the surface of the pathogen. It binds to complement receptor (CR) 1 on the surface of a macrophage; monocyte; B lymphocyte; polymorphonuclear leukocytes (neutrophil, eosinophil, basophil); and dendritic cells. The binding and activation of CR1 increase the ability of the phagocytes to carry out phagocytosis. C3b fragments are also important in the assembly of the C3 convertase.

There are a number of other CRs found on the surface of nonimmune cells that can bind to complement fragments. C4a fragment becomes a weak pro-inflammatory peptide. C5b fragment is involved in the assembly of the C5 convertase and formation of the MAC. MACs perforate cell membranes to create pores on the surface of pathogens, thus allowing an influx of extracellular fluid and causing osmotic lysis to destroy the pathogen.

Protection of Host Cell Against Damage From the Complement Cascade

The same complement system used to destroy pathogens can also endanger the host's own cells as well. We focus on the C3b fragment specifically. These fragments not only can opsonize pathogens, but also can form thioester bonds on the surface of the host cell and pathogen. Once attached, C3b fragments are able to initiate the formation of C3 convertases on the surface of the host cell. The result is the formation of MACs and the ultimate destruction of the cell. Therefore, host cells have evolved a number of mechanisms to protect themselves against activities of their own complement cascade.

Membrane-bound C3 convertase inhibitors include decay-accelerating factor (DAF), membrane cofactor protein, and CR1, while CR1 also performs the role of a C5 convertase inhibitor. Factor H, which is a soluble protein, is an additional C3 convertase inhibitor. Since IgMs (or IgGs) may have affinity to antigens on the surface of host cells, the C4 proteins may be recruited to initiate both the classical and MBL complement cascades. C4-binding protein (C4BP) is a soluble regulator that blocks the initiation of both of these complement cascades (Figure 2.9).

If somehow the complement cascade managed to evade the inhibitors discussed above and the C5 convertase is still active, membrane-bound CD59 (or protectin) and soluble-form vitronectin can still inhibit the formation of MAC.

Last, Factor I is an enzyme that can inactivate C3b, which is the most abundant complement protein fragment produced in these processes.

Adaptive Immune System

If the innate immune response is unsuccessful in clearing the pathogens, then adaptive immune responses will begin to dominate. However, the adaptive immune system is alerted from the moment the pathogens entered the mucous layer. That is, the mucosal immune system can act as an alarm system to alert the body of an invasion.

An important part of the adaptive immune system includes the multipotent hematopoietic stem cells (HSCs). Most immune cells originate from these stem cells found in the bone marrow. HSCs can undergo symmetric cell division, activated by signaling

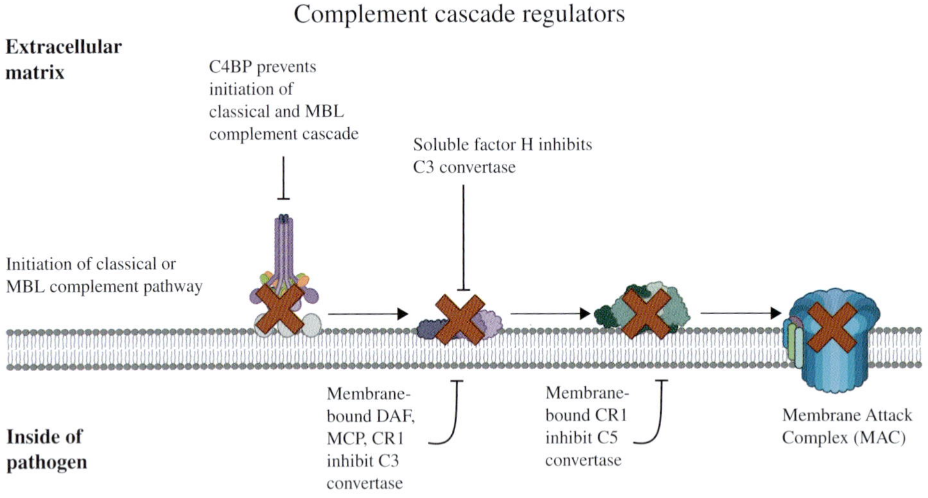

Complement cascade regulators

FIGURE 2.9   Protection of host cell against the complement cascade. DAF, MCP, and CR1 are membrane-bound inhibitors blocking the formation of C3 convertase. CR1 is also a C5 convertase inhibitor. Factor H is a soluble C3 convertase inhibitor. C4BP is a soluble inhibitor of C4. Membrane-bound CD59 and soluble vitronectin inhibit the formation of MAC. *Credit:* Figure partially created in BioRender.

molecules including IFN, Hedgehog, and Wnt ligand, for self-renewal to replenish their population. Daughter cells produced are not directed to undergo differentiation. Self-renewal may also be inhibited by the signals Tang1 and thrombopoietin, directing these cells to become quiescent (Figure 2.10).

Bone marrow stromal cells release IL-3, granulocyte-macrophage colony-stimulating factor, and macrophage colony-stimulating factor direct HSCs to differentiate into common myeloid progenitors, while IL-7 directs quiescent HSCs to differentiate into common lymphoid progenitors. Monocytes/macrophages, neutrophils, eosinophils, basophils, mast cells, erythrocytes, and platelets are all descendants of the common myeloid progenitor cell lineage.

The same bone marrow stromal cells also express IL-3, IL-4, IL-7, and stem cell factor to induce some common lymphoid progenitors to develop into mature B cells, again in the bone marrow. NK cells and a portion of the plasmacytoid dendritic cell population are also descendant of this lineage. Common lymphoid progenitors that are not affected by these signals would migrate to another primary lymphoid organ, called the thymus, and develop into mature T lymphocytes induced by IL-2, IL-7, and NOTCH released from thymic stromal cells. NOTCH is a critical signaling molecule that regulates downstream events of its receptor, NOTCH 1, to regulate cellular development, proliferation, differentiation, and apoptosis. Mature lymphocytes are then involved in both the adaptive immune response (directly) and innate immune responses (indirectly).

B- and T-Lymphocyte Maturation

The B lymphocytes undergo development and maturation in the bone marrow (Figure 2.10). They start out as common lymphoid progenitors that receive a signal to progress through the B-cell lineage. The major function of B-lymphocyte maturation is

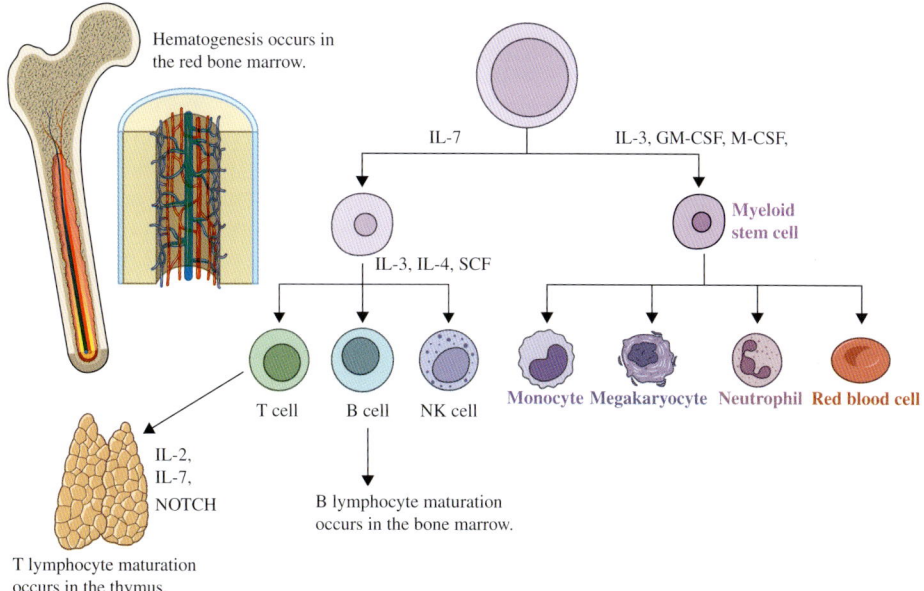

Hematogenesis occurs in the red bone marrow.

IL-7

IL-3, GM-CSF, M-CSF,

Myeloid stem cell

IL-3, IL-4, SCF

T cell   B cell   NK cell

Monocyte Megakaryocyte Neutrophil Red blood cell

IL-2, IL-7, NOTCH

B lymphocyte maturation occurs in the bone marrow.

T lymphocyte maturation occurs in the thymus.

FIGURE 2.10   Multipotent HSC replenishment and differentiation. HSCs divide in the bone marrow to replenish the stem cell population and to differentiate to produce functional blood cells. During mitotic cell division, HSC is directed to differentiate into either common myeloid progenitors or common lymphoid progenitors in the bone marrow. Each type of progenitor cell further divides and differentiates through different cell lineages, depending on the signals that are provided. Common lymphoid progenitors differentiate into B-cell or T-cell lineages. B-cell maturation occurs in the bone marrow, while T-cell maturation occurs in the thymus. *Credit:* Figure partially created in BioRender.

to carry out a process called gene rearrangement of the two gene loci, Ig heavy chain and Ig light chain, to encode a fully functional immunoglobulin (or antibody) folded in the immunoglobulin's quaternary structure.

The mechanism for gene rearrangement is somatic DNA recombination. At the beginning of their development phase and before being mature, these cells don't possess a complete gene that encodes the antibody molecule. Rather, these cells possess gene loci containing many gene segments that can be put together through different combinations via somatic DNA recombination to produce a full, active gene that can be expressed.

Gene rearrangement is a type of somatic DNA recombination where gene segments are recombined to form a full gene. This process affects the complementarity-determining regions within the antigen-binding region (or site) of the variable domain of the immunoglobulin, like constructing a structure with Lego pieces.

We do not describe here the process of somatic DNA recombination, but the result is the synthesis of a very diverse set of immunoglobulins that can recognize a wide variety of antigens that one's body encounters.

Since DNA recombination is permanent, each mature B lymphocyte will only express one particular version of the immunoglobulin. Before being fully mature, these developing B lymphocytes must undergo a process called central tolerance. During this process the system must ensure that the immunoglobulin produced by this cell exhibits self-tolerance by undergoing the process of selection. Cells that are positively selected express B-cell receptors (BCRs) that can recognize self-antigens without activating against them for destruction. Self-antigens are presented by specialized cells found in the bone marrow. Cells that have no strong reactivity (or cross-linking) to self-antigens exhibit self-tolerance and become mature B cells. The system is able to recognize self-entities but not destroy them. Mature B lymphocytes, but still naïve or not yet activated, are transported out of the bone marrow to a secondary lymphoid organ or tissue, like the spleen or lymph node, to be activated through interaction with Th2 lymphocytes. They are considered naïve because they have yet encountered an actual nonself-antigen.

Naïve, mature B lymphocyte expression membrane-bound immunoglobulins are called BCRs. The BCR complex is composed of surface immunoglobulin, both IgM and IgD, and is noncovalently bound to Igα and Igβ. Each immunoglobulin molecule contains multiple domains, two of which are the antigen-binding domain and effector domain.

There are two identical binding domains that bind the identical epitope on the same antigen. If activated, the effector domain of the BCR would, in turn, be activated, leading to the regulation of various signal transduction pathways. These BCRs, at this point, can bind to self-antigens weakly and thus would not be activated. The system would not be activated until bound more tightly to nonself-antigens at varying degrees of affinity. At this point, these cells are released from the bone marrow and circulate through the bloodstream as well as interstitial tissue.

Once bound to an antigen, the BCR is activated, which in turn induces the B lymphocyte to engulf the complex via receptor-mediated endocytosis. The cell breaks down the pathogen using a variety of digestive enzymes. Pathogenic proteins then undergo a process called peptide processing, where oligopeptides 12–24 amino acids long are complexed with the major histocompatibility complex (MHC) class II to create surface antigens to be recognized by TCRs on the surface of T lymphocytes. In this capacity, the B lymphocyte acts as a professional antigen-presenting cell (APC). A professional antigen-presenting cell is a specialized type of APC the is highly efficient at activating naïve T lymphocytes. This MHC class II + pathogenic peptide plays an important role in the B-lymphocyte activation.

The T lymphocytes also undergo development and maturation. They go through a similar process as B lymphocytes except in a different organ, the thymus. The major function of T-lymphocyte maturation is also to carry out gene rearrangement of the two gene loci, the α-T-cell receptor (α-TCR) and β-T-cell receptor (β-TCR), to encode a fully functional TCR folded in the quaternary structure. Common lymphoid progenitors that do not receive the multitude of signals to go through the B cell lineage migrate to the thymus to undergo maturation. Stromal cells in the thymal cortex express and secrete NOTCH 1, which induces the common lymphoid progenitors to progress through the T-cell lineage. Notch 1 is the receptor that is activated by NOTCH, which is a signaling molecule.

As with the process for B lymphocytes, the result of T-lymphocyte maturation is to produce a naïve, mature T lymphocyte that has been positively selected to express TCRs that can recognize self-antigens. Positive selection is achieved when the TCR molecule can recognize a self-MHC + self-peptide complex but binds loosely or weakly and therefore does not get activated by this interaction. This cell would have attained a self-tolerance property against self-antigens that are presented by specialized cells found in the thymus. Additionally, each mature T lymphocyte can express only one of the two co-receptors, either CD48 or CD4. We refer to these cells as either naïve, mature CD8-T lymphocytes or naïve, mature CD4-T lymphocytes.

Naïve, Mature T-Lymphocyte Activation

Naïve, mature B lymphocytes and T lymphocytes migrate away from their respective primary lymphoid organ and circulate throughout the body. Since B lymphocyte priming is directly regulated by the T lymphocyte, let's discuss T-lymphocyte priming first (Figure 2.11).

Naïve, mature CD4-T lymphocytes that had migrated out of the thymus make their way to secondary lymphoid tissues or organs. During the process of priming or activation, CD8-T lymphocytes can only differentiate into cytotoxic T lymphocytes (CTLs), which are also known as effector CD8-T lymphocytes. CTLs target abnormal cells, like virus-infected cells or cancer cells, for destruction by activating the extrinsic apoptotic pathway. Naïve, mature CD4-T lymphocytes can differentiate into one of the four effector CD4-T lymphocytes: Th1, Th2, Th17, or Treg.

Priming of naïve CD4-T lymphocytes starts with the plasmacytoid dendritic cell, the major type of dendritic cell. The most common professional APC is the plasmacytoid

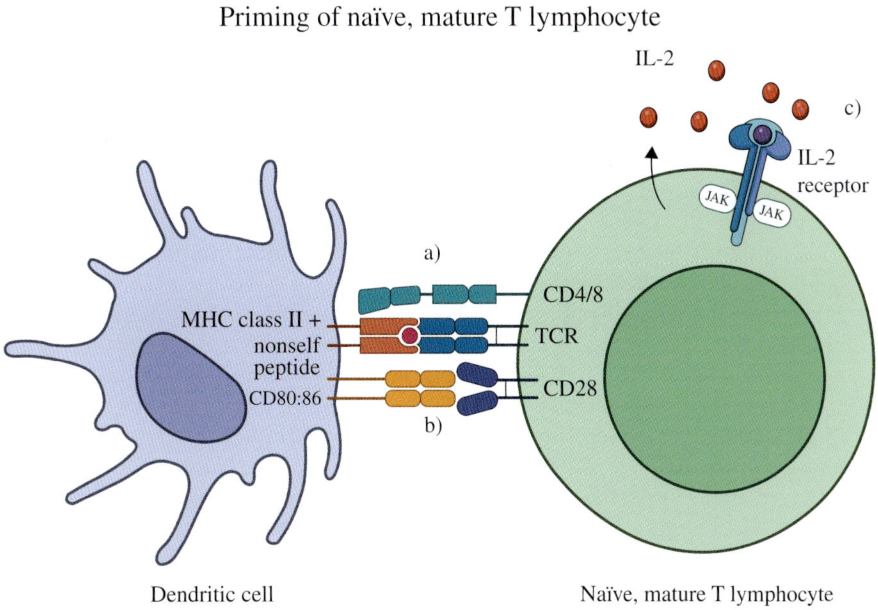

Priming of naïve, mature T lymphocyte

FIGURE **2.11**   T-lymphocyte priming. (a) The first step is lymphocyte activation, which is initiated by the interaction between the TCR:CD4 complex on the surface of the naïve T lymphocyte and the MHC class II:nonself-peptide complex on the surface of the dendritic cell. (b) Promotion of survival is regulated by the interaction between CD28 on the T lymphocyte and CD80:CD86 complex on the same dendritic cell. (c) Clonal expansion is stimulated by the binding of IL-2 to IL-2 receptor on the surface of the T lymphocyte. *Credit:* Figure partially created in BioRender.

dendritic cell. The dendritic cells are phagocytes that also perform the function of professional APCs required for T-lymphocyte priming. Dendritic cells are produced in the bone marrow, then released and circulate throughout one's body. They possess surface PRRs that recognize general PAMPs belonging to pathogens. Once bound by PAMP and activated, the phagocytic PRR induces phagocytosis of the pathogens. The pathogen is degraded and processed using the same mechanism as described for naïve B lymphocytes. The ultimate result is to present the MHC class II + nonself-peptide complex on the surface of the dendritic cell to be recognized by the naïve T lymphocyte for priming.

Dendritic cells that have encountered the pathogen and presented pathogenic peptides on their surface migrate into the T-cell zone of a secondary lymphoid tissue or organ. Priming requires three signals: (1) T-lymphocyte activation, (2) promotion of survival, and (3) clonal expansion.

1. Activation of the naïve T lymphocyte involves the interaction of TCR and CD4 complex on the surface of the naïve T lymphocyte with MHC class II + nonself-peptide complex on the surface of a dendritic cell. This signal directs the primed cell to undergo clonal expansion and cell differentiation.

2. Promotion of survival requires the interaction between CD28 on the T lymphocyte and CD80:CD86 complex on the same dendritic cell. The interaction provides a co-stimulatory response to promote survival of the T lymphocyte. Whereas the interaction between CTLA of the naïve, mature CD4-T lymphocyte and CD80:CD86 complex of the dendritic cell provides a co-stimulatory response to promote cell death to the T lymphocyte.

3. Clonal expansion is a large-scale cell proliferation of T lymphocytes that have received all the appropriate signals. Activation of TCR leads to the expression of IL-2, which possesses autocrine signaling activity, meaning that IL-2 that is secreted by the T lymphocyte undergoing priming activates the same cell to undergo clonal expansion.

The dendritic cell is crucial in directing cell differentiation of the T lymphocyte being primed. Interaction between the TCR + CD4 complex of the T lymphocyte and MHC class II + nonself-peptide complex of the dendritic cell activates not only the TCR, but also the MHC class II complex. Once activated, the dendritic cell is induced to express and secrete various signaling molecules to guide T-lymphocyte differentiation.

Dendritic cell secretes IL-12 to direct the naïve CD4-T lymphocyte to differentiate into Th1 cell, while secreting IFN-γ to prevent the naïve CD4 T-lymphocyte from differentiating into Th2 cell. The differentiated Th1 cell is induced to express the TS-TF T-bet, in turn, to regulate the synthesis of Il-2, IFN-γ, and the surface-bound CD40 ligand. IL-2 is secreted to induce clonal expansion of the Th1 cell itself. IFN-γ is secreted to continue to prevent differentiation into Th2 cells. CD40 ligand on the surface of Th1 interacts with CD40 (the receptor) on the surface of either the macrophage or CTL to provide additional activating signals for macrophages to destroy phagocytosed pathogens. This interaction also promotes increased CTL activity. Activation allows these effector cells to perform their functions more effectively and efficiently. Additionally, the Th1 cell can express and secrete IFN-γ to stimulate macrophages to secrete TNF-α to activate and/or prolong the inflammatory responses and secrete IL-12 to direct the differentiation of more naïve CD4-T lymphocytes into more Th1 cells. Th1 cells often continue to be produced in response to a prolonged viral infection.

Dendritic cells secrete IL-4 to direct the naïve CD4T-lymphocyte to differentiate into the Th2 cell, while secreting IL-10 to prevent the naïve CD4-T lymphocyte from differentiating into Th1 cells. The differentiated Th2 cell is induced to express the TS-TF GATA-3 (GATA binding protein 3) to regulate the synthesis and secretion of IL-2, IL-4, IL-5, and IL-10. GATA-3 is an important tissue-specific transcription factor that is involved in directing the differentiation of Th2 cells and regulating the expression of cytokines in response to an allergic reaction. Again, IL-2 induces clonal expansion of the Th2 cell itself to induce clonal expansion of itself as well. IL-4 and IL-5 are important in B-lymphocyte priming. IL-10 continues to prevent differentiation into Th1 cells.

Dendritic cells also secrete TGF-β to direct the naïve CD4-T-lymphocyte to differentiate into Th17 or Treg cell. The differentiated Treg cell is induced to express the TS-TF FoxP3 to regulate the synthesis and secretion of IL-10 and TGF-β. IL-10 suppresses the induction and proliferation of effector T lymphocytes when necessary. The differentiated T17 cell is induced to express the TS-TF RORγT (retinoic acid-related orphan receptor gamma t) to regulate the synthesis and secretion of IL-6 and IL-17. TS-TF RORγT is the master regulator of Th17 cell differentiation and IL-17/22 production. IL-6 regulates the balance between Th17 and Treg cell populations. Remember that IL-17 is a pro-inflammatory mediator and plays a role in defending against extracellular pathogens by recruiting neutrophils and macrophages to infected tissues.

Naïve, Mature B-Lymphocyte Priming

Naïve, mature B lymphocytes that had migrated out of the bone marrow make their way to secondary lymphoid tissues or organs, most importantly the spleen. During the process of priming or activation, B lymphocytes can differentiate into either plasma cells or memory B cells. Let's begin with the naïve B lymphocyte circulating and patrolling the body's tissues in search of nonself antigens. When these cells have identified and bound to the antigen with a certain level of affinity, the BCR is activated and activates phagocytosis of this entity (Figure 2.12).

The antigen, for instance, may be a capsid protein on the surface of a particular viral particle, in which case the virus would be phagocytosed. This activity is not virus-directed endocytosis but host-cell–directed phagocytosis. This means that the naïve B lymphocyte is well equipped to degrade the virus inside its phagolysosome. This cell carries out a process called peptide processing by which viral proteins would be digested and resulting peptides are complexed with MHC class II to be presented on the cell surface as antigens for TCRs. The B lymphocyte would now retreat to the B-cell zone of a secondary lymphoid tissue or organ to be primed. The purpose for the priming process is to produce a large number of similar active B lymphocytes that can destroy any entities that present this specific antigen. Once in the B-cell zone, this cell must wait to be selected by a specific Th2 cell.

Th2 cells that are produced as the result of T-lymphocyte priming can recognize and, in turn, prime B lymphocytes that express very specific antigens. The TCR + CD4 complex on the surface of the Th2 cell recognizes and binds to the MHC class II + nonself-peptide complex on the B lymphocyte. In fact, this MHC class II + nonself-peptide complex must be the same as that found on the surface of the dendritic cell that was involved in the priming of this particular T lymphocyte.

Priming of naïve, mature B lymphocyte

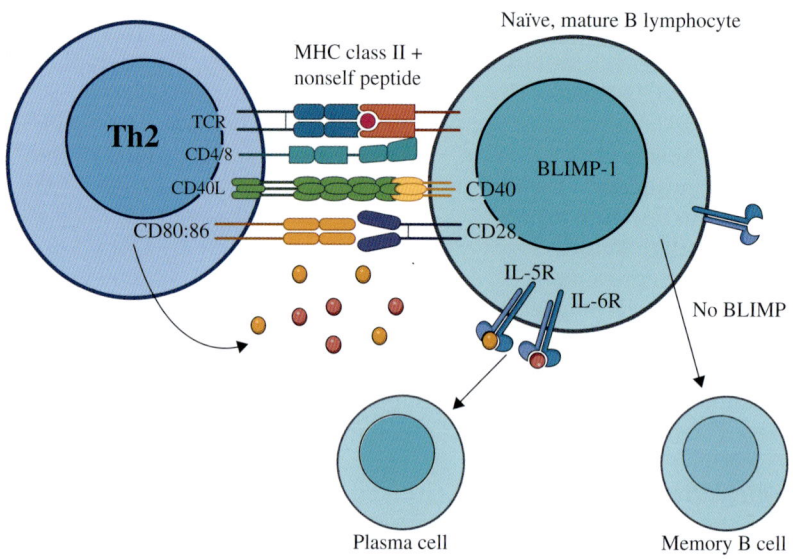

FIGURE 2.12   B-lymphocyte priming. Priming of naïve, mature B lymphocytes also involves activation, promotion of survival, and clonal expansion. Additionally, interaction between CD40L of Th2 and CD40 of the naïve B lymphocyte directs isotype switching. Last, the naïve B lymphocyte that receives the IL-5 and IL-6 signals can express BLIMP and becomes a plasma cell. The naïve B lymphocyte that does not receive the IL-5 and IL-6 signals becomes a memory B cell. *Credit:* Figure partially created in BioRender.

The MHC class II + nonself-peptide complex is referred to as the cognate antigen of this TCR + CD4 complex. Both the dendritic cell and naïve B lymphoid would have encountered the same pathogens from the same infection, more than likely in the same area of the body. Both of these cells would have engulfed the pathogen through phagocytosis.

The same nonself-peptide is then bound to an MHC class II molecule, and the cognate antigen is presented on the surface of both the dendritic cell and mature B cell. Because the TCR + CD4 complex must recognize its cognate antigen on both the dendritic cell and B lymphocyte, this concept is referred to as linked recognition by the Th2 cell. The interaction between the Th2 cell and B lymphocyte occurs within the border of the B-cell and T-cell zones.

The activation of the TCR + CD4 complex on the Th2 cell surface leads to the expression of a number of signaling molecules, including IL-4, IL-5, IL-6, and CD40 ligand. IL-4 is secreted and bound to IL-4 receptor on the surface of the B lymphocyte to induce multiple rounds of cell proliferation. B lymphocytes that are primed are also directed to differentiate into effector B lymphocytes, either plasma cells or memory B cells. IL-5 and IL-6 induce the expression of the TS-TF BLIMP-1 (B-lymphocyte-induced maturation protein-1) in the primed B cell. BLIMP-1 is a suppressor of gene expression that leads the primed B cell to differentiate into plasma cells.

Plasma cells produce almost exclusively soluble immunoglobulin and do not present MHC class II or CD40 on the cell surface. The absence of these two surface proteins means that these cells cannot undergo an additional priming process. They do not undergo further cell proliferation and are short lived at roughly 3 months. The majority of primed B lymphocytes are plasma cells. B lymphocytes that did not receive the IL-5 and IL-6 signals become memory B cells. Memory B cells express mainly membrane-bound immunoglobulins in the form of BCRs but are not necessarily just IgM. The BCRs are likely IgA, IgE, IgG, or IgM. They do express both MHC class II and CD40 and therefore can potentially be primed again. They may undergo cell proliferation but at a very slow rate. They are long lived at 10 years or longer and are important during a secondary infection of the same pathogen.

Affinity Maturation and Isotype Switching

The CD40 ligand expressed by the Th2 is another important component of B lymphocyte priming that leads the primed B lymphocyte to undergo affinity maturation and isotype (or class) switching. Affinity maturation is a process during priming in the B-lymphocyte to produce antibodies that are more effective at identifying and fighting pathogens. Isotype switching is a process to change the type of immunoglobulin produced by the primed B-lymphocyte, predominantly from IgM to IgG. The interaction of CD40 ligand and CD40 ultimately induces the expression of activation-induced cytidine deaminase (AID) in the primed B lymphocyte. AID induces hypermutation of the portion of the gene that encodes the existing antigen-binding site of the immunoglobulin. Hypermutation may result in a higher affinity, lower affinity, or unchanged affinity for the antigen. Of course, a fourth probable result is the production of a nonfunctional immunoglobulin. The system's goal is to generate an immunoglobulin molecule with a higher affinity for the antigen, which is produced as the result of the process of affinity maturation. Isotype switching also occurs during the priming of B lymphocytes. Instead of producing a BCR that is a combination of IgM and IgD, the primed B lymphocyte can be directed to express other immunoglobulin isotypes depending on the signals it received.

During the priming process, B lymphocytes are able to undergo a process called isotype switching. Each cell has the chance to switch from IgM/IgD to IgA, IgE, IgG, or IgM, depending on the type of pathogen and location of the infection. The five isotypes (or classes) of immunoglobulin or antibody are IgA, IgD, IgE, IgG, IgM. Immunoglobulin protein refers to the superfamily of CAMs expressed in a number of different cell types, while antibody is a type of immunoglobulin that's only produced by B lymphocyte and is involved in an adaptive immune response. All immunoglobulin isotypes recognize very specific antigens, but each is found in different locations in the body and confers different effector functions.

Immunoglobulin A, as discussed, is mainly a soluble molecule that's found predominantly in body external secretions, including mucus, saliva, tears, and breast milk. It is a part of the mucosal immune system, which is the first line of defense against invading microorganisms. IgA molecules opsonize pathogens to render them neutral. The neutralized pathogens are then physically removed from the body along with other waste products.

Immunoglobulin D is mainly a membrane-bound molecule on the surface of mature B lymphocytes. There are no known biological effector functions for IgD.

Immunoglobulin E is mainly a soluble molecule found in interstitial tissues. IgE molecules defend the body against parasitic infections involving eosinophils. They are also involved in hypersensitivity (or allergic) reaction types I and IV, which involve mast cell degranulation.

Immunoglobulin G is mainly a soluble molecule and the most abundant immunoglobulin isotype. These molecules are found in both interstitial tissues and bloodstream.

They are the major immunoglobulins produced during a secondary immune response and have many effector functions through opsonization of pathogens, including neutralizing the pathogen, increasing the effectiveness of phagocytosis by phagocytes and activating the classical complement cascade (but not as well as IgM). Furthermore, IgG molecules can be delivered from the mother across the blood-placenta barrier to provide temporary immunity for the fetus. They are also involved hypersensitivity types II and III, or drug allergy and serum sickness, respectively.

Finally, IgM is mainly a membrane-bound immunoglobulin found on naïve, mature B lymphocytes, acting as a BCR along with IgD. In their role as a BCR, these molecules are the first to be produced during B-lymphocyte maturation and are responsible for early stages of adaptive immunity. In the soluble form, IgM molecules are secreted into the blood and are potent activators of the classical complement cascade.

Summary of the Three Levels of Human Immune Systems Against a Viral Infection

Consider a sample scenario about how the host organism responds to a viral infection in the GI tract specifically. At this point, the virus is merely trapped in the mucous layer but an infection had not yet occurred. This invasion is considered an infection only at the moment immediately after the virus crossed or breached the epithelium into the connective tissue of the submucosa. In the mucous layer, the viral particles would be encountered mainly by IgA, which has some affinity to surface viral molecules, like capsomeres and spike proteins. If successful, the opsonization by IgA molecules would neutralize and allow viral particles to be eliminated from the host along with waste material.

Fragments of viral proteins can also be transcytosed by microfold cells into lymphoid tissues, like Peyer patches. In the lymphoid tissue, the transcytosed viral proteins can be phagocytosed by dendritic cells and naïve B lymphocytes. The likelihood of an infection depends on the dose of exposure and entry into the host organism. If these levels are low enough, IgA opsonization is more than sufficient to clear the pathogens and end this scenario. Also, not all viruses can cause an infection in this case; however, enteric viruses like poliovirus are very likely able to penetrate the epithelial lining to reach the underlying connective tissue.

Both the innate and adaptive immune responses are activated only after an infection occurred. Once in the submucosa of the mucous membrane, macrophages are generally the first type of leukocyte to encounter the pathogen. Macrophages can initiate the clearance of the virus through phagocytosis. They would also secrete pro-inflammatory mediators to activate the inflammatory system to recruit additional help. Neutrophils that are recruited would be the major phagocytic cells to attempt to clear this infection. Generally, the combined activities of all the innate immune responses should be sufficient to clear the infection. However, if the infection persists, the adaptive immune system would take effect.

Dendritic cells and naïve B lymphocyte can also phagocytose the viral particles as soon as they've crossed the epithelial barrier. There would be a certain number of effector T and B lymphocytes that were alerted by the transcytosis process of microfold cells. CTLs can identify and destroy host cells infected by the virus. Th1 can activated macrophages and CTLs to clear viral particles more efficiently through phagocytosis and destruction of infected cells, respectively. Th2 cells are involved in the priming of

B lymphocytes that had phagocytosed the virus. Once produced, plasma cells can produce a high amount of immunoglobulin, most likely IgG, to opsonize the virus and activate a number of pathways mentioned in the discussion of pathways, including the classical complement cascade.

Additional Virus-Specific Defense Mechanisms

There are a number of virus-specific defense mechanisms performed by host immunity. The infected host cell itself evolved several intracellular mechanisms that mount a defense against the virus once they've managed to enter the cell. We've seen the physiological mechanism of slowing down enzymatic activities by increasing body temperature, leading to a fever. Since most biological activities involve protein function, inhibiting protein synthesis is an effective means to disrupt pathogenic reproduction, therefore halting the spread of the infection.

Interferons Are Specific Against a Viral Infection

The TS-TFs IRF3 and IRF7 induce the expression of interferons in various cell types discussed next. Interferons are proteins known for their ability to "interfere" (hence the name) with viral replication in infected cells. In addition, these cytokines have a pyrogenic effect and antitumor and immunoregulatory functions. There are three types, type I (IFN-α, IFN-β), type II (IFN-γ), and type III (IFN-λ). Only type I is used as therapy. Type I and III are produced by a variety of cells (eg, leukocytes, dendritic cells, fibroblasts, and macrophages), whereas synthesis of type II is restricted to T lymphocytes and NK cells.

The presence of IFN-α, IFN-β, and even IFN-γ in the body indicates an immune response against a viral infection. IFN-γ is secreted mainly by NK and natural killer T (NKT) cells as a part of the host's innate immunity. They are also expressed in macrophages, Th1 cells, and CTLs once antigen-specific immunity has developed. IFN-γ performs a number of functions to help clear an infection. It promotes the activity of NK cells and differentiation of activated CD4-T lymphocytes into Th1 cells. Both of these activities are specific for viral infections.

Virus-infected cells are induced to express protein subunits to form immunoproteasomes. Immunoproteasomes are important in viral peptide processing and presentation onto MHC class I. IFN-γ also induces the expression of MHC class II in professional APCs. Last, IFN-γ induces activated B lymphocytes to undergo isotype switching to produce IgG2 and IgG3, both of which are powerful antibodies against a viral infection.

IFN-α and IFN-β are expressed in cells that come in contact with an infecting virus such as a variety of leukocytes, fibroblasts, and keratinocytes. IFN-α induces the expression of MHC class I whose function is to present viral peptides on the surface of virus-infected cells. The presentation of viral peptides by MHC class I is important for the clearance of virus by the adaptive immune response. Virus-infected cells can be identified by NK cells and CTLs. Both of these host cell types destroy virus-infected cells by activating the infected cell's extrinsic apoptotic pathway. Last, the complement cascades can also destroy viral particles by forming MACs on the surface of the enveloped virus.

Other proteins upregulated by IFN-α and IFN-β are 4E-binding protein (4E-BP) and eukaryotic initiation factor 4G (eIF-4G). EIF-4G interacts with the poly(A)-binding protein (PABP) to regulate translation initiation. EIF-4G dimerizes with eIF-4E to regulate cap-dependent translation initiation, while monomeric eIF-4G regulates cap-independent translation initiation. 4E-BP inhibits cap-dependent translation initiation, which is the main mechanism used for host protein synthesis. 4E-BP sequesters eIF-4E to prevent the formation of the eIF-4E:eIF-4G dimer, which is required for cap-dependent translation initiation. The monomeric eIF-4G, on the hand, promotes cap-independent translation initiation, which is the mechanism used to carry out the synthesis of critical host proteins when the cell is under stress (ie, during a viral infection).

Additionally, IFN-α and IFN-β induce the expression of PKR/DAI, which stands for protein kinase R/double-stranded RNA activated inhibitor of protein synthesis DAI. PKR/DAI regulates signaling pathways that are activated during stress-activated responses, such as a viral infection, by inhibiting translation and activating the intrinsic apoptotic pathway.

PKR/DAI phosphorylates guanosine diphosphate–bound eIF-2α in response to a viral infection to reduce all protein synthesis within the infected cell. In this event, the synthesis of both viral and host proteins is reduced with this mechanism. The host's strategy is to outlast the virus. Prompt and continuous viral protein synthesis is essential for virion replication. In other words, a halt in the production of viral proteins typically prevents virion replication.

On the other hand, the infected host cell can survive significantly longer, even though barely, and may be revived once the intracellular viral components are cleared. To reduce viral protein synthesis further, these interferons also activate the synthesis of RNase-L whose function is to degrade single-stranded RNA molecules in the cytosol, specifically messenger RNAs (mRNAs).

Last, IFN-α and IFN-β induce the cell to increase p53 activity to limit viral spreading by promoting the intrinsic apoptotic pathway.

RNA Interference

Virus-infected cells can provide protection for themselves as well as neighboring cells by carrying out RNA interference (RNAi), also referred to as posttranscriptional gene silencing. RNAi is activated by the presence of dsRNA and, in turn, induces the degradation and loss of corresponding, complementary mRNA (Figure 2.13). The dsRNA is not normally found in eukaryotes or at least not found in very large amounts. The dsRNAs are usually present during an infection by certain RNA viruses that replicate through dsRNA intermediates. The intended function of RNAi is to inhibit the replication of viruses by degrading their mRNAs. RNAi is also required to prevent certain transposons from transposing (or integrating) into the genome of the eukaryotic cell. This function is not related to viral infection.

Double-stranded RNA, like the ones produce by a number of viruses, are recognized and bound by many small endonucleases called Dicer. Dicer cleaves the full-length viral dsRNA into smaller dsRNA called small interfering RNA (siRNA). Dicer also possesses helicase activity, which is used to melt the ds-siRNA into two ss-siRNA. These two ss-siRNA molecules are the passenger ss-siRNA strand and the guide ss-siRNA strand. The passenger ss-siRNA is the sense strand and does not have a useful function; therefore, it is destroyed in the cytosol. The guide ss-siRNA is the antisense strand that finds and anneals with its complementary viral mRNA to initiate its degradation. The complex of mRNA and guide siRNA is recognized and bound by

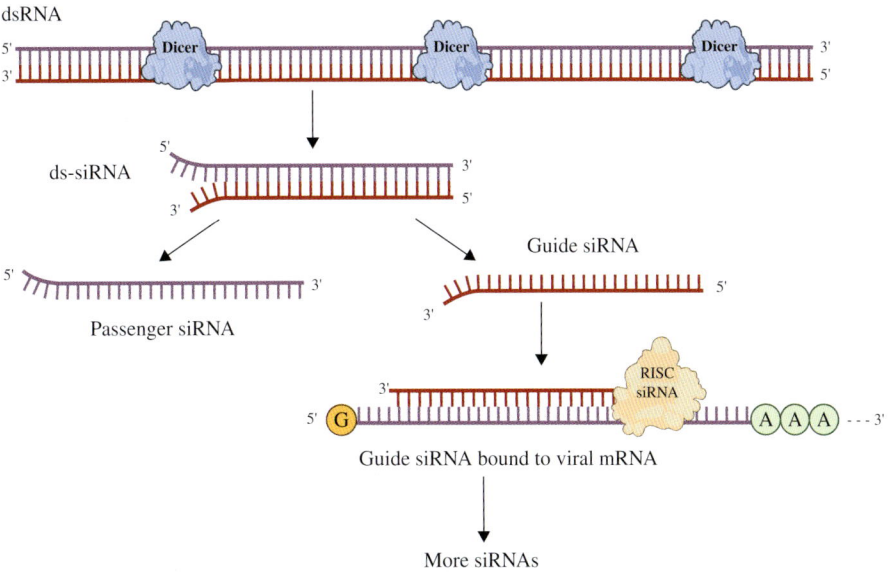

FIGURE 2.13 RNA interference for gene silencing. Viral dsRNA is cleaved by Dicer, producing siRNAs. The ds-siRNA is melted by Dicer into guide siRNA and passenger siRNA. Guide siRNA targets complementary viral mRNA. RISC binds and degrades the viral mRNA.
Credit: Figure partially created in BioRender.

an RNA-induced silencing complex (RISC), which possesses exonuclease activity to degrade viral mRNA, thus preventing further reproduction of the virus.

As a part of fighting against a viral infection in general, RNAi also possesses a mechanism to help neighboring cells that have not yet been infected fight against the viral infection more efficiently. The complex of mRNA and guide siRNA can also recruit an RNA-dependent RNA polymerase (RdRP), instead of RISC. RdRP catalyzes RNA synthesis using the guide siRNA as a primer and mRNA as the template. The double-stranded part is again recognized and bound by Dicer and the RNAi process continues. The process involving RdRP amplifies the production and accumulation of siRNAs inside the host cell. The excess siRNAs are secreted by the original viral-infected cell and internalized by neighboring cells. If this next cell is ever infected by a virus, viral mRNA degradation can be activated immediately by the binding of internalized siRNA without having to wait for viral dsRNA to be synthesized.

Bacterial Defense Against Bacteriophage

Restriction Digestion

In the early 1950s, Salvador Luria, Mary Human, Joe Bertani, and Jean Weigle were members of two teams of researchers who observed that some strains of bacteria were more resistant to specific bacteriophages than others. This observation led to the discovery of a groups of enzymes called restriction endonucleases (REases) or restriction enzymes by Werner Arber and Stuart Linn in the late 1960s. In 1970, Hamilton Smith purified a restriction endonuclease from *Haemophilus influenzae* serotype d, HindII (originally called endonuclease R). A year later, Dan Nathans and Kathleen Danna demonstrated the power of REases by digesting the SV40 genome into different size pieces of DNA by different combinations of REases. Drs. Arber, Smith, and Nathans shared the 1978 Nobel Prize in Physiology or Medicine for this discovery.

Restriction digestion is a defense mechanism used by bacteria against phage infection. The genome of bacteriophages is predominantly DNA. Restriction digestion is catalyzed by enzyme REases that cut phage DNA into many pieces to prevent virus replication. However, REases must be able to discriminate between invading DNA molecules and its own chromosomal and plasmid DNA. Host DNA is always methylated by the enzyme methyltransferase (MTase) to distinguish cellular DNA from the invading unmethylated viral DNA (Figure 2.14). MTases catalyze the transfer of the methyl group from *S*-adenosyl methionine to cytosine and adenine nitrogenous bases. Therefore, restriction-modification (R-M) systems are important defense mechanisms for bacterial cells that combine the activity of both enzymes.

The R-M systems are classified into four types referred to as type I-IV R-M system. Each type is classified based on its subunit composition: sequence recognition, cleavage position, cofactor requirements, and substrate specificity. REases are named according to the bacterial strain where they were initially isolated. For example, EcoRI was first isolated from *Escherichia coli* strain R. The "I" stands for the first ever enzyme extracted

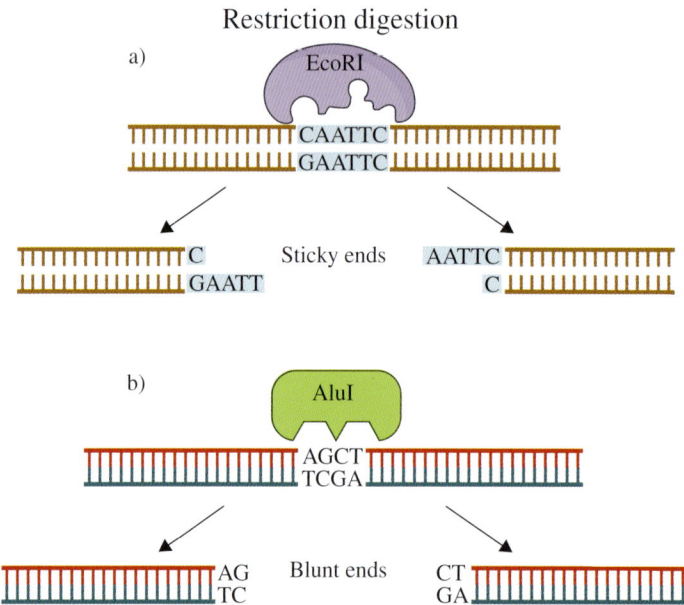

FIGURE 2.14 Restriction digestion. REase recognizes and binds the palindromic restriction site. (a) REase like EcoRI makes a staggered-end cut, leaving sticky ends. (b) REase like AluI makes a blunt-end cut, leaving blunt ends on DNA fragments. *Credit:* Figure partially created in BioRender.

from this particular strain. Thus, EcoRII was the second REase extracted from *Escherichia coli* strain R.

- Type I R-M system is a protein complex consisting of both REase and MTase. The DNA sequence or recognition sequence that is bound by these enzymes is hundreds to tens of thousands of base pairs away from the cleavage site. These enzymes require ATP hydrolysis and DNA translocation for their functions. Examples of type I R-M REase are EcoKI and EcoIq24I enzymes.

- Type II R-M system contains REase and MTase that do not form a complex with each other. The DNA recognition sequence of these enzymes is either within or near their cleavage site, generating consistent and predictable cleavage patterns. Examples of type II R-M REases are EcoRI, HindIII, and BamHI enzymes.

- The R-M system is a protein complex consisting of REase, MTase, and DNA-dependent nucleoside triphosphatase (NTPase) activities. The cleavage site of this complex is about 25 base pairs away from the recognition sequences. Cleavage requires the presence of two recognition sequences that are inversely oriented with respect to each other. Examples of type III REase are EcoP1I and EcoP15I enzymes.

- The type IV R-M system only recognizes and cleaves DNA sequences that contain methylated, hydroxymethylated, or glucosyl-hydroxymethylated. Examples of type IV REase are EcoKMcrBC and GmrSD. Examples of type IV MTase are MTase enzymes.

Currently, over 4000 REases have been identified that recognize more than 360 different recognition sequences from a wide variety of bacterial strains. Most REases recognize specific recognition sequences that are either long or short palindromic DNA sequences. Some REases, however, recognize asymmetric recognition sequences called nonpalindromic sequences. Once bound to these sites, REases cleave the dsDNA by making either a blunt-end cut or staggered cut. As the name indicates, a blunt-end cut produces a flat or blunt end at each end of the site of cleavage. A staggered cut produces two DNA ends with overhanging single strands, also known as sticky ends. These single-stranded overhangs or sticky ends can base pair with complementary single-stranded sequences found at the end of any other DNA fragments.

Restriction sites can be 4-bp (base pair), 6-bp, or 8-bp long. Of course, shorter recognition sequences have a higher probability of occurrence in any given viral DNA genome. For example, within a 5000-kilobase pair genome, a particular 4-bp sequence may be present 19,531 times, whereas within the same size genome a particular 8-bp sequence may be present only 76 times.

The following are examples of REases, their correlated recognition sequences, and site of cleavage (indicated by "/"): AcII recognizes AA/CGTT; HindIII recognizes A/AGCTT; AluI recognizes AG/CT; AscI recognizes GG/CGCGCC; EcoRI recognizes G/AATTC; BamHI recognizes G/GATCC; and SrfI recognizes GCCC/GGGC.

CRISPR

Bacterial cells have an intracellular mechanism to defend against the bacteriophage called clustered regularly interspaced short palindromic repeats (CRISPR). In 1987, while analyzing a gene responsible for alkaline phosphatase conversion in *Escherichia coli*, Yoshizumi Ishino discovered these unusual repetitive palindromic DNA sequences interrupted by noncoding spacers. These noncoding spacers were later found to be short viral DNA fragment inserts acquired during previous phage infections. Yoshizumi and his research team did not understand the biological function of these repeat sequences at the time. In 1993, Francisco Mojica characterized these repeat sequences and found that they shared a common set of features that are now known as the hall-marks of CRISPR sequences. He and Ruud Jansen coined the term CRISPR in 2002. Dr. Mojica further reported, in 2005, that these sequences act as a defense mechanism against bacteriophages.

The CRISPR system turns out to be an adaptive immune system used by bacterial cells to destroy invading phage DNA. This system targets the invading virus by comparing phage DNA to homologous sequences stored in their genome's CRISPR array, then degrades and destroys the phage DNA. The CRISPR sequences themselves are DNA fragments from bacteriophages that have previously infected this bacterial cell, which then are to identify and destroy phages with matching DNA sequences during future infections.

In 2005, Bolotin et al. reported an unusual CRISPR locus that encodes for the bacterial protein CRISPR-associated nuclease (Cas), usually located adjacent to CRISPR loci. In 2012, Emmanuelle Charpentier and Jennifer Doudna developed CRISPR into a gene-editing technique called CRISPR-Cas9. They further discovered that this system could be used to edit any desired DNA by just providing the right template or DNA fragment. CRISPR-Cas9 is a simple and cheap, yet precise, gene-editing technique that has a wide range of applications, including performing basic biological research, developing biotechnological products, and treating diseases. Gene or genome editing is a mechanism used to modify DNA sequences at specific locations in an organism's genome. Modification by the insertion, deletion, or substitution of desired DNA fragments results in either the inactivation of target genes, introduction of novel genetic traits, or repair of pathogenic gene mutations. The CRISPR-Cas9 technology has been applied to some basic biological research, biotechnological product development, and disease treatment, including studying gene function, understanding the pathogenesis of hereditary diseases, improving gene therapy, and breeding new agricultural crop varieties. In 2020, Drs. Charpentier and Doudna won the Nobel Prize in Chemistry for this work.

The CRISPR defense system protects bacteria from repeated viral attacks by capturing small fragments of previous infecting phage DNA called spacers. Spacers are then stored or integrated into the cellular chromosome or plasmid and are arranged in a palindromic fashion. The accumulation spacers interspersed throughout the host genome produce a CRISPR array, many of which are adjacent to the Cas gene. The number of spacers in a CRISPR array can range from 1 to several hundred, depending on the species. The CRISPR array functions as molecular memories for the bacterial cell when infected by the same bacteriophage.

The CRISPR-Cas9 system progresses through three basic steps: (1) adaptation, (2) crRNA (CRISPR RNA) biogenesis, and (3) target interference.

1. The adaptation step is also known as spacer acquisition. The bacterial host cell captures short DNA fragments belonging to the invading bacteriophage called protospacers. These protospacers are then integrated into the existing CRISPR array in the cellular genome to create a new set of palindromic sequences called spacers. This step allows the host cell to remember the genetic material of this particular phage. Think of this step as the production of memory B and

T lymphocytes during a primary viral infection in a human host or can be viewed as a "vaccination."

2. The crRNA biogenesis step involves the transcription of the repeat-spacer elements within the CRISPR array into precursor CRISPR RNAs (pre-crRNA), followed by cleavage yielding crRNAs. The crRNAs are a collection of small RNA fragments; each is complementary to a spacer acquired from a particular vaccination or infection event.

3. The function of the target interference step is to silence foreign nucleic acid sequences. This step occurs when the crRNA recognizes and base pairs with its complementary nucleic acid sequence from the current invading virus. The bacterial host then uses the Cas protein (ie, Cas9) to disable the invading virus by degrading its phage DNA.

Subversion Mechanisms of Pathogens

Viruses also have specific mechanisms to evade the body's immune system. Many of these mechanisms target the complement cascade. A number of CRs serve as entry receptors for various viruses. For instance, Epstein-Barr virus uses CR2, and possibly CR1, for its entry into the human B lymphocyte. Complement regulators, like decay-accelerating vactor (DAF), membrane cofactor protein (MCP), and protectin, have also been shown to act as a virus receptor for entry into host cells by members of the families Poxviridae and Orthomyxoviridae. These complement regulators are among the many factors that can protect host cells from being attacked and destroyed by the host body's own complement cascades. Binding to receptors on the surface of host cells induces receptor-mediated endocytosis of the virus. Binding of members of the families Flaviviridae and Retroviridae to soluble complement regulators, like factor H and vitronectin, has been shown to shield viruses from complement cascade attack.

Enveloped viruses acquire their envelope from the host cell membrane where they exited. This means that there are many host membrane components found in viral envelopes (ie, sialic acid). The soluble factor H complement regulator is attracted to sialic acid in the viral envelope and leads to evasion from the complement cascade. Host plasma membranes contain a high amount of membrane-bound complement regulators. This means that by acquiring their envelope from the targeted host cell, viral particles also exhibit these regulatory agents to thwart activities of the complement cascade. For example, protectin is found in the HIV envelope, which provides protection against the formation of MACs.

Further Readings

Agrawal P, Nawadkar R, Ojha H, Kumar J, Sahu A. Complement evasion strategies of viruses: An overview. *Front Microbiol.* 2017;8:1–19.

Aoshi T, Koyama S, Kobiyama K, Akira S, Ishii KJ. Innate and adaptive immune responses to viral infection and vaccination. *Curr Opin Virol.* 2011;1:226–232.

Asmamaw M, Zawdie B. Mechanism and applications of CRISPR/Cas-9-mediated genome editing. *Biologics.* 2021;15:353–361.

Bolotin A, Quinquis B, Sorokin A, Ehrlich S. Clustered regularly interspaced short palindrome repeats (CRISPRs) have spacers of extrachromosomal origin. *Microbiology.* 2005;151:2551–2561.

Cook-Mills JM, Marchese ME, Abdala-Valencia H. Vascular cell adhesion molecule-1 expression and signaling during disease: Regulation by reactive oxygen species and antioxidants. *Antioxid Redox Signal*. 2011;15(6):1607–1638.

Cornick S, Tawiah A, Chadee K. Roles and regulation of the mucus barrier in the gut. *Tissue Barriers*. 2015;3(1–2):1–15.

Elois M, Da Silva R, Pilati G, Rodriguez-Lazaro D, Fongaro G. Bacteriophages as biotechnological tools. *Viruses*. 2023;15(2):349.

Evans SS, Repasky EA, Fisher DT. Fever and the thermal regulation of immunity: The immune system feels the heat. *Nat Rev Immunol*. 2015;15(6):335–349.

Fahey E, Doyle SL. Cytokines and soluble mediators in immunity. *Front Immunol*. 2019;10:1–15.

Holmgren J, Czerkinsky C. Mucosal immunity and vaccines. *Nat Med*. 2005;11(4):545–553.

Hunt BJ. Endothelial cell activation. *BMJ*. 1998;316(7141):1328–1329.

Jinek M, Chylinski K, Fonfara I, Hauer M, Doudna J, Charpentier E. A programmable dual RNA-guided DNA endonuclease in adaptive bacterial immunity. *Science*. 2012;337(6096):816–821.

Lee GR. The balance of Th17 versus Treg cells in autoimmunity. *Int J Mol Sci*. 2018;19(3):1–14.

Linden SK, Sutton P, Karlsson NG, Korolik V, McGuckin MA. Mucins in the mucosal barrier to infection. *Mucosal Immunol*. 2008;1:183–197.

Loenen W, Dryden D, Raleigh E, Wilson G, Murray N. Highlights of the DNA cutters: A short history of the restriction enzymes. *Nucleic Acids Res*. 2014;42(1):3–19.

Morrison SF, Madden CJ, Tuponne D. Central control of brown adipose tissue thermogenesis. Front Endocrinol (Lausanne). 2012; 3(5): 1–19.

Sato S, Kiyono H. The mucosal immune system of the respiratory tract. *Curr Opin in Virol*. 2012;2(3):225–232.

Seth RB, Sun L, Chen ZJ. Antiviral innate immunity pathways. *Cell Res*. 2006;16:141–147.

Shmakov S, Utkina I, Wolf Y, Makarova K, Severinov K, Koonin E. CRISPR arrays away form *cas* genes. *CRISPR J*. 2020;3(6):535–549.

Smith KA. Louis Pasteur, the father of immnunology? *Front Immunol*. 2012;3:68.

Vasu K, Nagaraja V. Diverse functions of restriction-modification systems in addition to cellular defense. *Microbiol Mol Biol Rev*. 2013;77(1):53–72.

CHAPTER 3

Plus-Strand Single-Stranded RNA

Poliovirus

Historical Perspective

Poliovirus is known to be the causative agent of paralytic poliomyelitis and was discovered by Dr. Karl Landsteiner, a future Nobel Prize winner, in 1908. In 1931, two immunologists, Drs. Frank Burnet and Jean Macnamara, identified antibodies specific to the three different serotypes of this virus in the blood of infected individuals. In 1953, Dr. Jonas Salk developed an effective killed-virus vaccine against poliovirus at the University of Pittsburgh, known as the inactivated (or intramuscularis) poliovirus vaccine (IPV).

Evidence indicated that the effect of poliovirus infection existed as early as 1570 BCE in ancient Egypt. The mummy of the pharaoh Siptah shows telling deformity of one of his legs and foot. In the late 1700s, Dr. Michael Underwood in London described the first documented case of this paralytic disorder of infants in a medical textbook. However, it wasn't until the 19th century that known epidemics of this infection along with the ensuing disorder were observed in North America and Europe. The first outbreak appeared in Oslo, Norway, in 1868, where 14 cases were recorded, followed by another in Sweden in 1881, where 13 cases were recorded. The United States experienced its first outbreak in Vermont in 1894, with 132 cases recognized. The highest number of incidences recorded in the United States was 21,000 cases in 1952.

The Salk vaccine underwent a massive study trial from 1954 to 1955 and involved 1.8 million children across the United States. This trial, named the Francis Field Trial after Dr. Thomas Francis, who directed this program, was declared a success, and this vaccine was distributed to 450 million children. In 1961, Dr. Albert Sabin developed an oral poliovirus vaccine (OPV), known as the Sabin vaccine. The Sabin vaccine was approved in 1963 and was widely used in the United States and most other countries. In the United States, due to mass immunization campaigns, the average number of cases per year fell to 12 in the early 1970s compared to prevaccine levels.

While developed countries in North America and Europe had the resources to implement mass immunization of their citizens to tackle this health crisis, a global immunization campaign against poliovirus was initiated by the World Health Organization (WHO) in 1974 in less-developed countries. By 1989, roughly 67% of children were immunized through this very effective WHO campaign.

The effectiveness of the immunization campaign was evident by the beginning of the 20th century, when new cases of paralytic poliomyelitis worldwide dropped from hundreds of thousands of cases to 1000–2000. A number of countries were long declared polio-free. Even though polio was announced to be eliminated from the United States in 1979, a few cases have been identified occasionally. The last reported case occurred in the fall of 2022 when an unvaccinated individual became paralyzed due to a poliovirus infection; however, the source of transmission is unknown. Unfortunately, there are still countries that have new cases arising every year. Since this disease is linked to poor hygiene and unsanitary conditions, unfortunately countries that are plagued by poverty continue to see new cases. These are the same countries that do not receive proper distribution and storage of vaccines.

Classification and Structure of Poliovirus

Poliovirus belongs to the Picornaviridae family. This family includes five genera: *Enteroviruses* (eg, poliovirus and coxsackie viruses), *Rhinoviruses* (eg, common cold virus), *Hepatoviruses* (eg, hepatitis A virus), *Cardioviruses* (eg, encephalomyocarditis virus), and *Aphthoviruses* (eg, foot-and-mouth disease virus). There are three known poliovirus serotypes: PV1, PV2, and PV3.

Other related families of virus include Flaviviridae (eg, West Nile virus and hepatitis C virus), Togaviridae (eg, Semliki Forest virus and Ross River virus), and Coronaviridae (severe acute respiratory syndrome coronavirus 2 [SARS-CoV-2] and Middle East respiratory syndrome [MERS]).

The structure of the virus is a non-enveloped virion 18–30 nm in diameter (Figure 3.1). The single plus (sense)-strand RNA genome of 7440 bases is encased in an icosahedral capsid. Sense strand means that the viral genome has a similar nucleotide sequence as the mRNA (messenger RNA), which can sometimes be used directly as the template for translation. There's only a single open reading frame (ORF) encoding a polyprotein of 2209 amino acids, which is further cleaved into 11 viral translational products and 4 structural and 7 nonstructural proteins. A polyprotein, not to be confused with a polypeptide, is a stretch of amino acid sequence that contains a chain of covalently linked smaller proteins. Remember that a polypeptide is folded into a single protein in the tertiary structure. Let's stop briefly to appreciate the efficiency of this genome. Think about it: This is a large number of proteins encoded by a single, very small genome; therefore, the ORF is extremely compact and does not contain extraneous nucleotide bases. Even then, how can such a small genome encode for so many proteins? Some viruses, including poliovirus, have a clever strategy for translation processing, which is discussed further in the Infecting the Host Susceptible Cell, Viral Particle Replication, and Tissue Damage section of this chapter.

We can compare the poliovirus genome to the other virus that is discussed in this chapter, SARS-CoV-2, which also possesses 1 sense-strand RNA virus. SARS-CoV-2 is one of the largest RNA genomes compared to the poliovirus genome, which is one of the smallest. However, both viruses use the same strategy of synthesizing polyproteins that are eventually cleaved into smaller ones. When multiple entities evolve the same method to carry out a particular function, you know that this mechanism is a good one.

Poliovirus nucleocapsid structure

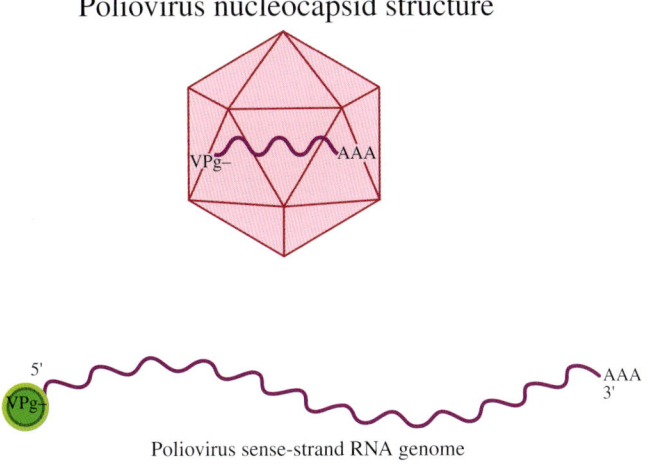

Poliovirus sense-strand RNA genome

FIGURE 3.1   Poliovirus structure and genome. Poliovirus is a non-enveloped virion. Its linear sense-strand RNA genome is encapsidated in an icosahedral capsid. *Credit:* Figure partially created in BioRender.

Host Range, Transmission, Tropism, and Susceptibility

Humans are the only known natural host for poliovirus, which has tropism for mainly epithelial cells of the digestive and respiratory tract, specifically microfold cells. It also has tropism for neurons in the central nervous system, especially the lower motor neurons in a person's spinal cord; however, neural dissemination is a rare event. Additionally, monocytes, macrophages, B lymphocytes, and follicular dendritic cells are susceptible cells.

The route of exit of poliovirus is through the elimination of host waste and is transmitted through the fecal-oral route via contaminated food and water (Figure 3.2). Here lies the problem with having unsanitary conditions aiding and propagating the transmission of this virus. In the external environment, these viral particles are resilient and survive quite well due to the compact and fortified structure of its capsid. Poliovirus can survive outside the body in soil, sewage, and infected water for up to 2 months. Poliovirus is extremely contagious and spreads by direct contact, usually entering the body through the mouth due to feces-contaminated food and water. It can sometimes spread via droplets that are expelled when the infected individual sneezes, coughs, or simply talks. The airborne droplets are then inhaled through the respiratory tract or land in one's mouth. Objects, like the wall or a table or toys, that are contaminated with the infected droplets are referred to as fomites and can act as a vector for virus transmission when the items are touched. They are quite resistant to common disinfectants but can be destroyed when exposed to a temperature of 50°C or higher.

The poliovirus can enter the oral or nasal cavity, then multiply locally in secondary lymphoid organs and tissues in the mucous membrane. The majority of the time the route of entry is through the mouth, mainly due to consuming food or water contaminated with feces. Once in the mouth, viral particles naturally make their way into the digestive tract as the result of host physiological activities. In the digestive system, this

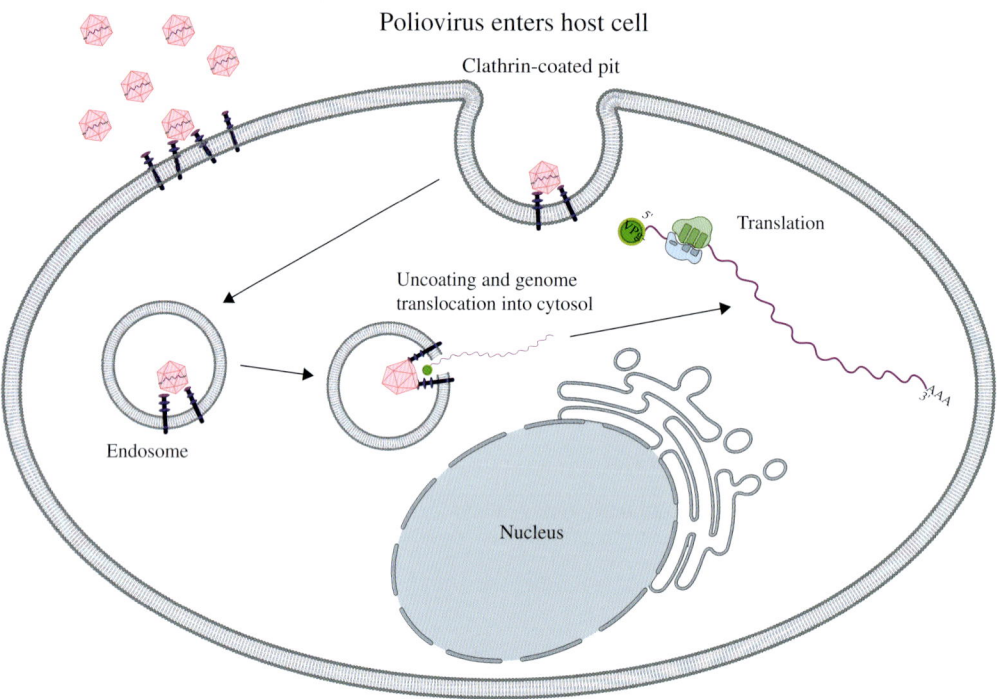

FIGURE 3.2 Poliovirus enters its host cell. Poliovirus is engulfed by its host cell via clathrin-dependent receptor-mediated endocytosis. Once inside the cell, the virion is localized in the endosome. The acidic environment in the endosomal lumen induces uncoating, leading to the release of the viral genome into the cytosol to be used immediately as a template for protein synthesis. *Credit:* Figure partially created in BioRender.

virus multiplies in the Peyer patches of the small intestine. The route of entry may also be through the nasal cavity, in which case poliovirus multiplies in tonsils and lymph nodes of the neck.

Susceptible host cells of poliovirus possess the receptor CD155 on their surface. CD155 is an adhesion protein belonging to the nectin/nectin-like family and immuno-globulin superfamily. This protein is exhibited on the surface of epithelial, endothelial, immune, and nerve cells and is involved in a number of physiological processes, such as cell polarity, cell proliferation and modulation of immune responses, cell differentia-tion, and cell adhesion and migration. It's worth noting the CD155 is expressed in a variety of cell types, but not all of them are known to be susceptible to poliovirus infec-tion; therefore, other components are required to determine tropism. Organ- or tissue-specific viral replication also correlates with the localization and types of translational modification required for viral protein synthesis.

Infecting the Host Susceptible Cell, Viral Particle Replication, and Tissue Damage

CD155 mediates the poliovirus binding inserting itself into the surface depression of the virus known as the *canyon*. The canyon surrounds each of the 12 axes of the icosahe-dral structure of the capsid. Entry of the virus occurs through clathrin-dependent, receptor-mediated endocytosis. Once inside the cell, the lower pH and acidic proteases in the lumen of the endosome induce uncoating of the virus. The viral genome is then translocated into the cytosol, where it is bound directly by ribosomes to carry out translation initiation. There is a protein called VPg attached to the 5′ end of the genome and the 3′ end is polyadenylated. Since the canonical m7G cap is absent, cap-independent or internal ribosome entry site (IRES)–mediated translation initiation is used for protein synthesis. The plus-strand RNA genome is used as the template by the host cell's machinery to synthesize a polyprotein that later undergoes proteolytic processing to produce multiple much smaller viral proteins (Figure 3.3). During cap-independent translation initiation, eukaryotic initiation factor (eIF) 4G binds to the IRES, which is located in the 5′ untranslated terminal region (UTR) of the plus-strand RNA, while poly(A)-binding proteins (PABPs) bind to the 3′-UTR as well as the poly(A) tail. In an effort to shut off cap-dependent translation of host cell proteins so that the machinery can be dedicated to making viral proteins, the viral 2A protease shuts down cellular cap-dependent translation initiation on the host cell's mRNA by digesting eIF-4F, a complex required for cap-dependent translation. Another viral enzyme, 3C^{pro}, has been shown to inhibit cellular transcription as well.

Viral proteins that are synthesized include 4 capsid proteins (VP1, VP2, VP3, VP4). The 2A protease shuts down cellular cap-dependent translation initiation, and 3CD protease performs cleavage of the polyprotein into smaller proteins VPg, and most importantly. 3D polymerase (3D^{pol}). The 3D^{pol} is the specific RNA-dependent RNA polymerase expressed by poliovirus. 3D^{pol} catalyzes the synthesis of more plus-strand RNA molecules to be packaged ultimately as the Polioviral genome. All these activities occur in

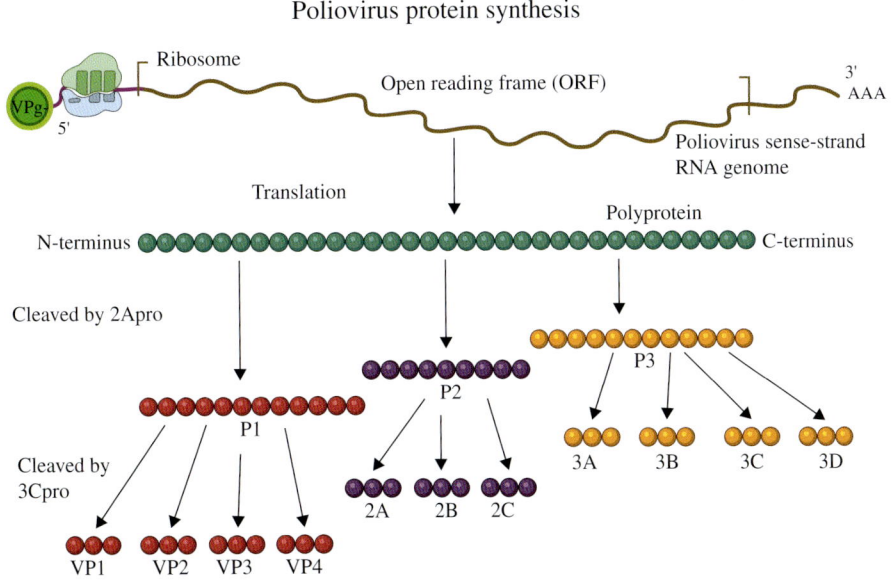

FIGURE 3.3 Polioviral protein synthesis. Translation from the single ORF produces 1 polyprotein. This polyprotein is first cleaved by 2A^{pro} into three smaller peptide fragments: P1, P2, and P3. Each fragment is further cleaved by 3CD^{pro} into a total of 11 functional viral proteins. *Credit:* Figure partially created in BioRender.

the cytosol of the host cell. The binding of 3D^{pol} also requires the interaction between the 5′ and 3′ ends of the viral plus-strand RNA to initiate the synthesis of the minus-strand RNA. VPg-pUpU (or VPg-oligonucleotide) is used as a primer for viral genome synthesis. The initial minus-strand RNA that is produced is used, in turn, as the template to synthesize many more plus-strand RNA to be assembled into virions (or viral particles). An important component of this minus-strand RNA template is that it contains a short polyuridine (poly-U) stretch at the 5′ end. The poly-U encodes for the poly(A) tail at the 3′ end of the plus-strand RNA viral genome. Assembly of virions occurs on the cytosolic side of vesicular membranes that are undergoing vesicular transport. Within about 4 hours of infection of the host cell, up to roughly 1 million virions can be produced. This massive number of virions in the cytosol that is produced so quickly causes the host cell to lyse, thus releasing the viral particles into host interstitial tissue for dissemination (Figure 3.4).

Dissemination in the Host Body

Once the initial primary poliovirus infection is established in the mucous membrane, viral particles that are released can spread to the bloodstream, known as hematogenous dissemination, then into the central nervous system (CNS), known as neural dissemination, targeting the lower motor neurons (Figure 3.5). There are two routes for neural

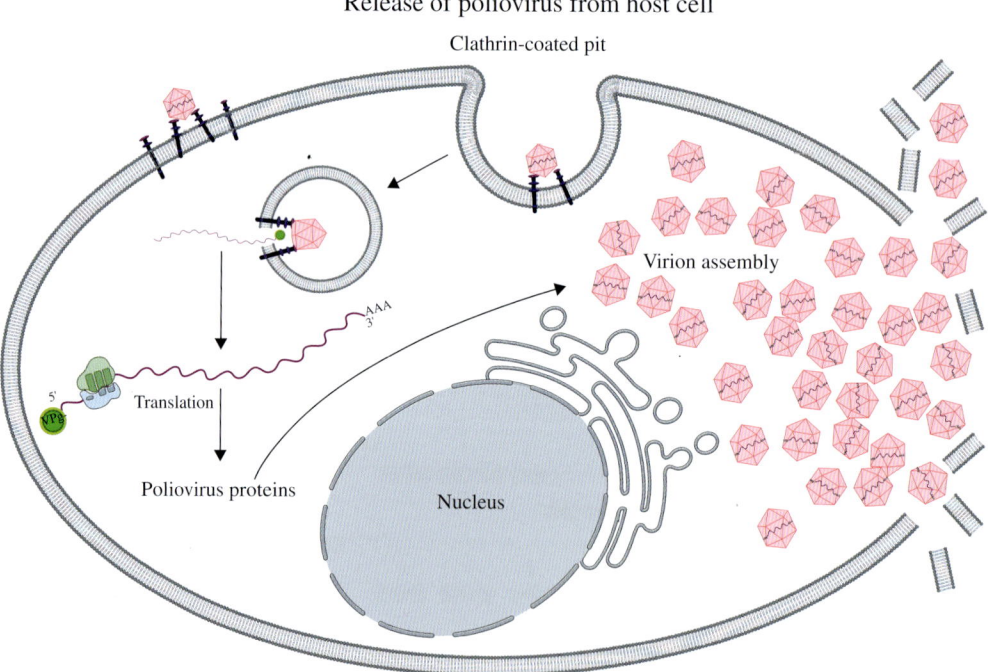

FIGURE 3.4 Release of poliovirus from host cell. Viral genome replication and protein synthesis occur in the host cell cytosol. Assembly also occurs in the cytosol. Virion production occurs at such a high rate and large amount that they cause the host cell to burst inducing cell lysis. *Credit:* Figure partially created in BioRender.

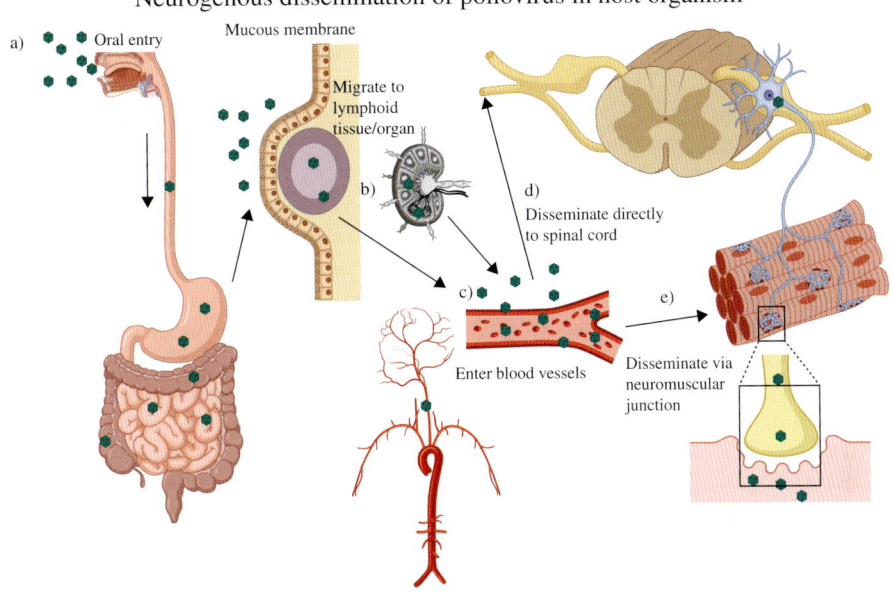

Neurogenous dissemination of poliovirus in host organism

FIGURE 3.5 Dissemination of poliovirus in host organism. (a) Poliovirus enters through oral cavity into the digestive tract. (b) After replication and release from the initial host cells, viral particles disseminate to lymphoid nodules and/or nodes. (c) Intravasation of viral particles into the bloodstream results in active viremia. Viral particles extravasate into the interstitial tissue, then disseminate either (d) directly to the spinal cord or (e) via the neuromuscular junction. *Credit:* Figure partially created in BioRender.

dissemination: (1) migrating from the bloodstream across the blood-brain barrier into the CNS and (2) invading lower motor neurons of the peripheral nervous system via the neuromuscular junction, then into the CNS. After neural dissemination, viral replication leads to the destruction of lower motor neurons located in the spinal cord. These nerve cells cannot regenerate; therefore, the absence of functional lower motor neurons prevents the infected individual from being able to innervate skeletal muscle cell. The inability to activate skeletal muscle contraction leads to lethargy, which in turn prevents muscle development resulting in underdeveloped muscles and a condition called acute flaccid paralysis (AFP).

Pathogenesis and Clinical Manifestation

Upward of 95% of individuals exposed to poliovirus are asymptomatic. During this time, viral particles would be shed and can be detected in stool and throat swabs. If hematogenous dissemination occurs, the active viremia can induce mild symptoms, such as sore throat, fatigue, muscle ache, nausea, headache, stomach pain, and fever. These symptoms usually dissipate after 2 to 5 days, a period referred to as the prodrome, thanks to our amazing innate immune responses. These symptoms should recede within a week, and complete recovery would ensue a few days later. However, in approximately 1% of cases, the virus may enter the CNS, causing aseptic meningitis, which is a common inflammatory disorder of the meninges.

The meninges are the protective layers around the brain and spinal cord. Symptoms of meningitis include severe muscle spasm of the neck, back, and lower limbs. The most severe outcome, which affects less than 1% of infected individuals, is AFP. AFP is the result of the destruction of lower motor neurons. Symptoms include excruciating incidences of pain in the back and lower limbs, followed by the appearance of asymmetrical paralysis of the limbs in children. AFP that is caused by a polioviral infection is known as paralytic poliomyelitis.

Spinal paralytic poliomyelitis is the most common cause of back paralysis, which involves muscle atrophy. Leg muscles are the most common tissue affected, leading to lower limb paralysis. More extensive paralysis can affect arm muscles and cause quadriplegia, which involves muscles of the thorax and abdomen in the most severe cases.

Another type of muscle atrophy is bulbar poliomyelitis, which is manifested by the infection and destruction of the upper motor neurons of the brain stem. This is the worst form of this disorder. Individuals with bulbar poliomyelitis have reduced breathing capacity, difficulty swallowing, and difficulty speaking.

Recovery from paralysis may occur and be complete in some patients; however, some experience post-polio syndrome (PPS), where the loss of motor functions persists beyond 12 months or so due to continued deterioration of motor neurons leading to progressive muscular weakness, joint deterioration, and increasing skeletal deformities, resulting in permanent disability. In fact, PPS may occur 2 or 3 decades after the initial prodromal period. Without respiratory support, fatality can result in extreme cases due to respiratory paralysis leading to asphyxiation. Muscular weakness, joint deterioration, and increasing skeletal deformities are symptoms of PPS.

Diagnosis, Treatment, and Prevention

Fatigue, even after minimal physical activity, and the presentation of sudden onset of limb weakness may be signs of this disorder. As discussed, flu-like, prodromal symptoms such as sore throat, fatigue, muscle ache, nausea, headache, stomach pain, and fever may also be an indication. Sudden onset of limb weakness is definitely a sign of the progression of this disorder. There are a number of laboratory tools that can be used for diagnosis. Diagnostic tools include sample collection of stools, blood, urine, and cerebrospinal fluid and throat samples by swabbing. Of these samples, poliovirus is most likely found in stool specimens, and polymerase chain reaction (PCR) is usually used for detection. Magnetic resonance imaging may be used to identify possible problems of the spinal cord.

In terms of treatment, unfortunately there is not yet a cure for paralytic poliomyelitis or a specific treatment. However, if implemented in the early stage of the disorder, management through physical and occupational therapy can provide release of joint contracture, limb strengthening, and reestablishment of muscle balance around a joint to prevent the progression of deformities. Replacement of joints and affected muscles through muscle transplant can be achieved through surgery. Surgery may also be used to reconstruct tendons and ligaments. There are devices that can assist individuals who have bulbar poliomyelitis breathe. The first of such devices was the iron lung, which is a negative-pressure ventilator first developed in 1928. Due to its enormous size, cost, and other issues, this machine is now obsolete. Modern devices are much smaller, cost relatively less, and are technically simpler to use.

Prevention of this infection begins with proper personal hygiene. This is a highly communicable disease, so a sanitary environment, which includes proper disposable of human waste, is definitely a must. Polio vaccine is, without a doubt, credited for the elimination of poliomyelitis from the United States and elsewhere. There are two types of vaccine that can prevent polio: IPV, introduced as an injection in the leg or arm, depending on the patient's age, and OPV, which is also known as the Sabin vaccine and, of course, is introduced orally. Only IPV has been used in the United States since 2000. The Centers for Disease Control and Prevention (CDC) recommend that children receive four scheduled doses of vaccine against poliovirus to protect their bodies and fight the virus. The vaccination schedule is at 2 months old, 4 months, 6–18 months, and 4–6 years old. Data showed that more than 99% of children who received the recommended doses of vaccine are protected from poliomyelitis.

Let's look at the comparison between these two most effective (really the only two known) defenses against poliomyelitis: IPV and OPV.

1. IPV is an inactivated or killed vaccine. The vaccine is made from viral particles that are inactivated with formaldehyde. Once the vaccine is introduced into the body through intramuscular injection, viral components are engulfed by phagocytic anticoagulant activated protein C (APC) through phagocytosis, processed, then displayed by major histocompatibility complex (MHC) class II molecule. Phagocytic APC is a natural anticoagulant whose functions include preventing blood clotting, reducing inflammation to reduce tissue damage, and stimulating neurogenesis, angiogenesis, and wound healing. Plasma cells synthesize both immunoglobulin (Ig) G and IgM. This vaccine is effective against active viremia and prevents mainly neural dissemination. There are almost no side effects except for redness and swelling at the site of injection, typical symptoms of the inflammatory response. This vaccine needs to be stored at 2–8°C and should be protected from light. Downsides of this vaccine include the fact that even though memory B cells are produced, no memory T cells are created. Viral components are displayed by MHC class II but not MHC class I since entrance of the virus is via phagocytosis, not endocytosis. This mechanism also means that there is no CD8 T-lymphocyte response; therefore, no CTLs (cytotoxic T lymphocytes) are produced. CTLs plays a major role in the adaptive immune response by seeking out and destroying abnormal cells such as cancer and pathogen-infected cells. Since no IgA is upregulated, IPV confers much weaker mucosal immunity than OPV; thus, viral particles that are ingested or inhaled have a higher probability of invading across the epithelial barrier and causing an infection.

2. Sabin vaccine (OPV) is a live-attenuated vaccine. This is a weakened, less-virulent version of the virus, although it can still replicate but does not cause illness. This vaccine creates a stronger and longer lasting immunological effects than IPV because the content of the vaccine is so similar to the virulent form of the virus. Once the vaccine is introduced into the body orally as drops, viral particles produce a local immune response and can infect susceptible cells, be processed, then be displayed by MHC class I. This vaccine can elicit CD8 T-lymphocyte response; therefore, CTLs and memory T lymphocytes are created. By the way, OPV, more so than IPV, is the main reason that poliomyelitis had been eradicated in the United States. Plasma cells synthesize and secrete IgA, which provides the body with strong mucosal immunity, especially in the gut, thus reducing the chance of an infection into the tissues. This vaccine needs to be stored at 2–8°C and should be protected from light.

The major downside of this vaccine is the fact that it is so similar to a live, virulent virus. Immunized individuals can experience the same symptoms of poliomyelitis up to a month after vaccination, causing a condition called vaccine-associated paralytic poliomyelitis (VAPP); however, this is an extremely rare problem. Because of the high rate of mutation during RNA genome replication, there is a high chance of reversion of live-attenuated virus back to wild type, called circulating vaccine-derived poliovirus (cVDPV). In this case, this vaccine can lead to paralytic poliomyelitis and not just VAPP. Individuals with immunodeficiency should never be given this vaccine.

Current Status

Even though paralytic poliomyelitis caused by wild-type poliovirus was eliminated from the United States in 1979, sporadic cases have been identified in this country. The last case was reported in July 2022, when an unvaccinated adult male was confirmed to be paralyzed as the result of a poliovirus infection in Rockland County, New York. Recently, poliovirus has actually been detected in sample wastewater taken from the sewage in multiple New York locations, which indicates that there is a local circulation of this virus. According to the CDC, a total of 100 positive samples of concern were found in New York State between March 9 and October 11 of 2022, including Sullivan, Rockland, Orange, and Nassau counties and New York City.

Currently, only IPV has been used in the United States since 2000, which confers 99%–100% protection from paralytic poliomyelitis. Some countries continue to use the OPV due to advantages, including low cost, ease of use, and high efficacy in halting an outbreak. Unfortunately, the live-attenuated virus can regain neurovirulence and circulate among the non-immunized populations. It's been proposed that this is what occurred in the July 2022 case in New York.

According to the European Center for Disease Prevention and Control, worldwide reported incidence of paralytic poliomyelitis decreased by more than 99% from an approximate 350,000 cases that occurred in more than 125 countries to 140 cases reported in only 2 countries in 2020. That's an impressive turnaround and cooperative success. The WHO certified the Americas polio free in 1994, Western Pacific Region in 2000, European Region in 2002, South-East Asia Region in 2014, and African Region in 2020. To date, the only reported outbreaks (1 case is now considered to be an outbreak) are proposed to arise from cVDPV originating from OPV and transmitted to non-immunized individuals. In 2016 cVDPV cases from 3 countries were reported, 378 cases from 20 countries were reported in 2019, and 106 cases from 27 countries were reported in 2020.

Scientific Significance and Discoveries

The study of poliovirus is central to the discoveries of a number of scientific advances, including the following:

1. The presence of uncapped mammalian mRNAs was discovered, which led to the identification of the IRES and the mechanism for cap-independent translation initiation.

2. The presence of a 5′-terminal genome-linked protein (VPg) and the 3′-terminal poly(A) sequence was identified.

3. The existence of polyproteins was discovered, which were later identified in a number of other RNA viruses.

4. The polioviral RNA molecule was the first animal virus RNA genome to be sequenced. Its structure, along with the human rhinovirus 14, was the first animal virus structure to be determined by x-ray crystallography.

SARS-COV-2

Historical Perspective

The CDC reported an appearance of a new strain of coronavirus in China in late 2019. The timeline of how this pandemic unfolded, its investigation, and responses to address this worldwide problem are outlined next.

On December 12, 2019, a number of individuals in the city of Wuhan, in China's Hubei Province, began to experience pneumonia-like symptoms, including shortness of breath and fever. However, standard known treatments were ineffective to reduce these symptoms. It wasn't until December 31, 2019, that the WHO Country Office in China was informed of this pneumonia outbreak of unknown etiology.

Since all initial reported cases appeared to be linked to the Huanan Seafood Wholesale Market, authorities closed this facility on January 1, 2020, to prevent additional spread of this pneumonia of unknown origin in fear of the similar outbreak of SARS-CoV-1 in late 2002. The SARS-CoV-1 (or SARS-CoV) outbreak lasted from 2002 to 2004 where 8,000 individuals were infected, resulting in 774 deaths globally. By January 5, 2020, the National Center for Immunization and Respiratory Diseases of the CDC was activated and began to investigate this novel disease by sequencing the genome of this atypical pneumonia virus, called Wuhan-Hu-1, and determined that the causative pathogenic agent was indeed a novel strain of coronavirus.

On January 7, 2020, the CDC begin their response against this novel coronavirus using the preparedness plans made for the Middle East respiratory syndrome coronavirus (MERS-CoV) in 2012. By January 11, 2020, the first death due to this disease was reported. The Thailand Ministry of Public Health confirmed the first case of the SARS-CoV-2 virus outside of China on January 13, 2020, followed by reports from the Japanese Ministry of Health, Labor, and Welfare on January 15, 2020. Evidence indicated possible human-to-human transmission of this virus, which was confirmed by the Chinese government on January 21, 2020.

It wasn't until January 17, 2020, that the CDC begin screening passengers on flights from Wuhan China to the United States for symptoms of this infection, called 2019 novel coronavirus, at that time. By January 19, 2020, there were 278 cases from China reported, along with 2 cases from Thailand, 1 case from Japan, and 1 case from the Republic of Korea. The first confirmed case in the United States, in Washington State, was reported on January 20, 2020. The CDC activated its Emergency Operations Center to respond to the emerging outbreak on the same day.

By January 22, 2020, the WHO's International Health Regulation Emergency Committee decided to monitor the situation instead of declaring the 2019 novel coronavirus a Public Health Emergency of International Concern (PHEIC). A citywide lockdown began in Wuhan, China, on January 23, 2020, due to this outbreak. By January 28, the CDC issued an advisory to avoid nonessential travel to China. Following the confirmation of the transmission of this virus between two individuals with no history of recent travel, the CDC issued 14-day quarantine orders to all US citizens who were recently repatriated from Wuhan, China.

On January 31, 2020, the WHO's International Health Regulation Emergency Committee declared the 2019 novel coronavirus outbreak a PHEIC. The Secretary of the Department of Health and Human Services also declared the 2019 novel coronavirus (2019-nCoV) outbreak a public health emergency.

On February 3, 2020, the CDC submitted an emergency use authorization (EUA), which is an authorization to allow the use of unapproved medical products or unapproved uses of approved medical products, to the Food and Drug Administration (FDA) to expedite approval for a SARS-CoV-2 diagnostic test kit, the CDC 2019-nCoV Real Time RT-PCR (reverse transcription–polymerase chain reaction). This request was approved by the FDA 1 day later. These test kits were used by February 5, 2020, on passengers arriving from Wuhan, China, at Travis Air Force Base in Sacramento, California. Others were used at a public health laboratory in Manhattan, New York City, New York. Unfortunately, initial test results in the field were deemed "untrustworthy results" by health professionals.

By February 10, 2020, more people worldwide died from the SARS-CoV-2 infection than the fatalities recorded during the SARS-CoV-1 outbreak between November 2002 and July 2003. On February 11, 2020, the WHO announced that the official name, COVID-19 (which stands for coronavirus disease 2019), designates the disease caused by the SARS-CoV-2 outbreak.

The first COVID-19–related death in the United States was reported on February 29, 2020, by the CDC and the Washington Department of Public Health. On March 11, 2020, the WHO declared COVID-19 a pandemic after more than 118,000 cases reported in 114 countries, resulting in 4291 related deaths. By March 13, 2020, the US government declared a nationwide emergency and finally started trying to solve the problem of controlling the pandemic.

By March 17, 2020, Moderna Therapeutics began the first human trials of its COVID-19 vaccine. The United States became the country with the most reported COVID-19 cases and deaths in April 10, 2020. The FDA issued an EUA to approve the use of remdesivir for treatment. On August 17, 2020, COVID-19 was reported to be the third leading cause of death in the United States.

On August 23, 2020, the FDA issued an EUA to approve the use of the liquid component of blood that contains antibodies from someone with COVID-19 (known as convalescent plasma) for treatment. More than 200,000 deaths from COVID-19 in the United States were reported in September 22, 2020, and more than 1 million deaths worldwide were reported in September 28, 2020.

The first Pfizer-BioNTech COVID-19 vaccine was administered on December 14, 2020, when the reported deaths surpassed 300,000 in the United States. On the same day, a new and more contagious SARS-CoV-2 variant, B.1.1.7, was detected, with the first case reported on December 29, 2020. Only 2.8 million individuals in the United States had received a COVID-19 vaccine dose by December 31, 2020. On January 22, 2021, an alpha variant was detected, and on January 25, 2021, a gamma variant was identified.

By January 26, 2021, there were 23 million COVID-19 vaccine doses administered in the United States, while more than 100 million cases were recorded worldwide. The first case of the beta variant was detected in January 28, 2021.

On March 29, 2021, the CDC confirmed that both Pfizer-BioNTech and Moderna COVID-19 vaccines were highly effective against SARS-CoV-2, reducing the risk of an infection by 90%. By April 21, 2021, more than 200 million vaccine doses had been administered in the United States.

A delta variant was identified in India on June 1, 2021, which became the dominant variant in the United States. Both Pfizer-BioNTech and Moderna COVID-19 vaccines appeared to be effective against all variants of SARS-CoV-2. On August 6, 2021, the CDC also released data showing that unvaccinated individuals were more than twice as likely to acquire COVID-19 compared to fully vaccinated individuals.

On October 6, 2021, the WHO identified the "post-COVID-19 condition" (also known as "long COVID"), with associated symptoms including fatigue, shortness of breath, and cognitive dysfunction that might persist for at least 2 months.

The omicron variant was detected in South Africa on November 19, 2021, and the first case was identified in the United States on December 1, 2021.

On December 15, 2021, more than 800,000 deaths from COVID-19 in the United States were reported, while the omicron variant was estimated to be the most transmissible variant thus far. The FDA approved Merck's molnupiravir for treatment.

The new year witnessed almost 1 million new, daily COVID-19 cases on January 3, 2022.

By March 5, 2022, the WHO reported that 10,704,043,684 COVID-19 vaccine doses had been administered worldwide, with roughly 56% of the world now fully vaccinated. Unfortunately, many regions still lacked access, especially on the African continent, where fewer than 20% of the total population were vaccinated.

By March 10, 2022, the WHO reported 450,229,635 confirmed cases and 6,019,085 confirmed deaths due to COVID-19. The US Census Bureau reported that deaths between 2019 and 2020 increased by 19%, which was the largest rise in mortality in a century.

By March 30, 2022, the United States reported 976,229 deaths and 79,853,683 total cases of COVID-19, and COVID-19 was the third leading US cause of death for the second year in a row. COVID-19–related deaths reached 1 million on May 12, 2022. By June 1, 2022, there were 84,145,569 COVID-19 cases and more than 1 million deaths recorded.

As of Fall 2024, everyone is encouraged to continue to get vaccinated, along with boosters, against SARS-CoV-2.

Classification and Structure of SARS-CoV-2

Severe acute respiratory syndrome coronavirus 2 belongs to the Coronaviridae family. This family includes four genera: *alphacoronavirus* (eg, feline and canine coronavirus); *betacoronavirus* (eg, SARS-CoV-1 and -2 and MERS); *gammacoronavirus* (eg, dolphin and bat coronavirus); and d*eltacoronavirus* (eg, swine and bird coronavirus). Two coronavirus outbreaks occurred earlier in the 21st century. Severe acute respiratory syndrome (SARS) was caused by SARS coronavirus (SARS-CoV) in 2002–2003. MERS was caused by MERS coronavirus (MERS-CoV) in 2012. All three species (SARS-CoV, SARS-CoV-2, and MERS) are known to be zoonotic viruses. SARS-CoV-2 is most genetically similar to SARS-CoV, with 86.85% nucleotide sequence and 77.2% amino acid sequence similarities. SARS-CoV-2 also shows genetic similarities to the other four common human coronaviruses (or HCoVs): HCoV-OC43, HCoV-HKU1, HCoV-229E, and HCoV-NL63. These four more common coronavirus strains cause mild upper-respiratory tract symptoms and are not as highly pathogenic as SARS-CoV, SARV-CoV-2, and MERS. Together, they account for 15-30% of "common colds" cases in human adults.

SARS-CoV-2 nucleocapsid structure

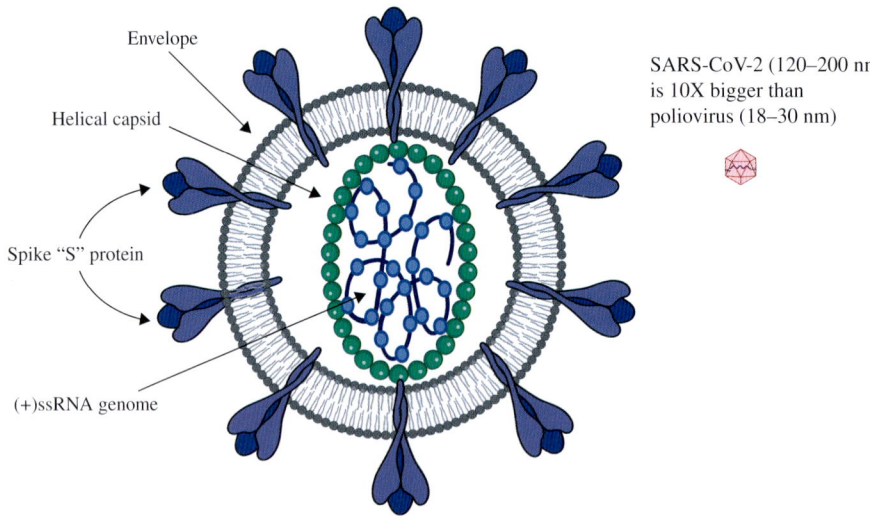

FIGURE 3.6   SARS-CoV-2 structure and genome. The single (+)ssRNA genome of SARS-CoV-2 is tightly encapsidated and surrounded by an envelope. Many spike "S" glycoproteins project outward and are important for the attachment and entry of the virion into the host cell. *Credit:* Figure partially created in BioRender.

The structure of the virus is an enveloped virion with a diameter of 120–200 nm. The single plus (sense)-strand RNA genome of 26–32 kilobases (kb), which is one of the largest genomes among RNA viruses, is encased in a helical capsid (Figure 3.6). A sense strand means that the viral genome can also be used directly as the template for translation, similar to poliovirus. Unlike the polioviral genome, where there is only a single ORF encoding 11 viral proteins, the SARS-CoV-2 genome is exceptionally large and has a polycistronic organization. There are two 5'-end overlapping ORFs, ORF1a and ORF1b, connected by a ribosomal frameshift site. The translation products are polyproteins designated pp1a and pp1ab. Each polyprotein is then cleaved into smaller proteins by the viral protease Mpro. These smaller proteins include 16 nonstructural proteins and 6 accessory and structural proteins. Mpro is an example of an accessory protein, and the envelope-embedded S glycoprotein is a structural protein that must undergo further translational modification and maturation during vesicular transport toward the plasma membrane.

Host Range, Transmission, Tropism, and Susceptibility

The Coronavirus family has a broad host range, and species members are known to cause a variety of lethal human and nonhuman animal diseases, targeting the respiratory and gastrointestinal tracts. SARS-CoV-2 itself is a known zoonotic virus, possibly having bats as their natural reservoir. To date, evidence supports that the masked palm civet, a small mammal indigenous throughout Asia, especially China and India, acted as an intermediate host for this virus before transmitting to mammals, like humans.

In the respiratory tract, goblet cells of the nasal passages and type II pneumocytes that line the alveoli of the lungs have been shown to be susceptible to SARS-CoV-2

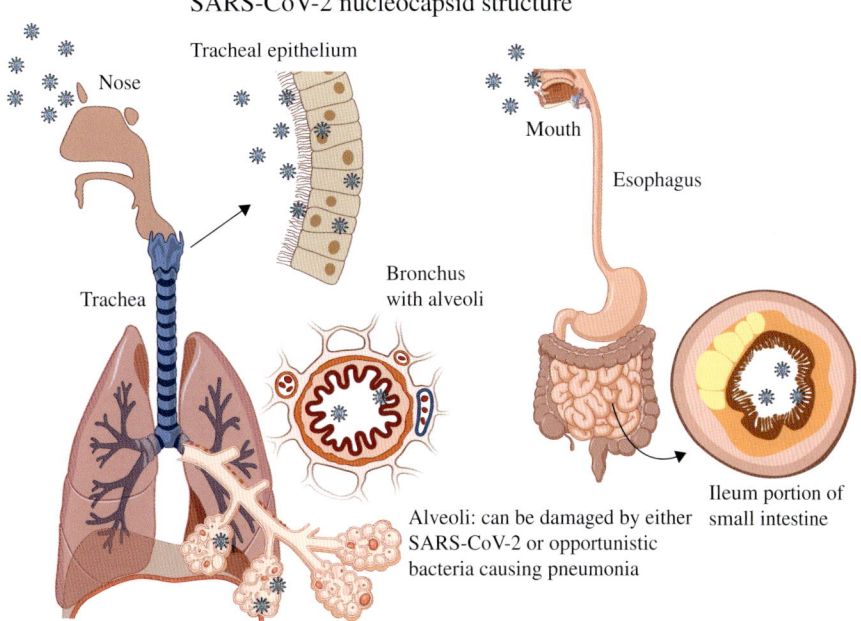

FIGURE 3.7 Mode of transmission and dissemination of SARS-CoV-2. SARS-CoV-2 is transmitted via respiratory droplets and enters the host organism through the nasal cavity, although this virus can also enter the host through the oral cavity. Along the respiratory tract, SARS-CoV-2 can infect the epithelium of the trachea, bronchus, and alveoli. In the digestive tract, this virus has been detected in the ileal lining of the small intestine. *Credit:* Figure partially created in BioRender.

(Figure 3.7). In the gastrointestinal tract, absorptive enterocytes of ileum are susceptible cells. As you can see, SARS-CoV-2 causes infection in both the upper and lower respiratory tract, which possibly explains the more efficient spread and worldwide pandemic.

The S "spike" glycoprotein on the surface of the viral particle contains a receptor-binding domain (RBD) that binds to a human receptor called angiotensin-converting enzyme 2 (ACE2) receptor. The ACE2 receptors are ubiquitous in human tissues and are especially overexpressed on intestinal epithelial cells; endothelial and smooth cells of blood vessels, brain, testis, and tubular epithelial cells of the kidney. Epicardia, adipocytes, fibroblasts, macrophages, myocytes, and coronary arteries of the heart as well as bronchial and tracheal epithelial cells, and type 2 pneumocytes of the lungs also over-express ACE2 receptor. The binding of spike S glycoprotein to ACE2 receptor along with other host surface proteins, principally transmembrane serine protease 2 (TMPRSS2), promotes entry of the viral particle into the susceptible cell and subsequent viral replication. Interestingly, ACE2 and TMPRSS2 co-express in nasal goblet cells, lung type II pneumocytes, and ileal absorptive enterocytes, which are the same major cell types found to be infected during testing. Type II pneumocytes produce pulmonary surfactant to prevent the collapse of alveoli. If these cells are destroyed, alveoli may likely collapse irreversibly, leading to the reduced ability to take in and distribute oxygen throughout the body.

Additionally, the neuropilin receptors (NRP1 and NRP2) were proposed to be another candidate class for the determination of tropism. This receptor is highly expressed in cells of the olfactory system, such as neuronal cells, which may be important in upper respiratory tract infection and may explain the neurological symptoms detected in some infected individuals.

The principal exit route for SARS-CoV-2 is the expelling of respiratory droplets containing viral particles during exhalation, for example, breathing, coughing, singing, and sneezing. There are two basic categories of respiratory droplets: larger droplets and smaller droplets; this designation is based on how long they can remain suspended in the air. Large droplets are visible and quickly fall out of the air within seconds of exhalation, causing them to be transmitted only a short distance, maybe 3 feet. Smaller droplets can remain suspended much longer, minutes to even hours; therefore, they can be transmitted through a much longer distance, especially when aided by air currents. When the small droplets evaporate, the viral particles can continue to be suspended in the air longer. How long depends on the size and density of the virion.

Note that the mode of transmission for SARS-CoV-2 is generally not through aerosol transmission since these viral particles are considered too big to be carried by such limited moisture. However, airborne transmission may still be possible under special circumstances, like an enclosed space or if tightly surrounded by a crowd where individuals are in closer proximity with inadequate ventilation. Once respiratory droplets are exhaled, viral particles are transmitted through the air. As time and distance pass, fewer droplets are moving toward the next host. Larger droplets would fall from the air more quickly than smaller droplets. Of course, other factors can certainly affect viral transmission (eg, humidity, temperature, airflow, pressure, even physical space). Viral particles may also fall on fomites.

The principal mode of entry is through inhalation of the droplets carrying SARS-CoV-2 directly into the respiratory tract. In this case, the risk of transmission is greatest when an individual is 3 to 6 feet from an infectious source while carrying out a conversation. That is to say, infectiousness increases as an individual comes closer to the infectious source. Viability of viral particles, which in turn affects the dose of exposure, is influenced by the amount of time spent in the external environment as well as environmental factors such as temperature, humidity, and sunlight. Other factors may influence the amount of dose of exposure. The dose of exposure increases in an enclosed area with inadequate ventilation and stagnant airflow, or an infectious individual tends to release more viral particles during physical activities.

Viral particles may also land onto exposed mucous membranes of the new host, such as the mouth, an exposed external injury, or even the exposed areas of the urogenital tissues. In this case, the viral load and the proximity of the two hosts affect the risk of transmission. Last, viral particles that are deposited onto one's skin through exposure to air droplets or contact with fomites can be delivered to the eyes, mouth, or nose. The viability of SARS-CoV-2 particles is 3 hours of experimental time in air droplets, 5.6 hours on stainless steel surfaces, 6.8 hours on plastic surfaces, and 72 hours on cardboard surfaces.

Infecting the Host Susceptible Cell, Viral Particle Replication, and Tissue Damage

The ACE2 receptor essentially marks susceptible cells and determines potential tropism and host range of SARS-CoV-2. However, permissibility can further be determined by various proteases, including TMPRSS2, discussed in the preceding section. Other proteases found in different cell types may also regulate viral entry, such as the following candidates: CD147, GRP78 (cell surface glucose-regulated protein 78), and heparan GRP78 is a molecular chaperone found in the endoplasmic reticulum (ER) that plays a number of roles including facilitating proper protein folding by stimulating the unfolded protein response (UPR) in the presence of misfolded proteins, sensing ER stress, regulating protein transport across the ER membrane, and protecting the cell from its apoptotic activities. ACE2 receptor polymorphisms can also affect susceptibility in different individuals. This may explain why some people are more susceptible to SARS-CoV-2 infection than others. The R_0 of SARS-CoV-2 was estimated to range from 1.4 to 6.49, with a mean of 3.28, which partially explains why this virus is so infectious. This value is higher than that of the SARS-CoV-1 outbreak in 2002, whose R_0 range was 2 to about 5. R_0 is defined as the average number of secondary infections produced by the infectious person. The designation of an epidemic by the Centers for Disease Control and Prevention is somewhat based on R_0. If R_0 is greater than 1, the number of infected cases would have increased exponentially and cause an epidemic or even a pandemic.

Attachment occurs when the ACE2 receptor dimer binds to two molecules of spike S glycoproteins on the viral surface (Figure 3.8). Viral entry via clathrin-dependent

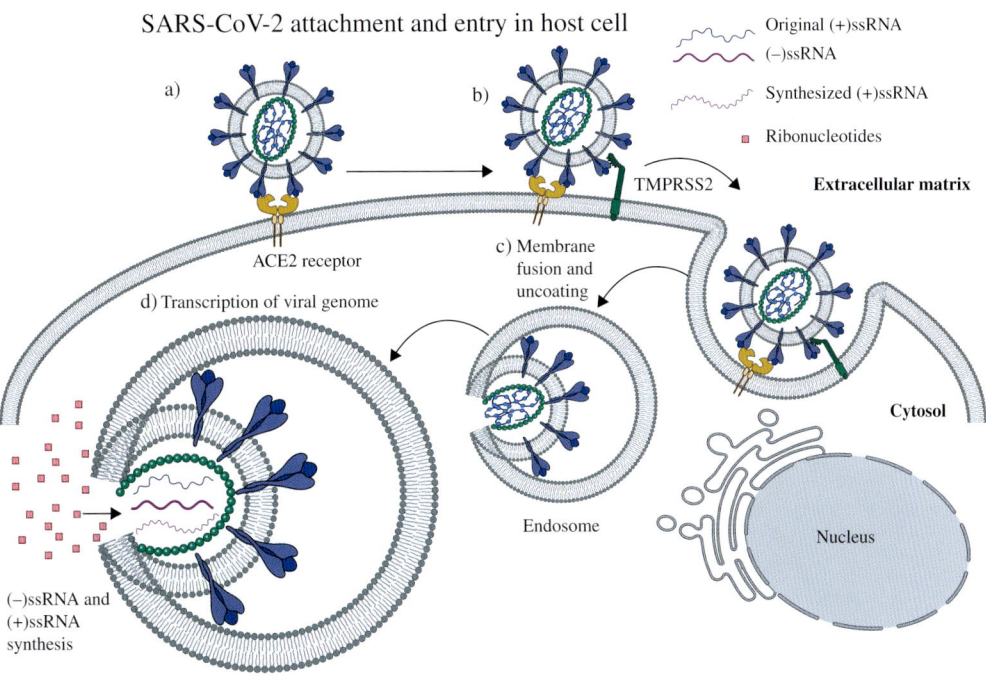

Figure 3.8 SARS-CoV-2 attachment, entry into host cell, and transcription. (a) SARS-CoV attaches to the host cell through the interaction of viral S protein and host cell ACE2 receptor. (b) Further interaction with host cell TMPRSS2 induces endocytosis of the viral particle. (c) Once in the endosome, membrane fusion and uncoating occur. (d) The viral genome is synthesized and translocated from the double-membrane vesicle into the host cell cytosol. *Credit:* Figure partially created in BioRender.

receptor-mediated and subsequent fusion of viral envelope to the endosomal membrane are aided by the cathepsin TMPRSS2 or other proteases. TMPRSS2 activates the spike S glycoprotein through acid-dependent cleavage at two different sites of the protein. Remember that the environment in the endosomal lumen is acid, with a pH of as low as 6. Cleavage of the spike S glycoprotein not only produces pores in the endosomal membrane, but also induces uncoating of the capsid, allowing a clear path between the nucleocapsid and host cell's cytosol.

The viral nucleoprotein RNA-dependent RNA polymerase (RdRP) regulates the synthesis of minus-strand RNA using the viral genome as the template within the double-membrane structure that resulted from the endocytotic process. All RNA synthesis occurs within this double-membrane vesicle, which means that molecules of ribonucleotides must also be transported into this structure to be used as substrates for transcription (Figure 3.9). The minus-strand RNA is, in turn, used as the template to produce many more plus-strand RNAs, which are used as the template for translation as well as packaged into virions. The plus-strand RNAs are transported into the cytosol where translation initiation occurs, while the minus strand remains within the double-membrane vesicle for further viral genome synthesis. Plus-strand RNAs in the cytosol are also encapsidated during virion assembly.

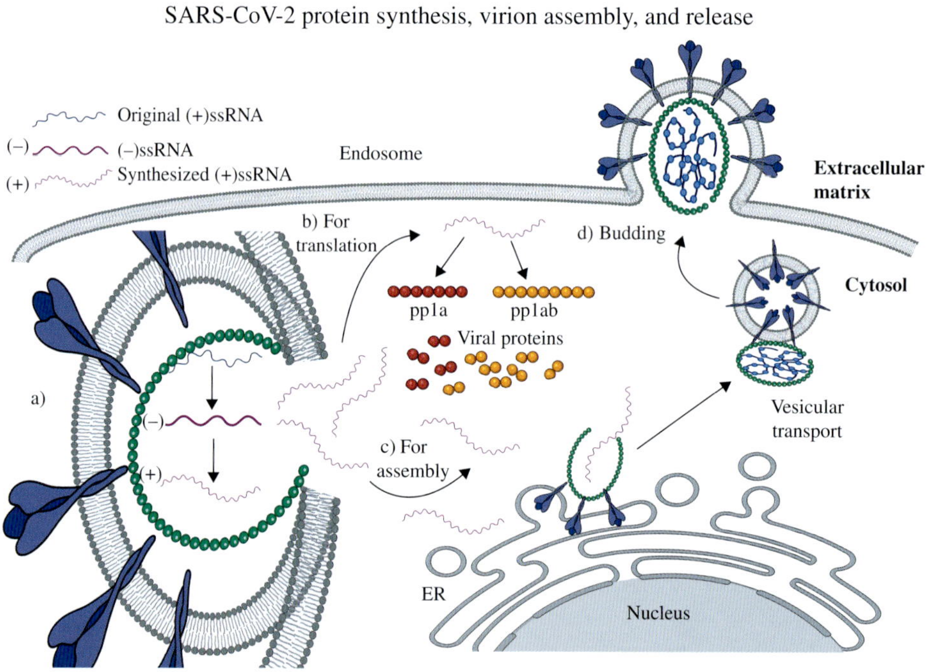

FIGURE 3.9 SARS-CoV-2 gene expression, assembly, and release from the host cell. (a) SARS-CoV-2 plus-strand RNA synthesis occurs within the double-membrane vesicle. (b) Some plus-strand RNA molecules are used as the template for viral protein synthesis. (c) The rest of the plus-strand RNA molecules are sent toward the endoplasmic reticulum (ER) membrane for assembly. (d) Release of virions from the host cell is via budding into the plasma membrane. *Credit:* Figure partially created in BioRender.

The RdRP of SARS-CoV-2 is reported to be faster than that of known coronavirus. This also means that this RNA polymerase is more prone to incorporate mistakes during elongation. By producing viral RNA genomic molecules that are highly variable, this virus can undergo a high rate of antigenic drift. Antigenic drift may result in selective advantages for SARS-CoV-2 strains by allowing them to evade the existing adaptive immune responses (ie, antibodies whose function is to prevent the reinfection of the same, initial viral strain) that were developed against the previous antigenic sites. An analogy for this process is that the virus simply takes off its original mask and puts on a new one. However, such a high rate and accumulation of errors, also called low replication fidelity, may also result in a high number of inactive viral genomic molecules, thus nonvirulent viral particles. To compensate for the low replication fidelity, SARS-CoV-2 exhibits proofreading activity in the form of nsp14, an exonuclease expressed by the SARS-CoV-2 genome. Nsp14 has both 3'-5' exonuclease activity and methyltransferase activity. Its function is greatly enhanced when dimerized with nsp10 cofactor, whose function is to stabilize nsp14 for substrate RNA binding to support its exonuclease activity.

As discussed in the classification and structure section, the SARS-CoV-2 genome contains two large overlapping ORFs that directly encode for two polyproteins, pp1a and pp1ab. These polyproteins are then cleaved into smaller proteins by the viral protease Mpro. There are 4 structural proteins:envelope (E), membrane (M), nucleocapsid (N), and spike (S). There are a number of nonstructural proteins: ORF3a is involved in the apoptotic pathway, interferon 1 antagonists (ORF3b, ORF6, ORF7a, ORF8), and suppressors of antiviral response (ORF9b and ORF9c). Proteins whose functions are yet to be known and awaiting description include ORF3d, ORF7b, ORF10. Structural proteins are synthesized and incorporated into the endoplasmic reticulum before being trafficked through the endoplasmic reticulum-Golgi intermediate compartment (ERGIC). Assembly of virions occurs near these vesicles that are being transported outward. Envelope-bound spike S glycoproteins are also synthesized in the host cell's cytosol, then embedded into the endoplasmic reticular membrane for further translational modification while being trafficked toward the plasma membrane via vesicular transport. After assembly, SARS-CoV-2 virions exit or are released through budding, resulting in virus particles containing an envelope that originated from the host cell's plasma membrane.

For some coronaviruses, including SARS-CoV-2, spike S glycoproteins that are not assembled into virions are exposed on the host cell surface. These surface-bound proteins can facilitate fusion between infected cells and adjacent, uninfected cells that contain ACE2 receptors on their surface. Fusion of infected cells to uninfected cells creates large, multinucleated cells, which allow the virus to spread within an infected organism without the detection or neutralization of virus-specific antibodies. The mechanism by which SARS-CoV-2 causes damage of the infected cell is still unknown. However, viral activities have been linked to the formation of the inflammasome and pyroptotic pathway.

Dissemination in the Host Body

Once the initial primary infection is established in the nasal mucosal membrane, a healthy individual may remain asymptomatic. SARS-CoV-2 virions can advance from the upper respiratory tract into the lower respiratory tract, infecting bronchial and alveolar epithelia. Along with the viral dissemination, opportunistic bacteria, which just happen to be present in the environment, may also enter the respiratory tract and make their way directly into the lower respiratory tract. Active viremia into the peripheral bloodstream through the lungs would likely ensue. Massive epithelial and endothelial tissue damage in the lungs may occur, which triggers exacerbated inflammatory responses and creates a cytokine storm. Cytokine storm syndromes (or hypercytokinemia) are disorders that are characterized by the uncontrolled overproduction of pro-inflammatory cytokines. This activity may prevent the immune system from functioning efficiently and may even cause dysfunction, such as acute respiratory distress syndrome (ARDS) and multiple organ failure (MOF).

Active viremia may lead to acute-phase pneumonia, which is the result of an overactive inflammatory response caused by a viral, opportunistic bacterial, or fungal infection. In this discussion, if the SARS-CoV-2 virions from the primary upper respiratory tract infection are disseminated and cause a lower respiratory tract infection, the disorder is referred to as secondary viral pneumonia. Opportunistic bacteria can cause secondary bacterial pneumonia as the result of a lung infection. Pneumonia is the buildup of fluid in one's lungs that is the consequence of edema during the inflammatory response.

Pathogenesis and Clinical Manifestation

Upward of 40% of SARS-CoV-2–infected individuals are asymptomatic. These data account for all known strains. Initial mild symptoms may include sore throat, congestion and/or runny nose, mild cough, and fatigue. Moderate and more serious symptoms include dry cough, shortness of breath, fever or chills, muscle and body aches, headache, loss of smell (anosmia) and loss of taste (ageusia), nausea and vomiting, and diarrhea. Life-threatening or fulminant symptoms include labored breathing (dyspnea); persistent pain or pressure in the chest; new confusion; hypoxia; thromboembolism; inability to wake up or stay awake; and pale, grey, or blue-colored skin and/or lips. These individuals should seek emergency medical attention immediately.

Severe tissue damage to various organs is a likely result of the high levels of inflammatory response, especially excessive cytokine storm, against SARS-CoV-2 infection. COVID-19–induced tissue damage may lead to health complications such as acute respiratory failure, pneumonia, ARDS, acute liver injury, acute cardiac injury, arrhythmias, secondary infection, acute kidney injury, and septic shock. MOF is the most common direct cause of death for COVID-19–related fatalities. Other major causes of death include respiratory failure and circulatory failure.

The median time from the initial mild symptoms to more severe symptoms, such as dyspnea, is between 5 and 8 days. The median time from disease onset to severe health complications is between 8 and 12 days, and to receiving intensive care in a hospital is 10 to 12 days. And, the median time from the onset of the disease to death is between 20 and 23 days. The average mortality rate varies from 0.7% to 8% based on different demographics.

The fatality rate is highest, at 10%–27%, among those 85 years and older. The rate for individuals between 65 and 84 years old is 3%–11%, between 55 and 64 years old is 1%–3%, and less than 55 years old is less than 1%. The fatality rate is higher for all age groups with underlying severe medical conditions (comorbidity), such as diabetes mellitus, cardiovascular disease, hypertension, chronic kidney disease, cancer, and chronic respiratory disorder. The rate is 10.5% for those with cardiovascular disease, 7.3% for those with diabetes, and 6% for those with chronic respiratory disorder or cancer.

Diagnosis, Treatment, and Prevention

Routine laboratory blood tests are a perfectly useful tool to diagnose positive cases of COVID-19. Other diagnostic methods, such as RT-PCR and imaging solutions like chest radiography (chest x-ray) and chest computed tomography (chest CT) offer better sensitivity but are also more expensive. RT-PCR is the gold standard for COVID-19 diagnosis but suffers from low sensitivity, long waiting time for the results, and costly equipment. Specimens for testing can be from nasopharyngeal swabs (nasal and throat), sputum, other lower respiratory tract secretions (bronchoalveolar fluid), blood, and feces. Chest x-ray and chest CT offer better sensitivity than RT-PCR, but raise multiple concerns of radiation safety, low specificity, and a high rate of false-negative results. Serological tests can detect lymphopenia (reduced lymphocytes in blood) and high levels of D-dimer (protein fragment from the breakdown of a blood clot). Abnormal levels of liver enzymes can also be detected such as C-reactive protein, ferritin, creatine phosphokinase, and lactate dehydrogenase. The presence of these enzymes in blood is generally increased during a high level of inflammatory response. However, antibodies are not used as a marker because the titers for IgM and IgG are low during the early phase of infection.

In terms of treatment, according to the CDC, the US Food and Drug Administration (FDA) had authorized and approved 3 antiviral drugs to reduce the risk of severe COVID19 symptoms and keep infected individuals out of the hospital. Paxlovid™ and Lagevrio™ (molnupiravir) are oral antivirals in pill form, while Veklury® (remdesivir) is an intravenous infusion antiviral. Paxlovid is appropriate for adults and children (12 years of age and older and weighing at least 88 pounds) who have mild to moderate symptoms. Paxlovid inhibits the viral protease Mpro from cleaving viral polyproteins into functional proteins, thereby preventing the replication of virions. Lagevrio is appropriate for adults only (18 years of age and older) who have mild to moderate symptoms but are at high risk for becoming severely ill from COVID-19. Lagevrio is a ribonucleoside analogue that is taken up by the RNA-dependent RNA polymerase (RdRP), thereby inducing viral RNA mutagenesis during viral RNA synthesis. The production of mutated complementary RNA strands affects infectivity and leads to the generation of nonfunctional virions. Veklury (remdesivir) is appropriate for mainly adults (18 years of age and older) but can be used children (28 days of age or older and weighing at least 7 pounds) under special circumstances. Veklury binds directly to the viral RdRP to inhibit viral RNA synthesis, thus terminating RNA transcription prematurely.

Convalescent plasma therapy is a type of immunotherapy that may be prescribed to COVID-19 patients. This therapeutic method uses blood from individuals who have recovered from COVID-19 themselves. Their plasma is likely to contain anti–SARS-CoV-2 antibodies that can act as an antiviral agent. Those who are also immunocompromised or those receiving immunosuppressive treatment are likely candidates for this therapy.

The National Institutes of Health recommends several general measures to prevent SARS-CoV-2 infection and COVID-19. COVID-19 vaccination is at the top of the list. Should I say that again? Vaccination. Vaccination is the most effective known means to prevent COVID-19. Unfortunately, there are a few misconceptions about what a vaccine is and its function. Dispelling the major misconception, the role of a vaccine is not to prevent the transmission of the virus into one's body. The function of vaccination is to prepare one's body to fight against the infection faster, more vigorously, and more efficiently. Through these defensive activities, an immunized individual who is infected by SARS-CoV-2 is much more likely to exhibit only mild to no symptoms and, most importantly, to avoid severe COVID-19 symptoms, including death.

There are simpler general treatments for individuals having mild to moderate symptoms as well. These strategies include bed rest, sufficient nutrition, and sufficient hydration. Vital signs such as heart rate, pulse, blood pressure, oxygen saturation, and respiratory rate should also be monitored if at all possible.

Since the transmission of SARS-CoV-2 occurs primarily through exposure to respiratory droplets, any means to avoid or reduce the chance of exposure is an effective prevention. Interventions include physical distancing, use of well-fitting masks in the community, and surface cleaning and disinfection. Avoid prolonged exposure to an infectious individual in poorly ventilated, enclosed spaces or in crowded spaces. Practicing good hand hygiene can prevent the transmission of virus through contaminated hands and surfaces. Avoid touching your eyes, mouth, nose, or any other mucous membranes with your hands just in case they're contaminated with the virus.

Current Status

As of July 2023, the WHO reported that the total worldwide number of COVID-19 confirmed cases was 773,511,195, with 766,488,068 recovered from illness and 7,023,127 related deaths. Pneumonia caused approximately 10% of those fatalities. A total of 13,474,348,801 vaccine doses have been administered worldwide. As of June 2023, the total number of confirmed cases in the United States was 103,436,829, with 102,709,647 recovered from illness and 1,127,152 related deaths. A total of 668,882,018 vaccine doses have been administered in the United States. The initial source of SARS-CoV-2 is still unknown. Even though the first cases of this outbreak were linked to the Huanan seafood market in Wuhan City, the exact original infectious organism is also unknown and could be any wild animals that were sold at the market, including birds, snakes, marmots, and bats.

There are definitely signs of improvement from the early days of this pandemic. By March 2022, COVID-19–related deaths substantially decreased, and this decline was sustained throughout the year in the United States. The various FDA-approved COVID-19 vaccines with booster doses continue to be effective in reducing the risk of fatality and the rate of mortality among all age groups. Adults aged 65 years or older, along with those with comorbidity, remain the most susceptible group to COVID-19–related mortality.

The CDC reports that individuals who recovered from COVID-19 may experience long-term effects known as long COVID (or post-COVID) conditions, which continue to exist as a complication of the initial SARS-CoV-2 infection. These conditions are mainly experienced by individuals with severe COVID-19 symptoms and can last anywhere from weeks to months to even years post-illness. Unvaccinated COVID-19 patients as well as individuals who are infected by SARS-CoV-2 multiple times are also at a higher risk for developing long COVID compared to vaccinated individuals. Remember that just because you never had symptoms or tested positive for the virus does not necessarily mean that you haven't been infected. Therefore, in some cases, long COVID had been the diagnosis in individuals who seemingly had never been affected (but not necessarily had never been infected) by SARS-CoV-2.

Scientific Significance and Discoveries

Thus far, a number of existing antiviral therapies have been approved by the FDA; however, none is undeniably effective against SARS-CoV-2. Most of these therapies were used during previous outbreaks (SARS and MERS) and repurposed to treat COVID-19. Antiviral and immunomodulatory drugs in use are Paxlovid, Lagevrio, and Veklury, which is used to prevent viral replication. Other therapies that are used to prevent virus entry include casirivimab (imdevimab) and convalescent plasma. Corticosteroids and agonists and antagonists of cytokine storm and interferons have also been used when an individual is hospitalized due to COVID-19. Many other drugs have been tested in a clinical setting to check for potency in the treatment of COVID-19. Unfortunately, none of these antiviral drugs is the definitive cure or suitable prophylaxis. Monoclonal antibodies (mAbs) have been shown to be a powerful and effective tool for both treatment against and detection of SARS-CoV-2 due to their high specificity and reliability. Unfortunately, current standard diagnostic tests and treatment approaches using mAbs lack the required efficiency to eradicate this virus.

In terms of detection, the CDC developed an assay based on reverse transcription–quantitative polymerase chain reaction (RT-qPCR) assay for SARS-CoV-2 that has been widely used for COVID-19 diagnostic testing on respiratory specimens. Unfortunately, this tool also has a number of drawbacks to the technique, including a high rate of false results, the lack of sensitivity, limited specificity, and the long wait time between testing and results. Other logistical limitations that require improvements include insufficient essential RT-qPCR infrastructure to handle large sample volumes and insufficient amounts of reagents and kits to be used during a pandemic.

Thus, the development of precise and effective COVID-19 therapeutics and SARS-CoV-2 detection is needed. Nanotechnology is probably the most promising tool used in the advancement of new diagnostics and therapeutics. This field deals with the study of very small materials (having a diameter of no more than 100 nm) and is applied to other fields, such as biomedicine, material engineering, chemistry, physics, engineering, computer engineering, and biological sciences. There are a number of potential benefits of nanotechnology to develop targeted therapies, to provide efficient novel drug candidates, as early diagnostic tools, and for patient-personalized medicines that are safer and more effective than the existing options. Benefits include increased bioavailability, reduced dosing regimen, improved chemical stability, enhanced detection system sensitivity and specificity, improved safety profile, improved ease of administration, increased patient compliance, enhanced immune response to improve overall treatment outcomes, added aid in the administration of water-insoluble pharmaceuticals, increased circulation of medicines to circulate in the body, improved drug efficiency and reduced side effects, and increased capacity to carry multiple drugs/vaccines simultaneously.

Further Readings

Flerlage T, Boyd D, Meliopoulos V, Thomas P, Schultz-Cherry S. Influenza and SARS-CoV-2: Pathogenesis and host responses in the respiratory tract. *Nat Rev Microbiol.* 2021;19(7):425–441.

Gao S, Zhang L. ACE2 partially dictates the host range and tropism of SARS-CoV-2. *Comput Struct Biotechnol J.* 2020;13:4040–4047.

Holmes E, Goldstein S, Rasmussen A, et al. The origins of SARS-CoV-2. *Cell.* 2021;184(19):4848–4856.

Jones A, Mourao A, Czarna A, et al. Characterization of SARS-CoV-2 replication complex elongation and proofreading activity. *Sci Rep.* 2022;12:9593.

Khorramdelazad H, Kazemi M, Najafi A, Keykhaee M, Emameh R, Falak R. Immunopathological similarities between COVID-19 and influenza: Investigating the consequences of co-infection. *Microb Pathog.* 2021;152:104554.

Li H, Liu S-M, Yu X-H, Tang S-L, Tang C-K. Coronavirus disease 2019 (COVID-19) current status and future perspectives. *Int J Antimicrob Agents.* 2020;55(5):105951.

Marjomaki V, Kalander K, Hellman M, Permi P Enteroviruses and coronaviruses: Similarities and therapeutic targets. *Exper Opin Ther Targets.* 2021;25(6):479–489.

Mehndiratta MM, Mehndiratta P, Pande R. Poliomyelitis: Historical facts, epidemiology, and current challenges in eradication. *Neurohospitalist.* 2014;4(4):223–229.

Mueller S, Wimmer E, Cello J. Polivirus and poliomyelitis: A tale of guts, brains, and accidental event. *Vir Res.* (2005);111(2):175–193.

Ohka S, Nihei C, Yamazaki M, Nomoto A Poliovirus trafficking toward central nervous system via human poliovirus receptor-dependent and -independent pathway. *Front Microbiol.* 2012;3:147–151.

Racaniello VR. One hundred years of poliovirus pathogenesis. *Virology.* 2006;344:9–16.

Rikan S, Azar A, Ghafari A, Mohasefi J, Pirnejad H. COVID-19 diagnosis from routine blood tests using artificial intelligence techniques. *Biomed Signal Process Control.* 2021;72(A):103263.

Shirbhate E, Pandey J, Patel V, et al. Understanding the role of ACE-2 receptor in pathogenesis of COVID-19 disease: A potential approach for therapeutic intervention. *Pharmacol Rep.* 2021;73(6):1539–1550.

Sofi M, Hanid A, Bhat S. SARS-CoV-2: A critical review of its history, pathogenesis, transmission, diagnosis and treatment. *Biosaf Health.* 2020; 2(4):217–225.

Solanki R, Shankar A, Modi U, Patel S. New insights from nanotechnology in SARS-CoV-2 detection, treatment strategy, and prevention. *Mater Today Chem.* 2023; 29:101478.

CHAPTER 4

Minus-Strand Single-Stranded RNA

Influenza A Virus

Historical Perspective

The "Russian flu" that occurred from 1889 to 1894 is considered the first pandemic of the industrial era. Originally, the influenza strain was identified as H2N2; however, more recently, researchers believe that it was actually H3N8. In the 20th century, three significant pandemics of influenza A occurred: in 1918, 1957, and 1968. All three were informally identified by their presumed sites of origin as the Spanish, Asian, and Hong Kong influenzas or flus, respectively.

The Spanish Flu

The 1918 pandemic, caused by the H1N1 influenza A virus, lasted from 1918 to 1920 and was the worst pandemic in the 20th century. Despite the name, this pandemic was thought to have not originated in Spain, although the total number of individuals who died of this infection in Spain were officially estimated to be 147,114 in 1918, 21,235 in 1919, and 17,825 in 1920.

Where did the 1918 pandemic originate? Most likely, it was the United States, where the earliest documented cases were among soldiers at Camp Funston in Fort Riley, Kansas, in March 1918. Other possible places of origin could have been among laborers in the Guangdong Province in southern China and among French, British, and German soldiers in World War I (WWI) on the Western Front during the winter of 1917.

Why was the 1918 pandemic, also known as the "Great Influenza Epidemic," called the Spanish flu in at least the United States? At various times and places, this pandemic was named the "Brazilian flu" in Senegal, the "German flu" in Brazil, and the "French flu" in Spain. No one claimed the notoriety of being the origin point except Spain inadvertently.

This pandemic began as WWI was winding down. Spain was one of the few European countries to remain neutral in the war. Wartime censorship and media blackout were rampant in countries that were participating in the war, leaving the Spanish media free to report all world events.

While other countries refused to admit having infection cases, the first reporting of the outbreak came from Madrid, Spain, which was then pegged to be the origin of

75

this pandemic. Increased coverage of the infected Spanish King, Alfonso XIII, solidified the infamy of Spain as the origin of this pandemic.

After 18 months of this pandemic, an estimated 500 million people worldwide had been infected in three successive waves: a mild outbreak in the spring of 1918, followed by deadlier outbreaks in the fall of 1918, and then the winter of 1919. That's roughly 40% of the global population at the time. The estimated number of deaths ranged from 17 to 50 million individuals but might be as high as 100 million. That makes this pandemic deadlier than WWI.

The Asian Flu

The first cases of the H2N2 influenza A virus were identified in the Guizhou Province of southern China, which spread to Singapore in February 1957 and then reached Hong Kong by April. The pandemic spread to the United States in the summer of 1957 and reached Europe by November 1957. This pandemic was known as the "Asian flu" since it was first identified in East Asia. The World Health Organization (WHO) estimated the number of deaths to be between 1.5 and 2 million worldwide. Approximately 20 million individuals in the United States were infected, and there were 116,000 deaths.

The Hong Kong Flu

The pandemic that occurred between 1968 and 1970 was caused by the H3N2 influenza A strain and resulted in 1 to 4 million deaths worldwide and 100,000 in the United States. This outbreak likely originated in the Guangdong Province of China but was widespread in British Hong Kong by July 1965. This outbreak spread quickly throughout Southeast Asia and then reached North America a few months later. By December 1968, this pandemic spread to Southeast Asia, India, Western Europe, Australia, Japan, Africa, and Central and South America.

Classification and Structure of Influenza A Virus

Influenza A belongs to the Orthomyxoviridae family. This family includes seven genera: *Alphainfluenzavirus* (eg, influenza A), *Betainfluenzavirus* (eg, influenza B), *Gammainfluenzavirus* (eg, influenza C), *Deltainfluenzavirus* (eg, influenza D), *Isavirus* (eg, salmon isavirus), *Thogotovirus* (eg, arbovirus Dhori thogotovirus), and *Quaranjavirus* (eg, arbovirus Quantifil quaranjavirus). Of these, only genera *Alphainfluenzavirus* and *Betainfluenzavirus* are clinically relevant for human hosts and diseases.

There are numerous influenza A virus strains, which are classified according to the combination of the glycoproteins hemagglutinin (HA) subtype and neuraminidase (NA) subtype. There are 18 known HAs (H1–H18) and 11 identified NAs (N1–N11) (Table 4.1). In the past several decades, highly pathogenic human influenza A strains have been identified, including: H1N1, H2N2, H3N2, H5N1, H5N2, H7N3, and H7N9. However, only strains H1N1, H2N2, and H3N2 have been responsible for human epidemics. Each virion exhibits approximately 300 to 400 HA "spikes" and 40 to 50 NA "spikes"; all are embedded into the viral envelope.

Other related families of virus include: Parmyxoviridae (eg, measles virus and respiratory syncytial virus [RSV]), Filoviridae (eg, Ebola virus and Marburg virus), and Bunyaviridae (eg, hantavirus and Rift Valley fever virus).

The structure of the virus is an enveloped virion that is 80–120 nm in diameter by 300 nm in length. The 8 segments of minus (antisense)-strand RNA genome (total of

A. Hemagglutinin (HA) subtypes and their presence in vertebrate species.							
Subtype	Human	Avian	Swine	Bat	Feline	Canine	Equine	Bovine
H1	yes	yes	yes	no	no	no	no	no
H2	yes	yes	yes	no	no	no	no	no
H3	yes	yes	yes	no	yes	yes	yes	no
H4	no	yes	yes	no	no	no	no	no
H5	yes	yes	yes	no	yes	yes	no	yes
H6	yes	yes	yes	no	yes	yes	no	no
H7	yes	yes	yes	no	yes	no	yes	no
H8	no	yes	no	no	no	no	no	no
H9	yes	yes	yes	no	yes	no	no	no
H10	yes	yes	yes	no	no	no	no	no
H11	no	yes	no	no	no	no	no	no
H12	no	yes	no	no	no	no	no	no
H13	no	yes	no	no	no	no	no	no
H14	no	yes	no	no	no	no	no	no
H15	no	yes	no	no	no	no	no	no
H16	no	yes	no	no	no	no	no	no
H17	no	no	no	yes	no	no	no	no
H18	no	no	no	yes	no	no	no	no

B. Neuraminidase subtypes and their presence in vertebrate species.							
Subtype	Human	Avian	Swine	Bat	Feline	Canine	Equine	Bovine
N1	yes	yes	yes	no	yes	yes	no	yes
N2	yes	yes	yes	no	yes	yes	no	no
N3	no	yes	no	no	no	no	no	no
N4	no	yes	no	no	no	no	no	no
N5	yes	yes	yes	no	no	no	no	no
N6	yes	yes	yes	no	no	no	no	no
N7	yes	yes	no	no	no	no	yes	no
N8	yes	yes	no	no	no	yes	yes	no
N9	yes	yes	no	no	no	no	no	no
N10	no	no	no	yes	no	no	no	no
N11	no	no	no	yes	no	no	no	no

TABLE 4.1   Influenza A Subtypes and Their Host Range

C. Identified Influenza A HN subtypes and their presence in vertebrate species.	
Species	**HN Subtypes**
Human	H1N1, H1N2, H2N2, H3N2, H5N1, H5N2, H5N3, H5N4, H5N5, H5N6, H5N8, H6N1, H6N2, H7N2, H7N3, H7N7, H7N9, H9N2, H10N3, H10N7, H10N8, H10N10
Avian	H1N1, H2N2, H3N1, H3N2, H3N3, H3N5, H3N6, H3N8, H4N2, H4N6, H5N1, H5N2, H5N3, H5N4, H5N5, H5N6, H5N7, H5N8, H5N9, H6N1, H6N2, H6N6, H7N1, H7N2, H7N3, H7N4, H7N5, H7N6, H7N7, H7N8, H7N9, H9N1, H9N2, H9N3, H9N4, H9N5, H9N6, H9N7, H9N8, H9N9, H10N3, H10N4, H10N5, H10N6, H10N7, H10N8, H11N2, H11N9, H12N2, H12N5, H12N8, H13N6, H14N5
Bat	H17N10, H18N11
Bovine	H5N1
Canine	H3N2, H3N8, H5N1, H5N2, H6N1
Equine	H3N8, H7N7
Feline	H1N1, H3N1, H3N2, H3N8, H5N1, H5N2, H5N6, H7N2, H9N1, H9N2,
Rodent	H1N1, H5N1
Swine	H1N1, H1N2, H1N7, H2N3, H3N1, H3N2, H4N6, H6N1, H6N2, H6N6, H9N2, H10N7

TABLE 4.1   Influenza A Subtypes and Their Host Range (*Continued*)

13,588 bases) are encased in a helical capsid (Figure 4.1). Antisense means that the single-stranded RNA is complementary to the messenger RNA (mRNA). Each of segments 1–6 encodes for a single nonstructural protein. Segment 7 encodes for 2 matrix proteins, and segment 8 encodes for another 2 nonstructural proteins, to produce a total of 10 viral proteins. The 10 influenza A viral proteins are HA; NA; matrix protein 1 (M1; proteins form a layer beneath the viral lipid envelope); matrix protein 2 (M2; ion channel); nucleoprotein (NP; transport of viral RNA); nonstructural protein 1 (NSP1; downregulates host cell mRNA by blocking polyadenylation, interferon [IFN] antagonist to inhibit innate immune responses, restrict cascade response of IFN, increase virulence of virus); nonstructural protein 2 (NSP2; nuclear export protein, NEP); polymerase acidic protein (PA); polymerase binding protein 1 (PB1; elongation); polymerase binding protein 2 (PB2; cap binding); and RdRP (RNA-dependent RNA polymerase). Each viral genome segment is bound by a number of viral RNA-binding proteins, called nucleoproteins, to form a ribonucleoprotein (RNP) complex, which is tightly packed inside the helical structure of the capsid. The RNA polymerase complex or RdRP is included in this complex.

Nucleic acid synthesis occurs at such a high rate that mutations are inevitably produced. The polymerization rate of DNA polymerase is 500–1000 nucleotides per second for prokaryotic cells and 50–100 nucleotides per second for eukaryotic cells, while the polymerization rate of known RNA polymerases, including RdRP, is 40–80 nucleotides per second. The estimated rate of error during the copying process is approximately 10^{-4} mistakes per base incorporated. To compensate for this high rate of error production,

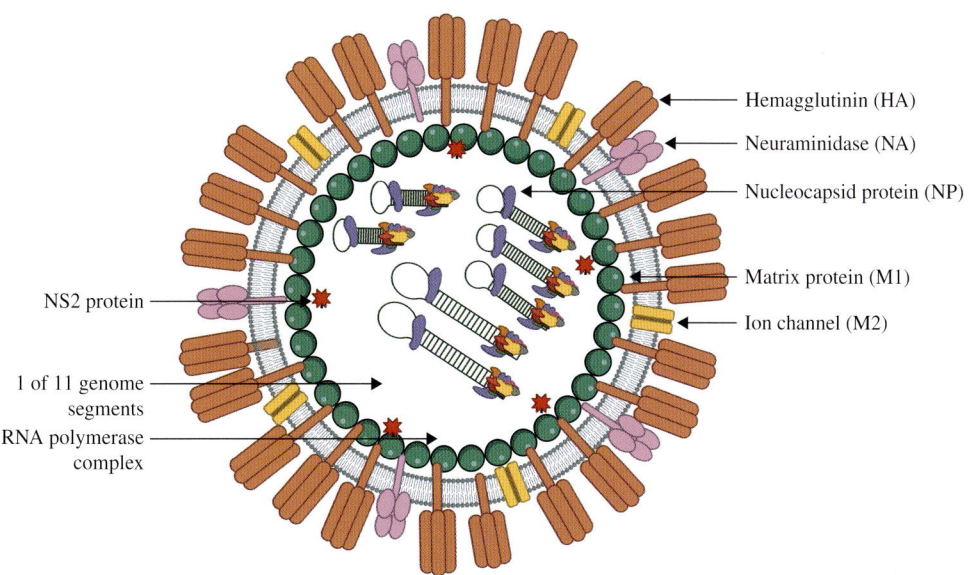

FIGURE 4.1   Influenza A nucleocapsid structure. This is an enveloped virion where the 8 segments of minus-strand RNAs are encapsidated by a helical capsid. The envelope exhibits about 500 molecules of HA and NA. The ratio of HA to NA is 45 to 1. Each genome segment is bound by an RNA polymerase complex and NP, forming an RNP complex. Several molecules of NS2 proteins are also packaged inside the capsid. *Credit:* Figure partially created in BioRender.

DNA proofreading mechanism and DNA repair systems reduce the resulting error rate to 10^{-9}.

However, in general, RNA proofreading mechanisms and repair systems are extremely rare in nature. Influenza A does not encode for such mechanisms; therefore, viral genome replication incurs high rates of error, thereby accumulating a high number of genomic mutations through successive replication.

The accumulated mutations that alter the antigenic portions of HA and/or NA is referred to as antigenic drift. Antigenic drift may result in selective advantages for influenza A strains by allowing them to evade the existing adaptive immune responses (ie, antibodies whose function is to prevent the reinfection of the same, initial viral strain) that were developed against the previous antigenic sites. Of course, antigenic drift should not interfere with the ability of viral particles to attach to the target host cells; otherwise, the new virus variation would become avirulent.

Antigenic shift may also occur when a single host cell is infected by 2 or more influenza A strains. During assembly of virions, a new HA subtype and/or NA subtype may be incorporated into the virion by the reassortment of genomic segments between two influenza strains. The result is a brand-new influenza A strain. Viral genera that acquire new strains frequently are referred to as emerging viruses where the new strain is a chimera of the parent strains. Remember that each influenza A virus contains 8 viral genomic segments. Severe acute respiratory syndrome coronavirus 2 (SARS-CoV-2) would never undergo antigenic shift since each virion only contains 1 molecule of the genome.

Host Range, Transmission, Tropism, and Susceptible Host Cell

There are many strains of influenza A virus that can infect a wide range of warm-blooded host animals, including human, swine, wildfowl, equine, bat, and other mammals (refer to Table 4.1). Aquatic birds have been found to serve as a natural reservoir for all known subtypes of influenza A virus.

Susceptible cells to human influenza A virus strains exhibit N-acetylneuraminic acid (Neu5Ac), which is the predominant form of sialic acid found on human cells. These residues are 9-carbon carboxylated monosaccharides that are bound to the outermost ends of glycans through $\alpha 2,3$, $\alpha 2,6$, or $\alpha 2,8$ linkages. Neu5Ac is synthesized in animals, but not plants, and is expressed in a cell-specific and species-specific manner.

For example, epithelial cells lining the human trachea express predominantly $\alpha 2,6$-linked Neu5Ac, but the entire respiratory tract does express $\alpha 2,3$-linked Neu5Ac (most prevalent in the lower respiratory tract, bronchioles, and alveoli), while epithelial cells lining the avian respiratory tract express predominantly $\alpha 2,3$-linked Neu5Ac. Furthermore, neuraminic acids are substrates for the NA enzyme. Different subtypes of NA identify and process different forms of neuraminic acids on the surface of viral particles.

Being able to infect a wide host range and having a broad tropism make influenza A viruses highly pathogenic and dangerous, with the potential to jump from an animal reservoir to humans. Some strains or subtypes of influenza A virus can also be zoonotic. Zoonosis is defined as a virus that can infect both human and nonhuman host. Human pandemics are most often caused by the emergence of novel influenza A viral strains that originated from a nonhuman animal and acquired the appropriate genetic variation to cross between species and be transmitted into the human organism.

As discussed in the Historical Perspective section, the three most significant pandemics of the 20th century were caused by zoonotic transmission of avian influenza A viral strains. The viruses that caused these outbreaks were shown to be originally avian influenza A viral strains whose HA genes accumulated relevant mutations. The new, mutated HA genes encode for glycoproteins that preferably bind to $\alpha 2,6$-linked Neu5Ac in humans, instead of $\alpha 2,3$-linked Neu5Ac in birds.

Common zoonotic strains are H5N1, H7N2, H7N3, and H9N2 from live poultry sold in the market; H1N1, H1N2, and H3N2 from swine, also from the market; H7N7 and H3N8 from equine; H6N1 from mice and ferrets; H5N6, H5N2, and H3N1 from canine and feline; and H10N7 from harbor seals.

Similar to SARS-CoV-2, the principal route of exit for influenza A virus is the expelling of respiratory droplets containing viral particles during exhalation. Unlike SARS-CoV-2, influenza A viral particles are transmitted primarily via large particle droplets. Large droplets do not remain suspended in the air for a long time and can only travel a short distance. Therefore, airborne or aerosol transmission is not a common route. Transmission most commonly occurs when an infected individual with a high viral load sneezes or coughs toward the recipient host; even then, the two individuals are usually very close in proximity. Direct contact with contaminated surfaces, either skin to skin or through fomites, is another mode of transmission. The viral particles are then delivered to the oral openings when an individual touches the face.

Entry of viral particles is generally through the mucous membrane of the nose or mouth to infect the upper respiratory tract first, then possibly the lower respiratory tract. Cells of the respiratory tract exhibit surface receptors that attract the influenza A virus. The eye has also been shown to be a portal of entry and to cause ocular disease in human. Viral particles adhere to the thin mucous membrane known as the conjunctiva that lines the inside of the eyelids and covers the white of the eye (sclera) . However, entry through the eye is rare.

Infecting the Host Susceptible Cell, Viral Particle Replication, and Tissue Damage

Attachment of the influenza A virion to its susceptible host cell occurs when the glycoprotein HA binds to Neur5Ac, which is the predominant sialic acid found in human cells (Figure 4.2). The glycoprotein NA may provide an important role in viral entry of certain host cell types; however, its precise role is currently not well understood. Some NA molecules have been reported to possess a binding site for Neur5Ac, known as the hemadsorption site, in addition to the NA catalytic domain. The interaction between HA and Neur5Ac triggers the internalization of the virus via clathrin-dependent, receptor-mediated endocytosis.

Once the endosome is formed, A host cell's adenosinetriphosphate (ATP)-dependent proton pumps, named vacuolar ATPases (v-ATPases), in the endosomal membrane pump protons from the cytosol into the endosomal lumen, reducing the pH to between 6.5 and 4.5. This acidic environment causes irreversible conformational change in HA, which in turn mediates membrane fusion between the viral envelope

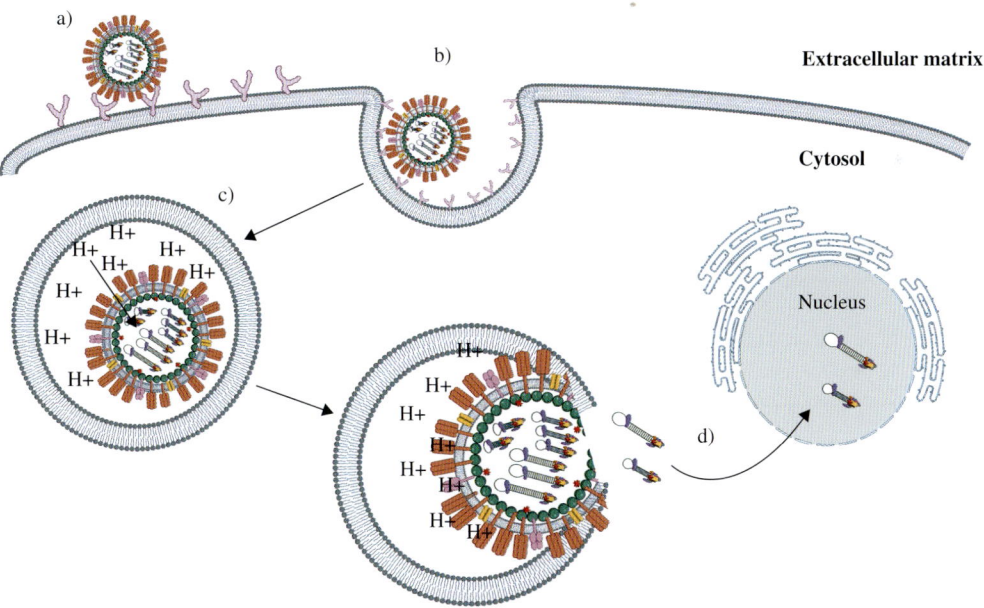

FIGURE 4.2 Influenza A enters the host cell via endocytosis. (a) Attachment of the virion to its host cell via the interaction between HA and cell surface Neur5Ac. (b) Influenza A enters the host cell via endocytosis. (c) Protons in the endosomal lumen diffuse into the virion through M2 ion channels. (d) The acidic environment inside the virion causes the viral envelope to fuse with the endosomal membrane and partial uncoating. This opening allows RNPs to migrate from the virion to the nucleoplasm. *Credit:* Figure partially created in BioRender.

and endosomal membrane. At the same time, the viral M2 ion channels pump protons from the endosomal lumen into the viral particle, enabling uncoating, where only a wide pore in the capsid is open to allow viral RNPs to be released into the cytosol and delivered toward the nucleus. Each RNP complex includes a viral genome segment bound by a number of viral RNA-binding proteins (nucleoproteins). The RdRP is among this group of nucleoproteins. The RdRP is composed of PA, PB1, and PB2. Viral RNPs then translocate into the nucleoplasm aided by members of the importin-a family.

Once in the nucleoplasm, the minus-strand RNA genome segments are used as the template for both viral mRNA and plus-strand RNA synthesis (Figure 4.3). The PA subunit of RdRP has endonuclease activity that cleaves the 5'-cap and oligonucleotide from the host cell's mRNAs, a process referred to as "cap snatching." This 5'-capped oligonucleotide is used as the primer for viral mRNA synthesis. Since an oligouridine or oligo(U) is found at the 5' end of each viral minus-strand RNA genome, the resulting viral mRNA contains a poly(A) tail at its 3' end. Having a 5' cap and 3' poly(A) tail allows viral mRNAs to exit the nucleus into the cytosol to be used as templates for cap-dependent translation initiation. The de novo primer is used for the synthesis of full-length viral plus-strand RNA and as the template for the amplification of viral minus-strand RNA genome segments.

Since viral protein synthesis uses cap-dependent translation, the predominant pathway used by the host cell, different inhibitory mechanisms are deployed by the

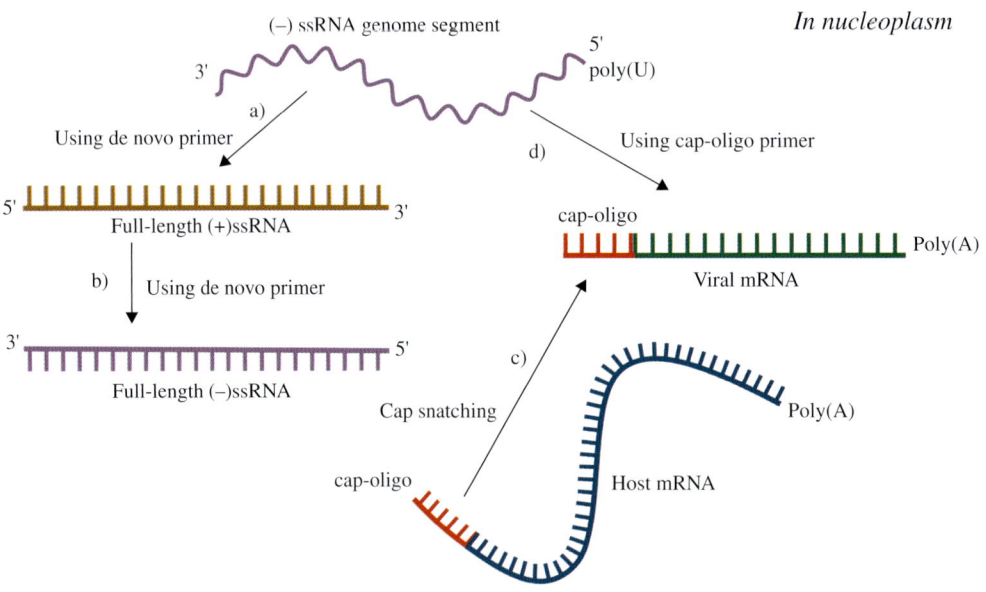

Figure 4.3 Influenza A genome replication and mRNA synthesis. (a) The (−) ssRNA genome is used as the template for full-length (+) ssRNA synthesis using de novo primer. (b) Full-length (+) ssRNA is used as the template for viral genome replication using de novo primer. (c) Cap-oligonucleotide is cleaved from host mRNA through a process called "cap snatching." The cap-oligonucleotide is used as the primer for viral mRNA synthesis. *Credit:* Figure partially created in BioRender.

influenza A virus to reduce host cell protein synthesis. One already mentioned mechanism, 5'-cap-free host mRNAs, caused by the activity of the viral endonuclease, cannot be used for cap-dependent translation, thus there are no host proteins. Additionally, the viral NS1 protein can also downregulate a host cell's mRNA processing by inhibiting polyadenylation of mRNA (Figure 4.4). Remember that viral mRNA already possesses a poly(A) tail without needing to undergo the polyadenylation process.

Following the synthesis of viral genome segments and their corresponding proteins, viral RNP complexes are assembled in the nucleoplasm. The 5' and 3' ends of each viral genome segment contain a high degree of complementarity to maintain molecule in the condensed form. RNPs are then exported into the cytosol, aided by viral NS2/NEP proteins that contain nuclear export signals.

All viral proteins are trafficked to the host cell's plasma membrane, where virion packaging occurs. Viral envelope proteins (e.g., HA, NA, and M2) are synthesized and processed in the endoplasmic reticulum (ER). HA and NA are glycosylated as they move through vesicular transport. All three membrane proteins are trafficked toward the cell's periphery and accumulate at lipid rafts in the plasma membrane.

Virion assembly occurs on the cytosolic side of the plasma membrane and then exits from the host cell via a budding process. A virion is not virulent or infectious unless it possesses one of each different viral RNA genome segment. Each RNA genome segment expresses different viral proteins. As described in the Classification and Structure

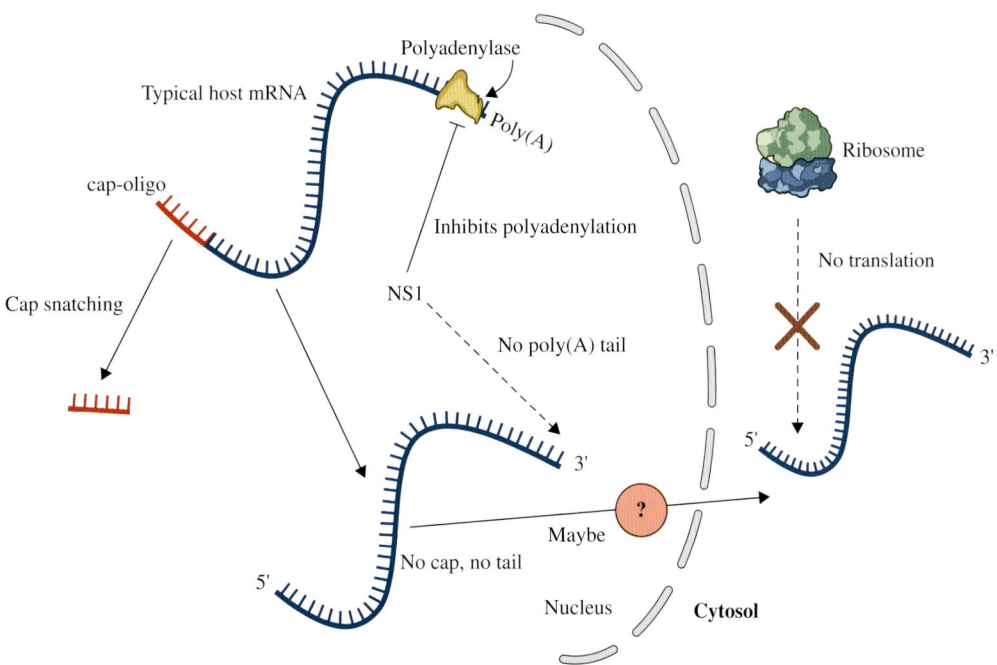

FIGURE 4.4   Downregulation of host protein synthesis. Cap snatching and the inhibition of polyadenylation by NS1 leaves the host mRNA without a cap or poly(A) tail. Uncapped mRNAs without a poly(A) tail rarely are exported into the cytosol. Even if this host mRNA somehow ends up in the cytosol, it cannot be recognized and bound by host ribosome. *Credit:* Figure partially created in BioRender.

of Influenza A Virus section, antigenic shift may occur if genome segments from different viral strains, which infected the same host cell, are assorted and assembled together
within a single capsid. Budding is initiated by an accumulation of enough M1 matrix
protein molecules at the cytosolic side of the lipid raft that contains HA and NA. The
viral envelope came from the host cell's own plasma membrane.

Once the virion is released from the host cell, NA is activated to remove sialic
acid from not only cellular receptors, but also HA and NA on the surface of newly
released virions (Figure 4.5). The removal of Neu5Ac from cellular receptors prevents newly released virions from reattaching to the same, now-dying host cell. The
removal of sialic acid from HA and NA prevents virions from aggregating to one
another to enhance infectivity and ensure efficient distribution of virions in the host
organism.

The replication and release of viral particles occur at a very high rate. As the host
cell is donating a portion of its cell membrane to each viral particle that is released, its
plasma membrane needs to be replenished through the delivery of components via
vesicular transport. However, if the rate of budding becomes higher than the host cell's
ability to replenish the components of its plasma membrane, then lysis of the host cell
may result, thus causing host cell death and tissue damage.

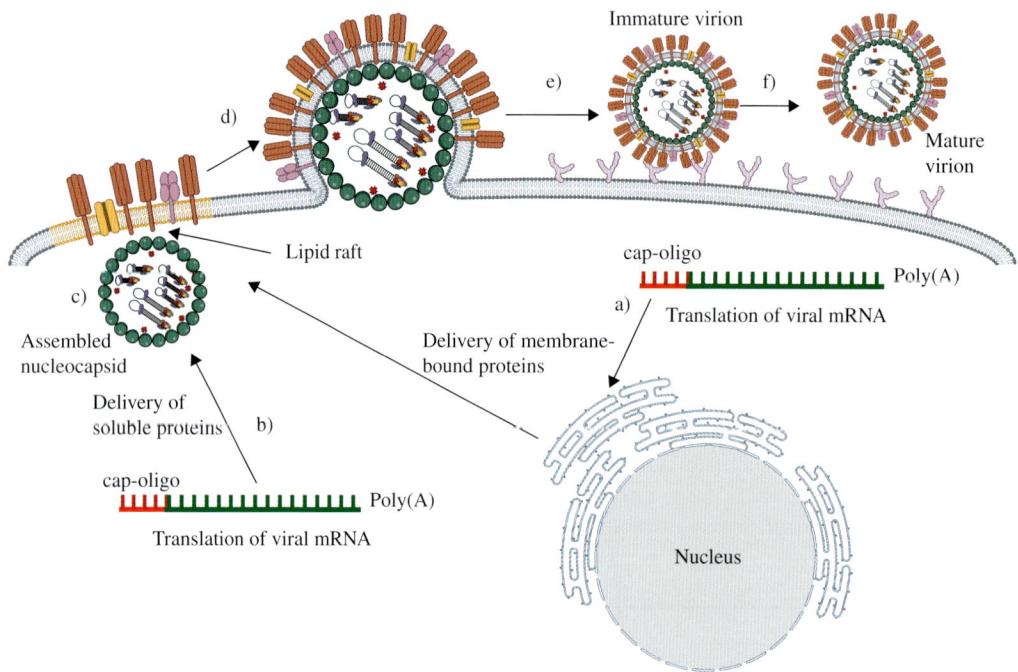

FIGURE 4.5 Assembly, release, and formation of mature influenza A. (a) Membrane-bound proteins
translated from viral mRNAs are sent to the ER for processing, then delivered to the plasma membrane
via vesicular transport. (b) Soluble proteins translated from viral mRNAs are trafficked toward the
plasma membrane. (c) Virion assembly occurs on the cytosolic side of the plasma membrane.
(d) Release of virions from the host cell is via budding. (e) An immature virion is immediately produced.
(f) Activity of NA allows the influenza A virus to mature and become a virulent viral particle.
Credit: Figure partially created in BioRender.

Dissemination in the Host Body

Influenza A virus replicates in both the upper and lower respiratory tract, but predominantly in the upper tract since α2,6-linked Neu5Ac is the prevalent form of sialic acid exhibited on epithelial cells in this region. This means that the initial primary infection is likely established in the trachea. Once released into the mucus layer, viral particles migrate toward the lower respiratory tract. To facilitate virus passage through the gel-like mucus layer, NA glycoproteins on the surface of the virus remove sialic acid from mucin, which is the major protein component of this respiratory secretion.

Once viral particles reach the lungs, influenza A virus can infect alveolar cells but not less likely since these cells do not express an overwhelming amount of α2,6-linked Neu5Ac. Consequently, a lower respiratory tract infection is unlikely unless the viral load passes a certain threshold. Acute-phase pneumonia may result due to an overactive inflammatory response caused by either a viral (other than influenza A), opportunistic bacterial, or fungal infection. If the influenza A virions from the primary upper respiratory tract infection are disseminated and cause a secondary infection in the lower respiratory tract, it is referred to as secondary viral pneumonia.

Opportunistic bacteria (most common are *Streptococcus pneumoniae/pneumococcus* and *Staphylococcus aureus*) can cause secondary infection in the lung as well, and this is referred to as bacterial pneumonia. A cytokine storm may be created to fight against the lower respiratory tract infection leading to pneumonia, which is the buildup of fluid in one's lungs that is the consequence of edema during the inflammatory response.

Pathogenesis and Clinical Manifestation

The pathology of influenza A viral infection is similar to that of SARS-CoV-2 infection. The "flu" caused by influenza A virus is an acute respiratory condition. The initial uncomplicated symptoms usually occur following an incubation period of 1–2 days. Primary symptoms include the sudden onset of high fever, chills, coryza (nasal discharge), sore throat and dry cough, headache, malaise, myalgia (muscle aches and pain) that leads to extreme physical weakness or emotional exhaustion, and inflammation of the upper respiratory tract. Malaise, headache, and myalgia are induced by the activity of IFN-α. Also, not all patients experience fever, especially elderly persons.

Fever is induced by pyrogens, either endogenous (like tumor necrosis factor alpha, interleukin 1b [IL-1b], IL-6, and IFN-γ secreted during a cytokine storm) or exogenous (like lipopolysaccharide and lipoteichoic acid from bacteria). Pyrogens activate hepatocytes and macrophages to synthesize and secrete prostaglandin E_2 (PGE_2). PGE_2 itself is considered a pyrogen when secreted at high amounts. PGE_2 in turn activates the hypothalamus to release hormones that regulate the degradation of multilocular adipose tissue (brown fat). Heat is released when adipose tissue is degraded, raising body temperature from 37° to as high as 39°. PGE_2 also regulates increased vasoconstriction, which leads to decreased loss of excess heat through the skin, therefore retaining the increased heat in the body. Fever is the body's defense mechanism to slow down viral replication and activity.

In most cases, these primary symptoms would dissipate within days to a week; however, some symptoms, like myalgia and malaise, may persist for weeks. Severe symptoms include hemorrhagic bronchiolitis, pneumonia, croup (swelling in the airway), and respiratory failure. Individuals who are at risk for severe symptoms associated with influenza A viral infection include infants, adults 65 years or older, and people with underlying medical conditions.

The underlying medical conditions can include chronic pulmonary disorders (eg, asthma), cardiovascular disorders, and diabetes mellitus. Such individuals are at very high risk of developing severe complications from influenza A viral infection. The complications can include cardiovascular disorders (eg, myocarditis); musculoskeletal disorders (eg, myositis, rhabdomyolysis); neurologic disorders (eg, encephalopathy, encephalitis); and multiorgan failure (eg, septic shock, renal failure, respiratory failure). Fatality may occur as little as 48 hours after the onset of the following symptoms: dyspnea (difficult and labored breathing), cyanosis (bluish discoloration of the skin due to low blood oxygen), hemoptysis (coughing up blood), and pulmonary edema leading to pneumonia.

An influenza A virus outbreak often occurs during fall into winter when the climate is temperate, meaning that the weather is warm to cool with humidity, not too hot or too cold. This is the perfect condition for viral particles to be transmitted via large droplets. Under these conditions, large droplets remain in the air longer, thus traveling farther without evaporating too quickly.

Diagnosis, Treatment, and Prevention

There are a number of diagnostic tests that can be performed to detect the presence of influenza A virus such as rapid diagnostic assays, serologic measurements, or a biopsy or autopsy tissue section analysis, confirmed by in situ hybridization or immunohistochemical techniques, and virus isolation by culture methods (rarely used). However, most influenza infections are diagnosed by their clinical symptoms, commonly known as the flu, and requires no laboratory tests. Physical examination and imaging are most often performed to detect pulmonary edema.

Diagnostic assays are generally implemented under special circumstances, such as diagnosing patients with severe symptoms as well as at-risk patients, like those with underlying conditions just discussed. These assays are also performed if their results are thought to be useful in planning the course of treatment or safeguard against possible complications. Rapid influenza diagnostic assays detecting the presence of viral components are the most widely used in hospital-based diagnostic laboratories. These assays can provide results quickly, within 30 minutes to an hour, and can detect viral components from samples of patients suspected to be infected.

Assay samples obtained by swabs or secretions of the nasopharyngeal area or throat secretions generally yield the highest sensitivity detection. Sensitivity is often higher at the onset of the disease, when a higher load of the virus exists. The "gold" standard for confirming influenza A virus infection is reverse transcription–polymerase chain reaction (RT-PCR) or viral culture. Some benefits for using the RT-PCR method of detection includes its ability to detect both viable and nonviable viral particles and the ability to be modified rapidly to adapt for the detection of novel targets (eg, new viral strains). A nonviable influenza A viral particle is one that is not infectious.

The US Food and Drug Administration (FDA) has approved several antiviral drugs to treat individuals who have a severe infection or are at higher risk for complications. Amantadine and rimantidine inhibit M2 ion channels, which in turn block membrane fusion and uncoating of the viral capsid. Oseltamivir (Tamiflu®), zanamivir (Relenza®), and peramivir (Rapivab®) inhibit NA activity. Baloxavir (Xofluza®) inhibits mRNA synthesis. A recent list of medications to treat symptoms of influenza A infection can be found in Table 4.2.

Most individuals who are not at high risk (eg, noninfants and adults younger than 65 years old and with no underlying medical condition) do not need the take

Medication	*Active Ingredients	Action of Medication
Alka-Seltzer Plus-D Multi-Symptom Sinus & Cold Liquid Gels (over-the-counter)	Acetaminophen, Dextromethorphan hydrobromide, Doxylamine succinate, Phenylephrine	Relieves headache, fever, body aches, stuffy nose, sinus congestion
Children's Tylenol Cold Plus Cough (over-the-counter)	Acetaminophen, Chlorpheniramine maleate, Dextromethorphan hydrobromide	Relieves fever, pain, cough, congestion
Comtrex Cold and Flu Maximum Strength Tablet (over-the-counter)	Acetaminophen, Chlorpheniramine maleate, Dextromethorphan hydrobromide, Phenylephrine	Relieves headache, fever, body aches, cough, sore throat, runny or stuffy nose, sneezing, sinus pressure
Coricidin HBP Cold & Flu (over-the-counter)	Acetaminophen, Chlorpheniramine maleate	Relieves fever, mild to moderate of headache and backache
Coricidin HBP Maximum Strength Flu (over-the-counter)	Acetaminophen, Chlorpheniramine maleate, Dextromethorphan hydrobromide	Relieves inflammation, runny nose, itchy nose, sneezing.
Decorel Forte Plus (prescription, over-the-counter)	Acetaminophen, Dextromethorphan hydrobromide, Guaifenesin, Phenylephrine	Relieves fever, pain, congestion, nasal secretions, coughs
Delsym Couth + Cold Daytime (prescription, over-the-counter)	Acetaminophen, Dextromethorphan hydrobromide, Phenylephrine	Relieves fever, pain, cough, chest congestion
Diabetic Tussin Night Time Formula (over-the-counter)	Acetaminophen, Dextromethorphan hydrobromide, Diphenhydramine hydrochloride	Relieves headache, fever, body aches, cough, runny nose, sneezing, sore throat
Mucinex Fast-Max Severe Congestion & Cold (prescription, over-the-counter)	Dextromethorphan hydrobromide, Guaifenesin, and Phenylephrine	Relieves cough, chest congestion, stuffy nose
Mucinex Fast-Max Severe Cold (prescription, over-the-counter)	Acetaminophen, Dextromethorphan hydrobromide, Guaifenesin, Phenylephrine	Relieves fever, pain, cough, chest congestion, stuffy nose

TABLE 4.2 Approved Drugs Used to Treat Influenza A Symptoms

Medication	*Active Ingredients	Action of Medication
Rapivab (prescription)	Peramivir	Prevents dissemination of mature viral particles
Sudafed PE Cold & Cough (prescription, over-the-counter)	Acetaminophen, Dextromethorphan hydrobromide, Guaifenesin, Phenylephrine	Relieves fever, pain, stuffy nose, sinus and ear congestion
Sudafed PI Pressure + Pain + Cold (over-the-counter)	Acetaminophen, Dextromethorphan hydrobromide, Guaifenesin, Phenylephrine	Relieves fever, pain, stuffy nose, sinus and ear congestion
Symmetrel (prescription)	Amantadine hydrochloride	Prevents dissemination of mature viral particles
Tamiflu (prescription)	Oseltaminvir	Prevents dissemination of mature viral particles
Theracaps Multi-Symptom Cough & Cold Reliever (over-the-counter)	Acetaminophen, Chlorpheniramine maleate, Dextromethorphan hydrobromide, Pseudoephedrine	Relieves fever, pain, sore throat, cough, nasal congestion
Triaminic Multi-Symptom Fever (over-the-counter)	Acetaminophen, Chlorpheniramine maleate, Dextromethorphan hydrobromide	Relieves fever, pain, sinusitis, bronchitis, watery eyes, itchy eyes/ nose/throat, runny nose, sneezing
Tylenol Cold (over-the-counter)	Acetaminophen, Chlorpheniramine maleate, Dextromethorphan hydrobromide, Phenylephrine	Relieves fever, pain, stuffy nose, sinus, ear congestion
Tylenol Cold & Flu Severe (prescription, over the counter)	Cetaminophen, Dextromethorphan hydrobromide, Guaifenesin, Phenylephrine	Relieves sore throat, cough, congestion, fever, minor aches
Tylenol Cold Multi-Symptom Severe (prescription, over-the-counter)	Acetaminophen, Dextromethorphan hydrobromide, Guaifenesin, Phenylephrine	Relieves fever, pain, headache, sore throat, nasal congestion, cough by loosening phlegm
Vicks DayQuil Severe Cold & Flu (prescription, over-the-counter)	Acetaminophen, Dextromethorphan hydrobromide, Guaifenesin, Phenylephrine	Relieves fever, cough, stuffy nose, body aches, headache, sore throat, sinusitis, bronchitis

TABLE 4.2   Approved Drugs Used to Treat Influenza A Symptoms (*Continued*)

Medication	*Active Ingredients	Action of Medication
(no trade name)	Rimantidine hydrochloride (prescription)	Prevents dissemination of mature viral particles
Xofluza (prescription)	Baloxavir marboxil	Prevents dissemination of mature viral particles
Zanamivir (prescription)	Relenza	Prevents dissemination of mature viral particles

*Mechanisms of ingredients:
- Acetaminophen inhibits Cyclooxygenase activity to relieve pain and reduce fever.
- Amantidine hydrochloride inhibits M2 ion channels to block membrane fusion and uncoating of the viral capsid. Less active than Rimantidine.
- Baloxavir marboxil inhibits PA endonuclease to prevent viral protein synthesis.
- Cetaminophen inhibits Cyclooxygenase activity to relieve pain and reduce fever.
- Chlorpheniramine maleate blocks histamine from binding to receptors on cells in the respiratory tract.
- Dextromethorphan hydrobromide is a synthetic analog of codeine that increases the cough threshold by depressing the cough center in the medulla oblongata through sigma receptor stimulation (acts as a cough suppressant).
- Diphenhydramine hydrochloride acts as an antagonist of the H1 receptor to reverse the effects of histamine on capillaries, thus reducing allergic reaction symptoms.
- Doxylamine succinate reduces nausea and vomiting by inhibiting histaminergic signaling to the vomiting center in the medulla.
- Guaifenesin selectively blocks polysynaptic reflexes in the spinal cord, reticular formation, and subcortical areas of the brain to act as an expectorant (or cough medicine) to help clear mucus from the airways to relieve congestion.
- Oseltaminvir inhibits neuraminidase activity to prevent virion maturation.
- Peramivir mimics cell surface sialic acid and acts as a neuraminic acid competitor to inhibit neuraminidase activity, thus preventing virion maturation.
- Phenylephrine is an agonist of α_1-adrenoceptors in the arterioles of the nasal mucosa and has nasal decongestant action.
- Pseudoephedrine directly stimulates beta-adrenergic receptors and indirectly stimulates alpha-adrenergic receptors to cause the release of norepinephrine (NE) from neurons, leading to reduced swelling of blood vessels in the nose for nasal congestion relief.
- Relenza inhibits neuraminidase activity to prevent virion maturation.
- Rimantidine hydrochloride inhibit M2 ion channels to block membrane fusion and uncoating of the viral capsid. More active than Amantidine.

TABLE 4.2   Approved Drugs Used to Treat Influenza A Symptoms (*Continued*)

antiviral drugs. Additionally, taking any of these drugs within the first 24 hours of appearance of mild symptoms is expected to result in the greatest positive benefit. The most common reported side effects are nausea and vomiting for oseltaminvir, difficulty breathing with wheezing for zanamivir, and diarrhea for peramivir.

To treat secondary bacterial infection, macrolide antibiotics, like azithromycin and erythromycin, have been shown effective against bacterial pneumonia for adults. To treat bacterial pneumonia in children, amoxicillin is commonly used.

The simplest strategy to prevent influenza A virus infection is to avoid close contact with known infected individuals, especially in small, unventilated spaces. Regular handwashing and avoiding touching eyes, nose, and mouth; avoiding direct contact with infected individuals; and practicing good health habits such as proper nutrition, enough sleep, and not smoking are all thought to reduce the chance of infection.

Vaccination, leading to immunization, is an effective measure to prevent influenza A infection or at least reduce the symptoms associated with an infection. Vaccination is the act of introducing a vaccine into the body, while immunization is the process by which one's body becomes protected against a disease through vaccination. Unfortunately, a vaccine may provide protection for only a short period of time, usually months, not years.

Additionally, due to antigenic drift, a flu vaccine for 1 year might no longer be protective the following year. Therefore, the Centers for Disease Control and Prevention (CDC) recommend getting vaccinated annually, especially for those in high-risk populations. Vaccinated individuals can still get infected but are less likely to come down with severe symptoms since the body is better able to clear invading viral particles more efficiently at a higher capacity.

Influenza A flu vaccines have a 70%–90% efficacy rate in preventing influenza A virus–induced illness among study groups of healthy young adults and only slightly lower efficacy in individuals 65 and older. From recent studies, the CDC reported that vaccinated individuals spent 4 fewer days in the hospital, once admitted, and were 26% less likely to be admitted to the intensive care unit than those who are unvaccinated. Additionally, those who were vaccinated had a 31% lower risk of fatality from influenza A virus–induced illnesses.

Many of the antiviral drugs mentioned have not only therapeutic effects but also prophylactic (preventive) effects. To take advantage of the full protective effect of these drugs, they must be administered continuously when the viral load is high. Both of the NA inhibitors, zanamivir and oseltamivir, are effective in preventing influenza when administered prophylactically.

Current Status

Annually, between 5% and 20% of the population in the United States contracts the flu. From 2010 to 2019, deaths resulting from influenza infection and pneumonia in the United States were 12.3 to 15.7 per 100,000 population. According to the WHO, between 290,000 and 650,000 deaths occur annually due to influenza-related pneumonia and respiratory illnesses worldwide. These US and worldwide death rates are higher when including pneumonia-related deaths caused by all secondary diseases that stem from the initial influenza infection, such as opportunistic bacterial pneumonia.

Unfortunately, the exact numbers cannot be tracked due to a number of difficulties and flaws in collecting these data. For instance, individuals who are infected and even have serious symptoms might not go to the hospital or visit a physician. Related illnesses and deaths may be recorded without any reference to an influenza infection since the etiology was not known.

To provide a better estimate of these data to include unknown and unreported cases, the CDC and other public health agencies in the United States and around the world use statistical models to estimate the annual number of seasonal influenza-related cases, hospitalizations, and deaths. This projection is known as the burden of flu disease, which is an annual estimate of the number of influenza-induced illnesses, medical visits, hospitalizations, and fatalities. This estimate is used to make informed policy, guide communications about the risk of infection, and make preparations for the given flu season. This estimate is also used to gauge the impact of flu vaccination and provide an idea of variations that occur from season to season.

The projected range of the average estimated annual burden of flu disease in the United States from 2010 to 2022 was up to 41 million illnesses; 140,000–710,000 hospitalizations; and 12,000–52,000 fatalities. The actual numbers reported by the CDC in the 2021–2022 influenza season were considerably lower at 9 million illnesses, 4 million medical visits, 10,000 hospitalizations, and 5000 deaths. As you can see, the number of hospitalizations and deaths are drastically different.

Scientific Significance and Discoveries

The 1918 pandemic caused by H1N1 influenza A virus was the worst pandemic in the 20th century in terms of deaths caused. The search for causes and cures of this tragedy greatly contributed to a number of research tools and clinical technologies that are still being used today. The most significant innovation, a chick-embryo system, was developed at Vanderbilt University in 1931 by Ernest William Goodpasture and his colleagues. At this time, viruses had to be grown on living tissue, either recovered from a living infected host or grown in tissue culture. Both techniques were not satisfactory for viral research for different reasons. Growing a virus in a living host was expensive and difficult to control. Tissue culture was susceptible to contamination without added antibiotics, which were not yet developed at this time.

The chick-embryo system involved meticulously cutting a small pore in an eggshell while avoiding damaging the vitelline membrane to help preserve the sterility of the egg. A fertile egg provided a cheap living environment for viral growth. An isolated strain of influenza virus was then injected through the membrane into the egg. Finally, the eggshell opening was sealed with a piece of glass. This chick-embryo system was reproducible and produced consistent pure cultures of a variety of viruses. This technique was so successful that it was used to produce viral particles in large enough quantities to enable the production of an influenza vaccine a few years later by Dr. Thomas Francis.

Ebola Virus

Historical Perspective

On July 20, 2014, Patrick Oliver Sawyer, a Liberian American lawyer, traveled from Liberia and arrived at the international airport in Lagos, Nigeria. He exhibited symptoms before, during, and after the flight, where he vomited midflight. Upon arrival at the airport, he was driven to a private hospital in a private car. He told the hospital staff that he might have malaria but denied that he was in contact with any individuals who'd be infected with the Ebola virus when he had actually been under observation in a hospital in Monrovia, Liberia, for possible Ebola virus infection. Sawyer came into contact with his sister, who was infected with the Ebola virus. He visited her in the hospital, then later attended her funeral and burial ceremony. He developed a fever and other symptoms but left the hospital against medical advice. He was transported to a private hospital, where he was reported to have fever, vomiting, and diarrhea. He died 5 days after arriving in Lagos. An investigation identified 894 individuals with whom he may have come into contact. The police officer who transported him to the hospital, along with 4 of the 9 doctors and nurses who came into contact with him, died within a few days. Sawyer was recorded to be the index case (or patient zero) for the introduction of the Ebola virus into Nigeria during the West African Ebola epidemic in 2014.

There have been many outbreaks of Ebola virus through the decades. The first reported outbreak in human population was in 1976 when a 44-year-old school teacher in Zaire (now the Democratic Republic of the Congo), near the Ebola River valley, was recorded as the index patient. As with the above 2014 outbreak, the initial diagnosis was malaria. However, his symptoms became much more serious, starting with violent vomiting, then bleeding from his mouth, nose, and anus. He died roughly 2 weeks later. More than 300 individuals were admitted to the hospital with similar symptoms, and most died.

The last reported outbreak started on September 20, 2022, in Mubende District, Uganda. This was the sixth confirmed Ebola virus outbreak in Uganda. There were 164 reported (142 confirmed and 22 probable) cases, with 55 deaths during this outbreak. Ugandan officials declared the end of this outbreak on January 11, 2023.

By far the largest Ebola outbreak was reported by the WHO on March 23, 2014, which started in Liberia, as discussed at the beginning of this section. There were 28,610 cases and 11,308 deaths. It spread through Guinea, Liberia, and Sierra Leone and is known as the West African Ebola epidemic.

Suffice it to say, Ebola virus is one of the most contagious and virulent human viruses, with a high rate of fatality and an extremely short incubation period.

Classification and Structure of Ebola Virus

Ebola virus belongs to the Filoviridae family. This family includes six genera: *Ebolavirus* (eg, Ebola virus), *Marburgvirus* (eg, Marburg virus and Ravn virus), *Thamnovirus* (eg, Huángjiāo virus), *Striavirus* (eg, Xīlǎng virus), *Cuevavirus* (eg, Bombali virus), and *Dianlovirus* (eg, Měnglà virus). Four of the 5 known Ebola virus species cause disease in humans: Zaire ebolavirus, Sudan ebolavirus, Taï Forest ebolavirus, and Bundibugyo ebolavirus. The fifth, Reston ebolavirus, causes disease in nonhuman primates but not in humans.

The structure of the virus is an enveloped virion that is filamentous in shape with a length of 1–2 μm and diameter of 80 nm. The single minus (antisense)-strand RNA genome of 18,960 bases is encapsidated by a helical complex of capsid proteins (Figure 4.6). The genome contains 7 open reading frames (ORFs) that encode for 10 viral proteins: nucleoprotein (NP), viral protein 35 (VP35), VP40, VP30, VP24, RNA polymerase ("large" L protein), and glycoprotein (GP). The polymerase L protein is packaged inside the viral envelope along with the polymerase co-factor VP35, and transcription factor VP30. Molecules of the matrix protein VP40 form a layer between the helical capsid and viral envelope. The IFN-response antagonist VP24 is integrated into the VP40 layer. GP undergoes further post-translational processing to produce three smaller products as well as the full-length protein: soluble secreted glycoprotein (sGP), soluble small secreted glycoprotein (ssGP), enterotoxin Δ-peptide, and full-length

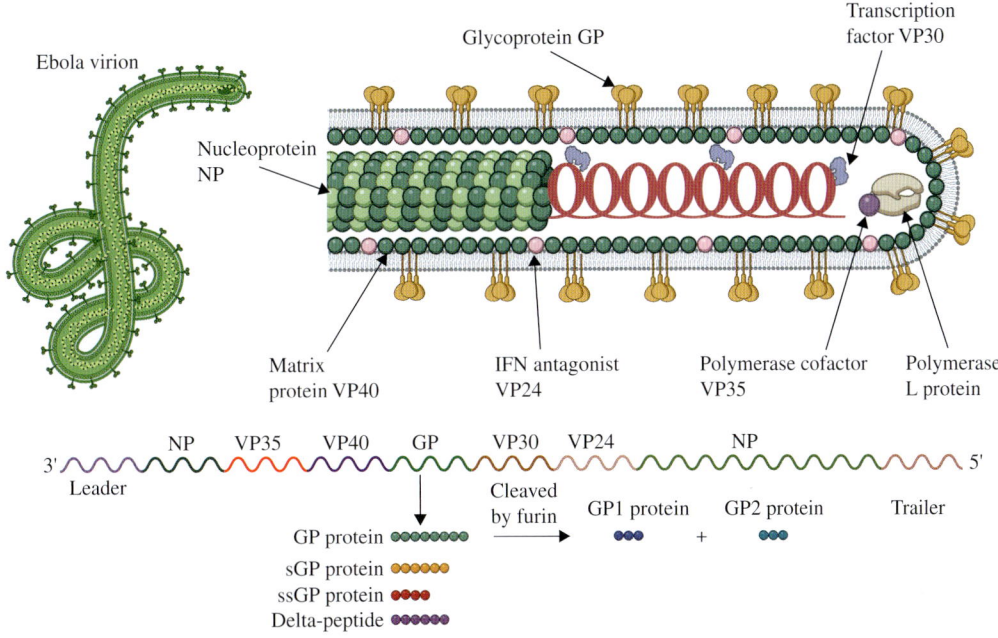

FIGURE 4.6   Ebola virus capsid structure and genome. (a) The Ebola viral particle is an enveloped virion surrounding a nucleocapsid. The envelope itself presents numerous molecules of the GP used for attachment to the host cell plasma membrane. There is a protein layer composed of matrix VP40, interlaced with VP24, between the envelope and the capsid. The helical capsid encapsidates the viral genome that is complexed with several protein to form an RNP complex. (b) The Ebola genome consists of a single, linear minus-strand RNA containing 7 ORFs that encode for 11 functional proteins. *Credit:* Figure partially created in BioRender.

transmembrane spike GP. The full-length GP is further cleaved by host cell protease, furin, into GP1 and GP2. Large L protein performs a number of activities, including RNA synthesis, capping, and polyadenylation.

Host Range, Transmission, Tropism, and Susceptible Host Cell

Even though Ebola virus infection is rare in the human population, it can be deadly, resulting in life-threatening conditions such as hemorrhagic fever, intravascular coagulation, and liver dysfunction. It is a known zoonotic virus with a broad host range, including humans, monkeys, gorillas, chimpanzees, and bats. Strains of Ebola virus that infect humans most likely come from bats.

When infecting primates, Ebola virus is known to be pantropic, which means that it affects a spectrum of host cells without showing affinity for any particular one. This virus is able to invade almost all human cells because the viral envelope GP is able to interact and attach to a variety of receptors for entry into the host cell. However, Ebola viruses most often target hepatocytes, endothelial cells, adrenal cells, dendritic cell, monocytes, macrophages, and fibroblasts.

Susceptible host cells exhibit one or more of the following entry receptors: human folate receptor α (HFR-α), β1 integrins, TYRO3 (protein tyrosine kinase) receptor tyrosine kinase family members, T-cell immunoglobulin, and mucin domain 1 (TIM1), dendritic cell-specific intercellular adhesion molecule-3-grabbing nonintegrin (DC-SIGN), liver/lymph node-specific ICAM-3 grabbing nonintegrin (L-SIGN), and human macrophage galactose and *N*-acetyl-galactosamine-specific C-type lectin (hMGL). There is a link between the life-threatening conditions listed above and the list of susceptible cells exhibiting surface receptors for Ebola viral particles. In addition, Niemann-Pick C1 (NPC1; cholesterol transporter) is required for a susceptible cell to be permissive as well.

Ebola virus is transmitted through direct contact with contaminated bodily fluids. In an infected individual, viral particles can be found mostly in blood, but also in bodily fluids such as urine, saliva, sweat, feces, vomit, breast milk, ocular fluid, cerebrospinal fluid, and semen. The virus can also be present in organs like the skin. Therefore, viral particles can easily exit the infected host via a number of portals.

Once exited into the extracellular environment, the virus can be transmitted via direct person-to-person contact (ie, shaking hands, through sexual intercourse, or even breastfeeding). Transmission can easily occur via brief or even slight contact with objects (fomites) such as clothing, beddings, needles, hospital walls, and more where viral particles can survive up to nearly 15 days on these dry surfaces. Since Ebola virus is a known zoonotic species, transmission to a human host is possible via contact with blood or body fluids from infected nonhuman animals. Ebola virus is thought to have been introduced into the human population through direct contact with fruit bats, chimpanzees, gorillas, monkeys, forest antelope, or porcupines found ill or dead or in the rainforest. Transmission can occur even if an individual prepares, cooks, or eats an infected nonhuman animal, which can likely and often happen in some parts of the world.

Ebola virus can easily enter a new human host via a number of routes of entry. Viral particles can enter when touching one's eyes, nose, or mouth or simply being sprayed with contaminated fluids while talking, exercising, caring for an infected patient, or even direct contact with the deceased during burial ceremonies. Ebola is not known as an airborne virus; however, contaminated fluids may enter through any opening in the body, even broken skin.

Infecting the Host Susceptible Cell, Viral Particle Replication, and Tissue Damage

Attachment occurs when the viral envelope GP is recognized and bound by one of the entry receptors on the surface of the susceptible host cell. This interaction induces the internalization of the virus via not only mainly micropinocytosis but also clathrin-mediated endocytosis or caveolin-mediated endocytosis, depending on the shape of the viral particle (Figure 4.7). Once in the endosome, GP is proteolytically modified by the host cell's cathepsin B and cathepsin L. Cleaved GP fragments are now able to bind to NPC1, leading to the fusion of viral envelope to endosomal membrane. Membrane fusion and uncoating are aided by the acidic environment of the endosomal lumen. The viral RNP complex can now be released into the cytosol.

Ebola and influenza A viruses not only share some similar infection mechanisms but also differ in others. Despite the influenza A viral genome being composed of 8 segments while the Ebola viral genome is composed of 1 segment, both encode for the same number of viral proteins. Another difference is the location of viral genome and

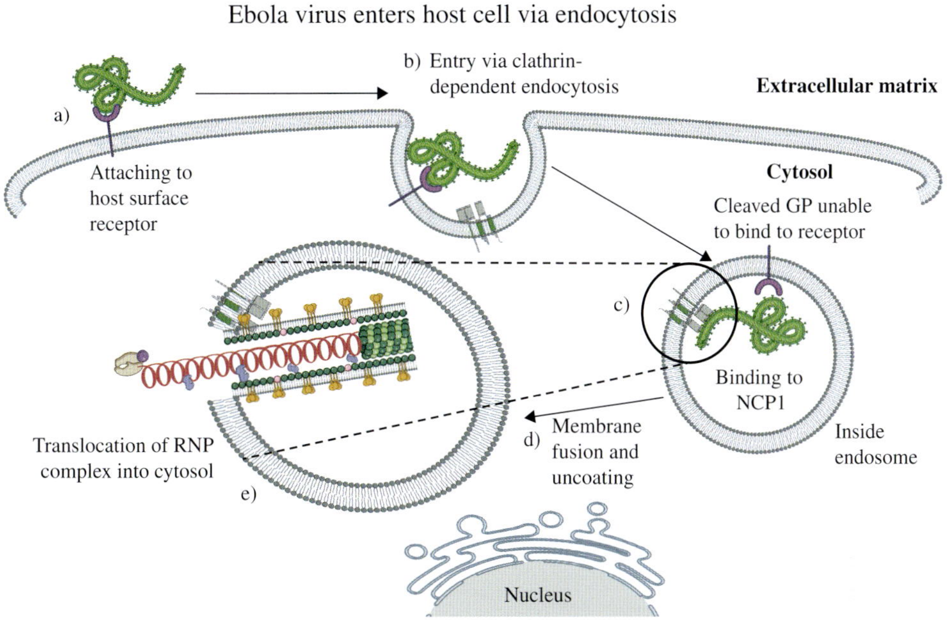

Ebola virus enters host cell via endocytosis

FIGURE 4.7   Entry of Ebola virus into the host cell. (a) Ebola virion attaches to a host cell receptor. (b) Host cell internalizes the viral particle via endocytosis. (c) Cleavage of viral GP enables the binding to NPC1. (d) The interaction of the virion to NPC1 induces membrane fusion between the viral envelope and endosome. This interaction also leads to uncoating of the capsid. (e) The RNP complex is delivered to the cytosol. *Credit:* Figure partially created in BioRender.

mRNA synthesis in the host cell. Whereas influenza A RNA synthesis occurs in the nucleoplasm, Ebola viral RNA synthesis occurs in the cytosol.

Viral genome replication begins with the synthesis of the complementary plus strand in the cytosol of the host cell (Figure 4.8). The full-length plus-strand RNA is then used as the template for the synthesis of more minus-strand viral RNAs to be assembled into virions. Both activities are catalyzed by the viral RdRP. Viral protein VP30 is also crucial for the regulation of RNA transcription initiation. VP30 activity depends on its ability to bind zinc ions, interaction with NP, and phosphorylation status.

For viral protein synthesis, RdRP also uses the minus-strand viral RNA as the template to synthesize 7 different mRNAs. The 5′ end of each mRNA is capped by the large L protein via a noncanonical pathway, which is different from the capping mechanism used by eukaryotic host cells. Polyadenylation is also regulated by the large L protein using the polyadenylation site at the 3′ end of each mRNA. Thus, viral protein synthesis is performed by host cell machinery through cap-dependent initiation of translation.

Several nonstructural viral proteins are important in viral replication and host immune response evasion. NP binds to various host cell proteins to help direct and facilitate viral RNA synthesis and virion assembly.

VP35 binds dsRNA, thus preventing the recognition of viral dsRNAs from being recognized by the host cell's intracellular pattern recognition receptors, such as retinoic acid–inducible gene I (RIG-I) and melanoma differentiation-associated protein 5 (MDA-5).

Ebola viral RNA and protein synthesis

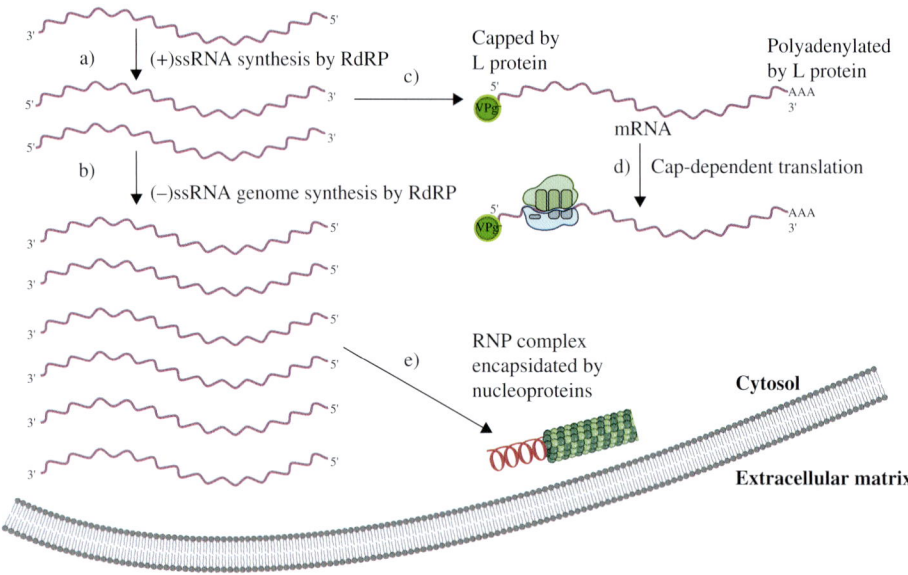

Figure 4.8   Genome transcription and gene expression. (a) The (−) ssRNA genome is used as the template for the synthesis of complementary (+) ssRNAs. (b) The (+) ssRNA is used as the template for the synthesis of viral genome molecules. (c) The (+) RNA is capped and polyadenylated by viral L proteins, producing an mRNA. (d) Viral protein synthesis occurs via cap-dependent translation. (e) Viral genome and proteins are transported toward the plasma membrane for virion assembly. *Credit:* Figure partially created in BioRender.

VP35 can also inhibit the activation of the tissue-specific transcription factor (TS-TF) interferon-regulatory factor 3 (IRF-3), which is a critical step in the host interferon response. Last, VP35 can inactivate protein kinase R (PKR; an antiviral protein), thus enabling continuous viral protein synthesis.

Soluble secreted glycoprotein (sGP) has been shown to inhibit pro-inflammatory cytokine production from noninfected macrophages. They can also impair chemotaxis of activated macrophages, thus reducing the number of phagocytes at the site of infection. A large concentration of sGP molecules is found in the blood at the early stages of the infection. This leads researchers to believe that sGP may contribute to host immune evasion by acting as a decoy for anti-GP antibodies.

VP24 can also inhibit IFN responses by blocking the p38MAPK (mitogen-activated protein kinase) pathway and inhibiting the activation of nuclear factor–κB (NFκB), which is a major TS-TF of the IFN response pathway. VP24 may also play a role in initiating the budding process of virion release.

Once produced, sGP molecules are trafficked to the plasma membrane via vesicular transport where they undergo glycosylation and proteolysis and are targeted to lipid rafts (Figure 4.9). Other viral proteins as well as the RNP complex are transported toward the plasma membrane for assembly along cytoskeletons. Viral nucleocapsids

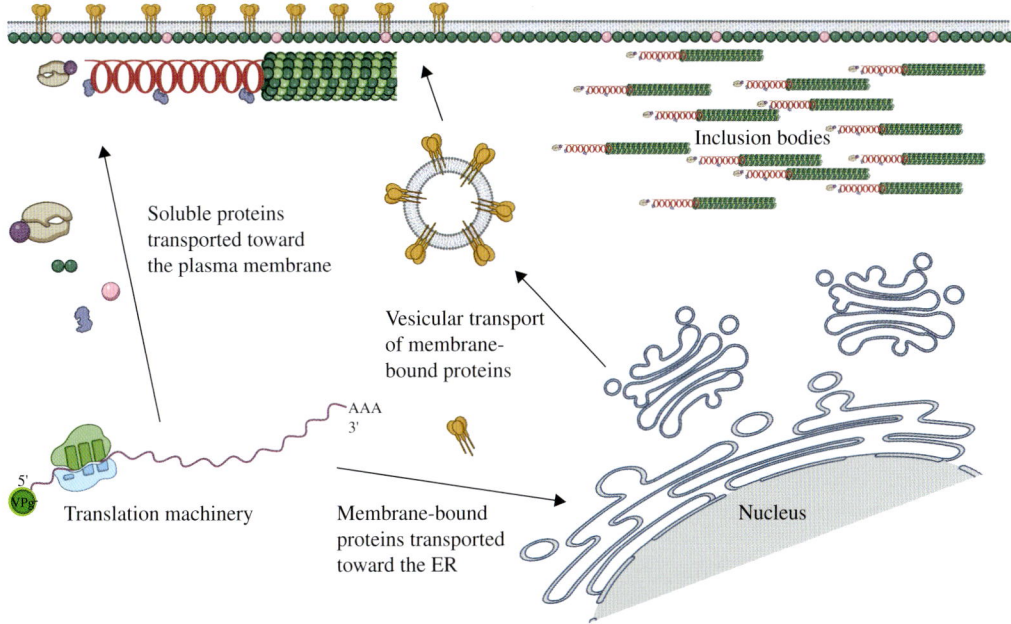

Inclusion bodies

Soluble proteins
transported toward
the plasma membrane

Vesicular transport
of membrane-
bound proteins

AAA
3'

5'
VPg   Translation machinery

Membrane-bound
proteins transported
toward the ER

Nucleus

FIGURE 4.9   Viral protein transport, virion assembly, and inclusion bodies. Viral proteins are synthesized in the cytosol. Soluble proteins are transported toward the plasma membrane along with genomic RNA molecules. Assembly of virions occurs near the cytosolic side of the plasma membrane, producing inclusion bodies. Ebola viral particles are released from the host cell via budding, where they acquire their envelope from the plasma membrane. *Credit:* Figure partially created in BioRender.

are assembled in the cytosol of the host cell and are identified as inclusion bodies under the light microscope at low magnification.

The release (or exit) of virions is through budding; therefore, the viral envelope comes from the host plasma membrane. Similar to the effect of influenza A virus, Ebola virus undergoes a high rate of replication and release from the host cell. If the rate of budding becomes higher than the host cell's ability to replenish the components of its plasma membrane, then lysis-caused death of the host cell may result, thus causing tissue damage.

Dissemination in the Host Body

Since Ebola virus has a wide range of tropism, dissemination from the primary infection and the establishment of secondary infections almost always occurs. Hematogenous dissemination is the major mode of spreading viral particles throughout the infected individual. While in the bloodstream, this virus can infect and cause damage to not only blood cells (eg, monocytes), but also endothelial cells. Dissemination into the lymphatic system is also extremely likely, leading to infection of lymph nodes and lymphatic organs. Neurogenous dissemination can lead to a number of neurologic complications, such as seizures, memory loss, headaches, cranial nerve abnormalities, and tremor.

Pathogenesis and Clinical Manifestation

Early symptoms of an Ebola virus infection can be sudden and may include nonspecific flu-like symptoms such as fever, malaise, myalgia, headache, and sore throat. The infection often gets resolved by the body's immune system within 14 to 21 days; however, as the infection progresses, the individual may exhibit more serious symptoms, including vomiting, diarrhea, rash, and coagulation abnormalities. Severe external bleeding as well as gastrointestinal bleeding may also occur, along with hematological abnormalities such as lymphopenia (low lymphocyte counts) and neutrophilia (high neutrophil counts). The latter stage of the infection, if remaining untreated or not treatable, involves exaggerated inflammatory responses due to cytokine storm. The sGP, also known as shed GP, binds to and activates entry receptors on the surface of noninfected dendritic cells and macrophages to secrete massive amounts of pro-inflammatory cytokines, leading to a cytokine storm.

Once in the host organism, Ebola virus most often targets macrophages and dendritic cells first. If host entry is via passive viremia where viral particles are transmitted directly into the host bloodstream, monocytes are a likely first target. Ebola virus causes problems for many host tissues in different ways.

- They attack a number of leukocytes, thereby reducing the effectiveness of the host immune system. Their rapid replication induces heightened innate immune responses, including a cytokine storm, which tend to cause great tissue damage.

- By attacking the spleen and kidneys, the body is unable to regulate its fluid and chemical balance satisfactorily and cannot produce sufficient clotting factors to deal with blood vessel damage.

- Liver damage means that the body is unable to produce adequate amounts of plasma proteins, some of which are important components of the innate immune system,-such as complement proteins and acute-phase proteins.

- The combination of liver damage and heightened viremia can lead to intravascular coagulopathy, which is the presence of blood clotting in one's blood vessels. The destruction of the endothelium and vascular integrity contributes to hemorrhagic fever symptoms, hypertension, and capillary leaks. Diffuse bleeding and hypotensive shock, leading to progressive multiorgan failure, account for a high number of fatalities during the terminal stages.

- Dysfunction of the gastrointestinal tract causes vomiting and diarrhea.

In summary, the extraordinarily high rate of Ebola virus replication overwhelms the host immune defenses. Different from many other viral infections already discussed, Ebola virus can only be transmitted from an infected individual with symptoms. Meaning that infected individuals who are asymptomatic are not known to be contagious. The incubation period for Ebola virus infection is 2 to 21 days. Therefore, if an individual had been in contact with a known infected patient but does not exhibit symptoms within 21 days, then chances are this individual is either not infected or not contagious.

Also, unlike many other viruses, Ebola virus can be transmitted even after the infected host had died. According to the WHO, an individual who had survived an Ebola virus infection may continue to experience symptoms for 2 or more years, referred to as post-Ebola virus disease (EVD) syndrome. Common sequelae in survivors include feeling tired, headache, arthralgia, myalgia, eye pain and vision problems, weight gain, abdominal pain and loss of appetite, hair loss and skin problems, trouble sleeping, memory loss, hearing loss, and possibly depression and anxiety.

Diagnosis, Treatment, and Prevention

Symptoms for an Ebola virus infection are very similar to those of other infectious diseases like malaria, typhoid fever, and meningitis; therefore, it can be quite difficult to obtain a specific diagnosis of Ebola virus infection clinically. Most often, symptoms are attributed to malaria, especially in regions of the world where malaria is prevalent. There are, however, a number of differential diagnostic assays to detect and confirm Ebola virus infection, including antibody-capture enzyme-linked immunosorbent assay, antigen-capture detection tests, serum neutralization test, RT-PCR assay, electron microscopy, and virus isolation by cell culture.

It is important to note that early care and treatment greatly improve one's chance of surviving an Ebola virus infection; therefore, an individual showing symptoms should seek medical care immediately.

For treatment, the WHO recommend two monoclonal antibody therapies: ansuvimab (formerly known as mAb114, whose brand name is Ebanga™) and REGN-EB3 (brand name Inmazeb®), developed by Regeneron Pharmaceuticals (Regeneron) to treat Ebola virus infection. The REGN-EB3 is a cocktail of three fully-human monoclonal antibodies that includes atoltivimab (REGN3470), maftivimab (REGN3479), and odesivimab (REGN3471). Amsuvimab is a neutralizing antibody that is introduced intravenously into patients who have Ebola virus–induced diseases. Blood transfusions are given to treat diffuse bleeding. The patient should receive supportive care with rehydration and treatment with medicine to eliminate or control symptoms such as pain, nausea, vomiting, and diarrhea.

Vaccines are the primary and most important means to prevent most viral infections and to control spread of the virus. Recombinant vesicular stomatitis virus–Zaire ebolavirus (rVSV-ZEBOV), under the brand name Ervebo®, is an FDA-approved vaccine to prevent Ebola virus–related illnesses for individuals aged 1 year and older. Ad26.ZEBOV (brand name Zabdeno®) and MVA-BN-Filo (brand name Mvabea®) are not FDA-approved vaccines, but they are approved for medical use in the European Union. Two other vaccines not approved by the FDA are Ad5-EBOV (approved in China) and rVSV/Ad5 (approved by the Russian Federation).

People can protect themselves from Ebola virus transmission through simple but effective preventive measures like washing hands thoroughly, avoiding contact with the body fluids of infected individuals, and not coming in direct contact with deceased bodies of infected individuals.

The WHO has established effective interventions, such as case management, surveillance and contact tracing, good laboratory service, safe burials, and social mobilization to provide successful control of an outbreak. The WHO is assisting countries to maintain surveillance for and prevent Ebola virus outbreaks by managing preparedness, alert, control, and evaluation protocols. They have procedures in place to respond to an outbreak by supporting community engagement, disease detection, contact tracing, vaccination, case management, laboratory services, infection control, logistics, and training and assistance with safe and dignified burial practices.

Health care workers are in the front line and are very susceptible to Ebola virus infection when caring for patients. Those who care for suspected or confirmed infection must be extra vigilant in protecting themselves by using personal protective equipment to shield from direct contact with fluids or other contaminated materials, good hand and respiratory hygiene, and safe injection practices. Clinical laboratory technicians should also use proper equipment and procedures to protect themselves from contaminated samples.

Current Status

The latest outbreak ended in Uganda on January 11, 2023, as declared by Ugandan officials. Overall, the average Ebola virus case fatality rate is roughly 50%, varying from 25% to 90% between 1976 and 2023. The effectiveness and swiftness of the response greatly influence the rate of fatality.

Many lessons were learned from these past outbreaks, enabling the partnership between the CDC and WHO to developed global policies to monitor, report, and mount swift and effective response to control future Ebola virus outbreaks. Ebola virus vaccines remain the most effective means to prevent and fight against possible future outbreaks.

Therefore, a major effort to battle Ebola virus diseases and control outbreaks is currently dedicated to the development of several types of vaccines that demonstrate efficacy, potency, durability, and cost-effective methodologies in development. A number of vaccines against the Ebola virus are under development and testing. The following are strategies used by researchers for vaccine development:

Virus-Like Particles Vaccine: A baculovirus-based expression system regulated in insect cells is used to amplify the production of virus-like particles (VLPs) containing matrix protein VP40 and GP (and sometimes NP). Multiple vaccines containing VLPs were successfully developed, which have been shown to trigger innate, B-cell–mediated, and T-cell–mediated immune responses. This type of vaccine is effective and safe but at present quite expensive.

DNA Vaccine: Vaccines based on plasmids containing either GP or NP are used to develop DNA vaccines. There are many benefits to the use of plasmids. Plasmids are noninfectious, simple to produce in vast quantities, and can be modified and adaptable as the pathogen undergoes its own genetic alteration. This type of vaccine can induce both B-cell–mediated and T-cell–mediated immune responses.

Recombinant Whole-Virus Vaccine: Whole virus, whether live attenuated or inactivated, is generally the most immunogenic since the host immune system can respond broadly (against multiple viral components) and vigorously to a complete viral particle. Even though 100% of animal study subjects survived, this is the most uncertain type of vaccine from the point of view of safety. Since Ebola virus can generally cause the host immune system to overrespond, whole-virus vaccine may also cause symptoms related to overactive immune responses.

Replication-Incompetent Vaccine: The replication-incompetent vaccine contains viral particles that had lost their ability to replicate and do not release shed GP to upregulate innate immune responses. The induction of the B-cell–mediated immune response is the primary target of this vaccine.

Replication-Competent Vaccine: A replicating-competent vaccine induces the production of a high titer of neutralizing antibodies. This vaccine is based on the introduction of a recombinant virus that contains one select Ebola viral gene: GP, NP, polymerase cofactor VP35, or VP40. Ervebo is a replication-competent vaccine. This vaccine is durable and immunogenic, but there are safety concerns.

Scientific Significance and Discoveries

Learning from the lessons of previous outbreaks throughout the past 6 decades, numerous worldwide organizations have developed and expanded their experience and expertise to guide a quick, effective response to future outbreaks.

Following the 2014 outbreak that reached the United States, a national network of treatment centers was established to support national preparedness against outbreaks from the Ebola virus and other special pathogens. There are 13 Regional Emerging Special Pathogens Treatment Centers across the country. These centers are equipped with biocontainment units and are staffed by highly trained health care professionals to manage patient needs.

Health care professionals at these centers provide resources, education, training, research, and assessments of operational readiness to health care workers at other medical facilities. They continually participate in training with the biocontainment units and devise ways to incorporate technology and acquired knowledge to respond to outbreaks of infectious diseases. They also communicate with international organizations to exchange information and best practices.

No new methods were invented, but traditional laboratory diagnostic technology was greatly improved in light of the many Ebola virus outbreaks. For instance, more accurate molecular assays and faster and more efficient diagnostic methodologies had been developed. Field diagnostic laboratories are more widely deployed, and diagnostic testing has also been made simpler and more sensitive, yet more efficient.

Further Readings

Bouvier N, Palese P. The biology of influenza viruses. *Vaccine*. 2008;26(4):D49–D53.

Broadhurst M, Brooks T, Pollock N. Diagnosis of Ebola virus disease: Past, present, and future. *Clin Microbiol Rev*. 2016;29(4):773–793.

Du R, Cui O, Rong L. Competitive cooperation of hemagglutinin and neuraminidase during influenza A virus entry. *Viruses*. 2019;11(5):458.

Falasca L, Agrati C, Petrosillo N, et al. Molecular mechanisms of Ebola virus pathogenesis: Focus on cell death. *Cell Death Differ*. 2015;22(8):1250–1259.

Flerlage T, Boyd D, Meliopoulos V, Thomas P, Schultz-Cherry S. Influenza and SARS-CoV-2: Pathogenesis and host responses in the respiratory tract. *Nat Rev Microbiol*. (2021;19(7):425–441.

Jain S, Martynova E, Rizvanov A, Khaiboullina S, Baranwal M. Structural and functional aspects of Ebola virus proteins. *Pathogens*. 2021;10(10):1330.

Khorramdelazad H, Kazemi M, Najafi A, Keykhaee M, Emameh R, Falak R. Immunopathological similarities between COVID-19 and influenza: Investigating the consequences of co-infection. *Microb Pathog*. 2021;152:104554.

Long J, Mistry B, Haslam S, Barclay W. Host and viral determinants of influenza A virus species specificity. *Nat Rev Microbiol*. 2019;17:67–81.

Malik S, Kishore S, Nag S, et al. Ebola virus disease vaccines: Development, current perspective & challenges. *Vaccines*. 2023;11(2):268.

McAuley J, Gilbertson B, Trifkovic S, Brown L. Influenza virus neuraminidase structure and functions. *Front Microbiol*. 2019;10:39. doi: 10.3389/fmicb.2019.00039.

Moghadami M. A narrative review of influenza: A seasonal and pandemic disease. *Iran J Med Sci*. 2017;42(1):2–13.

Mostafa A, Abdelwhab E, Mettenleiter T, Pleschka S. Zoonotic potential of Influenza A viruses: A comprehensive overview. *Viruses.* 2018;10(9):497.

Munoz-Fontela C, McElroy A. Ebola virus disease in humans. Pathophysiology and immunity. *Curr Top Microbiol Immunol.* 2017;411:141–169.

Sullivan N, Yang Z-Y, Nabel G. Ebola virus pathogenesis: Implications for vaccines and therapies. *J Virol.* 2003;77(18):9733–9737.

Takada A. Filovirus tropism: Cellular molecules for viral entry. *Front Microbiol.* 2012;3:34.

Taubenberger J, Morens D. The pathology of Influenza virus infections. *Annu Rev Pathol.* 2008;3:499–522.

CHAPTER 5

Double-Stranded RNA (dsRNA)

Rotavirus

Historical Perspective

On April 25, 2005, a 2-year-old boy developed a moderate fever of 38.5°C accompanied by diarrhea and vomiting. He remained at home, but since his bouts with nonbloody diarrhea increased throughout the day, the child was given Idravita Humana, which is a supplement containing mineral salts and sugar, to drink to rehydrate his body. He was admitted into the hospital 2 days later due to his worsened condition, with additional symptoms of hyporeactivity (underresponsive to sensory input) and asthenia (lack of energy). He was intubated for ventilation, but it was discontinued after 30 minutes when he was pronounced dead.

An autopsy was performed, and the cause of death was determined to be tonsillar herniation, characterized by the descent of the cerebellar tonsils through the foramen magnum, which is a large opening in the occipital bone of the skull. This condition would have caused severe cerebral edema (or swelling of the brain). Indeed, histological analysis showed that there was significant inflammation and blood stasis, which is the slow pooling of blood, in the central nervous system. Neurological damage was likely triggered by rapid dehydration and consequent electrolyte imbalance. The small intestine showed signs of severe mucosal damage in multiple areas and loss of villous epithelium, but the intestinal wall was not affected. Lymphoplasmacytic lymphoma (LPL) and neutrophil infiltration were detected in the lamina propria of both the small and large intestine. LPL is a slow-growing cancer caused by reproduction of B lymphocytes, mainly in the bone marrow but sometimes in the lymph nodes and spleen as well. Neutrophil infiltration is the accumulation of neutrophils, which can contribute to inflammation in inflammatory bowel disease. Rotavirus was found in postmortem samples but not detected in peripheral blood. All these conditions arose within a few days from the initial detection of symptoms. The incubation time for rotavirus is roughly 2 days.

Severe gastroenteritis (commonly referred to as stomach flu), with acute diarrhea as a major symptom, is the major cause of infant mortality for ages 5 and under worldwide, including in the United States. The causative agent was identified as a reovirus-like pathogen, later named rotavirus, in May 1973 by Dr. Rudge Townley, director of gastroenterology at the Royal Children's Hospital in Melbourne,

Australia, and his team. Historically, before a vaccine was introduced in the United States, the disease caused dozens of deaths in this country and up to 500,000 fatalities worldwide. A live-attenuated oral vaccine was finally developed in 2006 for global use.

An earlier rotavirus vaccine was actually developed and briefly in use in 1998. Unfortunately, its use was discontinued due to a rare adverse effect called intussusception, which is a life-threatening condition where a portion of the intestine (usually small and rarely large) slips inside another segment, causing obstruction of the passage and absorption of food. Even though the introduction of vaccines reduced the number of deaths, rotavirus is still the leading pathogen causing gastroenteritis, and severe diarrhea results in more than 200,000 fatalities annually, especially in low-income countries. However, vaccine developers and researchers do not clearly understand why vaccines are much more effective in high-income countries than in low-income countries.

Classification and Structure of Rotavirus

Rotavirus belongs to the Reoviridae family. This is the largest family of double-stranded RNA (dsRNA) viruses and includes 15 genera and 75 virus species: *Aquareovirus* (eg, Chinook salmon reovirus); *Cardoreovirus* (eg, *Eriocheir sinensis reovirus*); *Coltivirus* (eg, Colorado tick fever coltivirus); *Cypovirus* (eg, *Dendrolimus punctatus* cypovirus); *Dinovernavirus* (eg, *Aedes pseudoscutellaris* reovirus); *Fijivirus* (eg, maize rough dwarf virus); *Idnoreovirus* (eg, idnoreoirus 1-5); *Mimoreovirus* (eg, *Micromonas pusilla* reovirus); *Mycoreovirus* (eg, mycoreovirus 1); *Orbivirus* (eg, bluetongue virus); *Orthoreovirus* (eg, piscine orthoreovirus); *Oryzavirus* (eg, rice ragged stunt virus); *Phytoreovirus* (eg, wound tumor virus); *Rotavirus* (eg, rotavirus A); and *Seadornavirus* (eg, Banna virus). However, only 4 of these genera are known to infect humans and animals: *Coltivirus*, *Orbivirus*, *Orthoreovirus*, and *Rotavirus*.

Other related families of virus include Birnaviridae (eg, rotifer birnavirus), Chrysoviridae (eg, *Penicillium chrysogenum* virus); Cystoviridae is the only known dsRNA virus that infects prokaryotes (eg, *Pseudomonas* virus phi6); Hypoviridae (eg, alphahypovirus); Partitiviridae (eg, *Penicillium stoloniferum* virus); Picobirnaviridae (eg, human picobirnavirus); Quadriviridae (eg, *Rosellinia necatrix* quadrivirus 1); and Totiviridae (eg, *Giardia lamblia* virus). There are 8 species of rotavirus, named *Rotavirus A–H*; however, only species A, B, and C have been detected in humans. *Rotavirus A* causes 90% of human infections and is the leading cause of disease in infants and young children.

The structure of the Rotavirus is a non-enveloped virion. Eleven segments of dsRNA genome totaling 16,500–21,000 base pairs are encased in a triple-layer particle, which is a capsid consisting of 3 layers of icosahedral shells: outer, intermediate, and inner shell (Figure 5.1). The dimension of the virion is 85 nm in diameter (100-nm spike to spike). The capsid is composed of several distinct viral proteins (VPs). The inner core shell is made up of VP2 molecules; the intermediate shell is made up of VP6 molecules; and the outer shell is made up of VP7 molecules that are embedded with VP4 spike proteins. Furthermore, VP4 spike proteins are anchored onto the intermediate shell and then protrude through the outer shell to mediate attachment to susceptible host cells. The VP7 outer shell itself is associated with a trace level of Ca^{2+}; this layer's stability is calcium dependent.

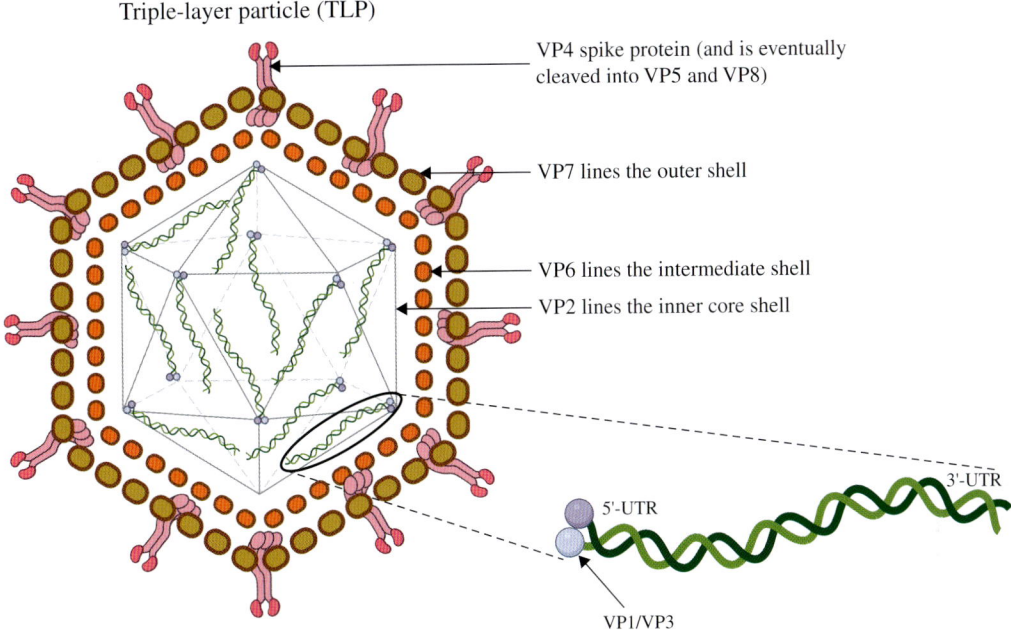

FIGURE 5.1   Rotavirus capsid structure and genome. The rotavirus 11 RNA genome segments are encapsidated by three layers of icosahedral shell, termed TLP. Each dsRNA segment is attached near a vertex of the inner core shell by the VP1/VP3 dimer. *Credit:* Figure partially created in BioRender.

Each of the 11 dsRNA segments is associated with a number of nucleoproteins to form an Ribonucleoprotein (RNP) complex that includes RNA and RNA-binding proteins that are involved in functions such as viral transcription and translation, RNA metabolism, and genome engineering. The nucleoproteins include VP1 (which is the RNA-dependent RNA polymerase [RdRP]) and VP3 (which is the RNA capping enzyme). The VP1/VP3 pair anchors each RNP complex to the area located near each vertex of the icosahedral structure of the inner capsid, known as the 5-fold axis. All 11 genome segments are used as the template for de novo messenger RNA (mRNA) synthesis. Each segment contains an open reading frame (ORF), which together encode for 6 structural VPs (VP1, VP2, VP3, VP4, VP6, and VP7) and 6 nonstructural proteins (NSPs; NSP1, NSP2, NSP3, NSP4, NSP5, and NSP6). Additionally, VP4 is eventually post-translationally cleaved into 2 proteins, VP5 and VP8. VP5 is the portion that is anchored to the intermediate capsid shell constructed with VP6 molecules. VP8 is the hemaglutinin spike that attaches to the epithelial cells in the gut during an infection. NSP1 is an antagonist of the innate immune response. NSP4 is a glycoprotein that acts as the viral enterotoxin. NSP4 has both intracellular functions when bound to endosomal membranes and extracellular functions when secreted. Each ORF encodes for a 5′-untranslated terminal region (UTR) and a 3′-UTR in the mRNA.

Host Range, Transmission, Tropism, and Susceptible Host Cell

Reoviridae is the largest known family of dsRNA viruses. The 75 discovered species have been isolated from a wide host range, including mammals, birds, reptiles, fish, crustaceans, marine protists, insects, ticks, arachnids, plants, and fungi. Because of the wide host range and the presence of multisegmented genome leading to possible genetic reassortment, rotavirus is classified as zoonotic; however, since they are so species specific, the incidence of zoonotic transmission has been found to be extremely low.

Susceptible host cells for rotavirus exhibit surface-bound receptors and co-receptors such as ganglioside (sialic acid–containing glycosphingolipid), histo-blood group antigen (HBGA), integrin, heat shock cognate protein (hsc70), and junctional adhesion molecule A. The various surface molecules aggregate within lipid rafts in the plasma membrane and act as receptors and/or co-receptors for different strains of rotavirus. The main target cells are terminally differentiated intestinal epithelial cells located at the tip of villi, primarily of the ileum and jejunum.

Since the primary site of infection for rotavirus is the epithelial lining of the small intestine, its portal of exit is the anus via the elimination of host waste, and it can be transmitted through the fecal-oral route and hand-to-mouth contact, mainly by close person-to-person interaction. Large amounts of viral particles are shed in diarrheal stools of infected patients. Direct transmission through the oral cavity is very likely when proper, good hygiene is not practiced or there are unsanitary conditions. The tough, triple-layer capsid of rotaviral particles renders them quite stable in the environment; therefore, transmission can also occur through ingestion of contaminated food or water, as well as contact with contaminated fomites, especially in care facilities and hospitals. In a temperate climate, like that in the United States, epidemics usually occur between January and June of every year. Infants and children younger than 5 are most susceptible to the illness, although adults can also acquire the infection but exhibiting milder disease symptoms.

Infecting the Host Susceptible Cell, Viral Particle Replication, and Tissue Damage

Many mechanisms for rotavirus replication are still not well understood, and this is an area of ongoing research.

The infectious form of rotavirus is a full triple-layer particle (TLP); however, viral particles must be modified to render them competent for entry into the host cell. The maturation step involves the cleavage of VP4 spike proteins into two fragments, VP5 and VP8, by trypsin-like proteases found in the lumen of the intestine (Figure 5.2). Note that VP4 cleavage is not required for cell attachment since the exposed terminal domain can still interact with the host cell receptor. However, the presence of VP5 is required for entry into the host cell since it regulates membrane penetration. Attachment occurs when the VP8 domain of VP4, produced after cleavage, binds to ganglioside or HBGA receptors on the surface of the target host cell. Interaction with various co-receptors, which are specific to the host organism and tropism, as postattachment interaction is also required for efficient internalization of the virion through membrane invagination.

Internalization occurs via either direct cell membrane penetration or endocytosis; however, different strains adopt different pathways. Human rotavirus has been shown to carry out mainly clathrin-mediated endocytosis (Figure 5.3). Strains that infect many

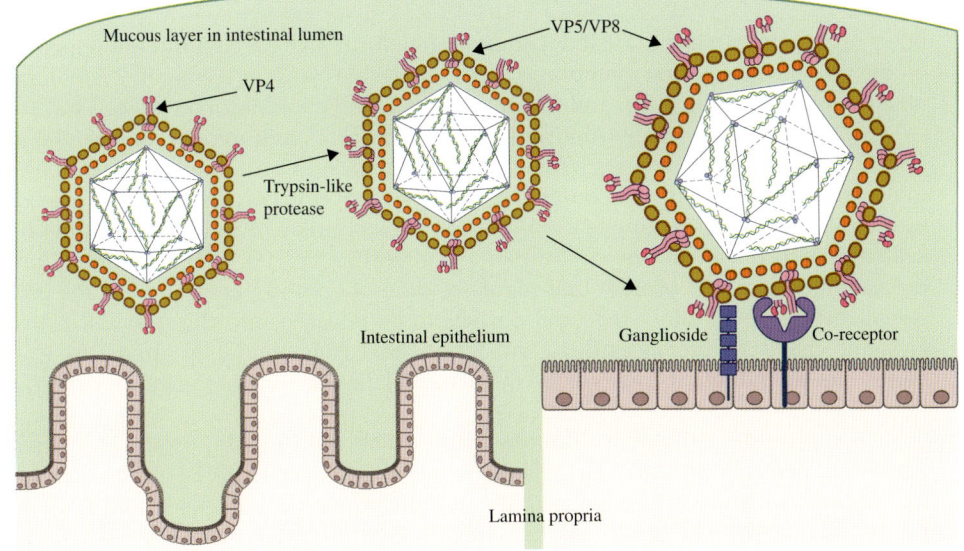

FIGURE 5.2 Preparing the rotavirus for internalization. Trypsin-like proteases in the intestinal lumen cleave VP4 spike proteins into VP5 and VP8. The VP8 fragment then interacts with host cell surface gangliosides for attachment of the viral particle. Further interaction with a co-receptor on the host cell surface induces internalization of the TLP. *Credit:* Figure partially created in BioRender.

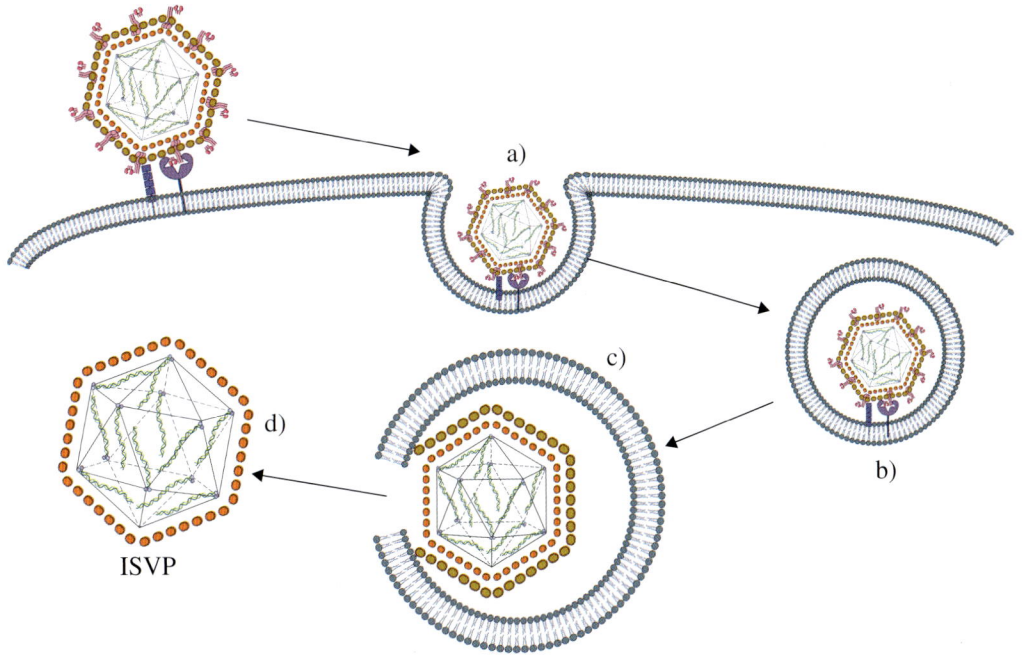

FIGURE 5.3 Rotavirus entry into host cell by endocytosis. (a) Internalization of rotavirus is through clathrin-mediated endocytosis. (b) NSP4 induces loss of Ca^{2+} ions from endosomal lumen, leading to the shedding of VP7 molecules. (c) The shedding of VP7 outer shell occurs within 10 minutes after entry into the host cell. (d) The DLP is transported from the endosomal lumen into the cytosol. This DLP is referred to as the ISVP. *Credit:* Figure partially created in BioRender.

107

other organisms appear to be independent of clathrin and caveola but are dependent on the presence of plasma membrane-bound molecules such as dynamin 2, small guanosinetriphosphatases RhoA and Cdc42, actinin-4, and cholesterol on the cell surface.

Once in the endosome, viroporin NSP4 in a low pH environment induces the loss of Ca^{2+} ions from the incoming virus-containing vesicle to promote the release of VP7 molecules, thus removing the outer capsid layer. Viroporins are ion channels encoded by a viral gene. The shedding of the VP7 outer shell occurs within 10 minutes after host cell entry, which leads to the conformational alteration in VP5. VP5 is the membrane-permeabilizing capsid protein that regulates disruption and penetration of the endosomal membrane to allow the viral particle to be released into the host cell cytosol. Cysteine cathepsin proteases B, L, and S delivered from the Golgi apparatus to the endosome are thought to be involved in the partial disassembly of the intermediate and inner shells in the endosomal lumen.

The transcriptionally active double-layer particle (DLP) with porous intermediate and inner shells referred to as the infectious subvirion particle (ISVP) is then deposited into the cytosol. The ISVP contains all the machinery needed for plus-strand RNA synthesis and capping. The removal of the outer shell permits conformational changes that activate VP1 and VP3 molecules to carry out transcription. However, nucleotides must still be diffused into the ISVP from the cytosol for nucleic acid synthesis. It is proposed that, in addition to its role as an RdRP, VP1 activates NSP4 to form a distinct tunnel to allow nucleoside triphosphates to enter the particle and synthesized plus-strand RNAs to exit.

The dsRNA molecules remain in the ISVP, where each genome segment is used as the template for repeated transcription of plus-strand RNA molecules by the RdRP activity of VP1 polymerase using de novo initiation (Figure 5.4). The capping enzyme VP3 places a cap structure on the 5′ end of each plus-strand RNA molecule, which would eventually serve as the mRNA templates for VP synthesis.

Capped viral mRNA molecules synthesized in the ISVP are released into the cytosol to act as templates for cap-dependent translation. These mRNA templates are not poly-adenylated whose role is to determine the stability of the translation initiation complex to regulate the amounts of proteins to be synthesized. In the absence of the poly(A) tail, the amount of protein molecules produced correlates with the length of each mRNA. Since the poly(A) tail is missing in these mRNAs, specialized structures in the 3′-UTR are required to aid translation initiation.

NSP3 binds to the 3′-end of the mRNA in the cytosol to serve as a binding site for the 5′ cap-associated eukaryotic initiation factor (eIF-4G), in place of cellular poly(A)-binding proteins (PABPs) that arerequired for cap-dependent translation.

Also, since rotaviral mRNAs do not possess a poly(A) tail, the actual length of the viral mRNA affects the amount of protein synthesis. More VPs are synthesized from shorter viral mRNA due to the quicker rate of synthesis, while lower amounts of VPs are synthesized from longer viral mRNA.

Once VPs are synthesized in the cytosol, genome assortment and the assembly of viral particles, up to the formation of the DLPs, occurs mostly in a region of the cytosol referred to as the viroplasm, which is a membrane-free structure that is nucleated and assembled by NSP2 and NSP5.

The interaction between NSP2 and NSP5 is important for the formation of the viroplasm. Additionally, the NSP4-induced elevation of cytosolic Ca^{2+} concentration contributes to the regulation viroplasm formation. NSP2 is also involved in viral genome

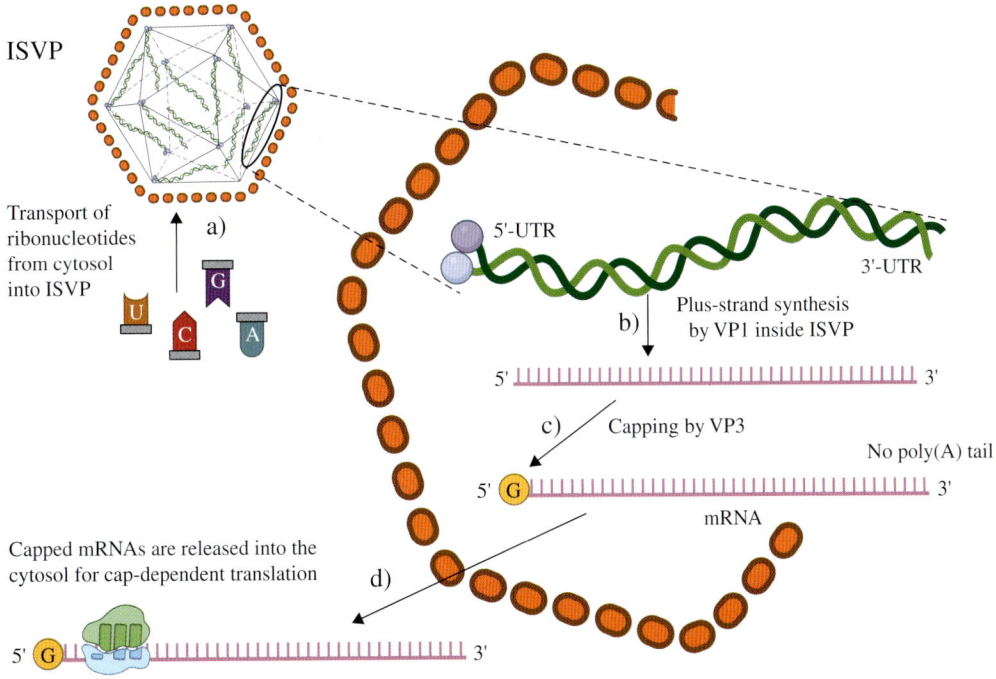

ISVP

Transport of ribonucleotides from cytosol into ISVP

a)

5'-UTR

3'-UTR

b) Plus-strand synthesis by VP1 inside ISVP

c) Capping by VP3

No poly(A) tail

mRNA

Capped mRNAs are released into the cytosol for cap-dependent translation

d)

FIGURE 5.4 Genome transcription from the ISVP. (a) Nucleotides diffuse through the porous intermediate and inner shells into the ISVP. (b) VP1 catalyzes plus-strand RNA synthesis. (c) VP3 places a cap on the 5'-end of the RNA, forming a capped mRNA without a poly(A) tail. (d) Capped mRNAs are released from the ISVP into the cytosol to be used for cap-dependent translation of viral proteins. *Credit:* Figure partially created in BioRender.

replication by binding to the initial plus-strand RNA to ensure equimolar pregenome RNA assortment. In this site, viral structural proteins VP1, VP2, VP3, and VP6 interact with one another to drive viral particle assembly.

Groups of 11 selected plus-strand RNAs are packaged into assembled or assembling VP2 inner layer. Currently, the mechanism by which rotavirus acquires 1 of each of its 11 genome segments is poorly understood. However, this assortment process is believed to be mediated by RNA-RNA interactions among the single-stranded transcripts, in addition to the activities of nonstructural VP.

It has been proposed that structures in the 5'- and 3'-UTRs are important in directing genome assortment, and this is initiated by the binding of VP1 (RdRP) and VP3 (capping enzyme) molecules to viral mRNA segments. NSP2 also helps organize the plus-strand RNAs for packaging and genome replication by removing secondary structures in the 5'- and 3'-UTRs.

Next, VP2 molecules form the inner shell around ribonucleoprotein complex, producing a single-layer particle (SLP), then dsRNA synthesis is initiated by VP1 using each plus-strand RNA as the template (Figure 5.5). This SLP is believed to expand in size as dsRNA synthesis continues. Once dsRNA genome synthesis is completed, VP6 molecules form the intermediate shell to produce a complete DLP, referred to as a replicase particle.

Rotavirus assembly and genome synthesis

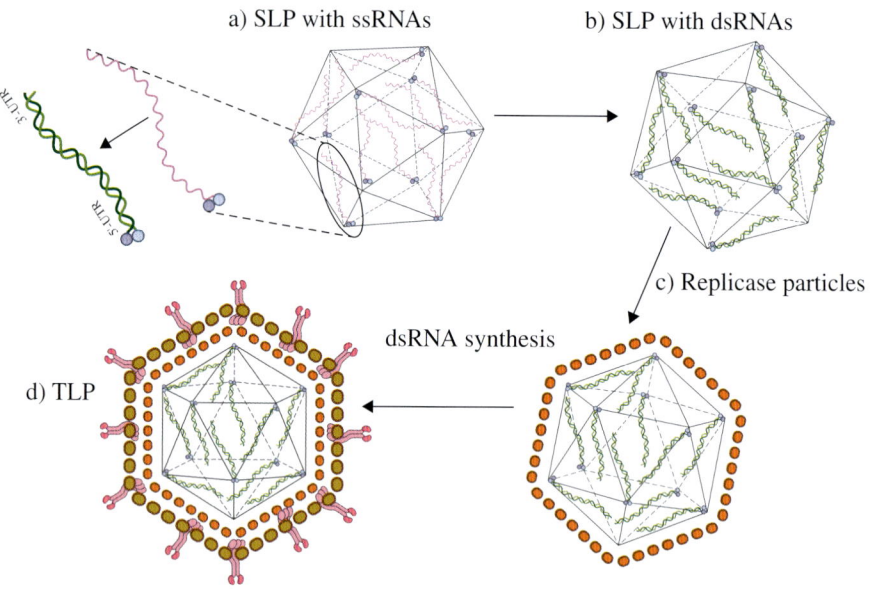

FIGURE 5.5   Assembly of the three layers of rotavirus capsid. (a) The inner core shell encapsidates 11 viral ssRNA segments producing a SLP. (b) Synthesis of dsRNA occurs within the SLP using each of the 11 ssRNA molecules as a template producing an SLP with dsRNAs. (c) The intermediate shell forms around the SLP, producing a DLP, also known as the replicase particle. (d) The outer shell forms around the DLP, producing a TLP. *Credit:* Figure partially created in BioRender.

In the replicase particle itself, the 11 dsRNA segments serve as templates for the synthesis of additional plus-strand RNAs. The interaction between VP1 and VP2 triggers the polymerase to undergo a conformational change, thereby activating its enzymatic activity to allow for the initiation of RNA synthesis. The majority of the synthesis of both mRNAs and plus-strand genomic RNAs occurs in replicase particles. This process is known as secondary transcription, which greatly amplifies the production of RNAs for VP synthesis as well as viral replication.

The final step of infection is the formation of the TLP, which is composed of all three capsid shells. Replicase particles are transiently localized at the endoplasmic reticulum (ER), where VP7 molecules are located. The internalization of the DLP from the cytosol into the ER lumen is necessary for the acquisition of the outer layer, resulting in TLP assembly (Figure 5.6). Even though still unknown, it is proposed that the formation of a complex including NSP4, VP4, and VP7 is involved in this mechanism. The release of the TLP from the ER lumen is also not known. It is proposed that NSP4 reprises its role to perforate the ER membrane, similar to its activity on the endosomal membrane.

Very early during the infection, rotavirus selectively downregulates host cellular protein synthesis while advancing vigorous VP synthesis (Figure 5.7). It has been proposed that NSP3 causes host mRNAs to be localized and accumulate in the nucleus by blocking their export to the cytosol. Translation machinery is then dedicated to VP

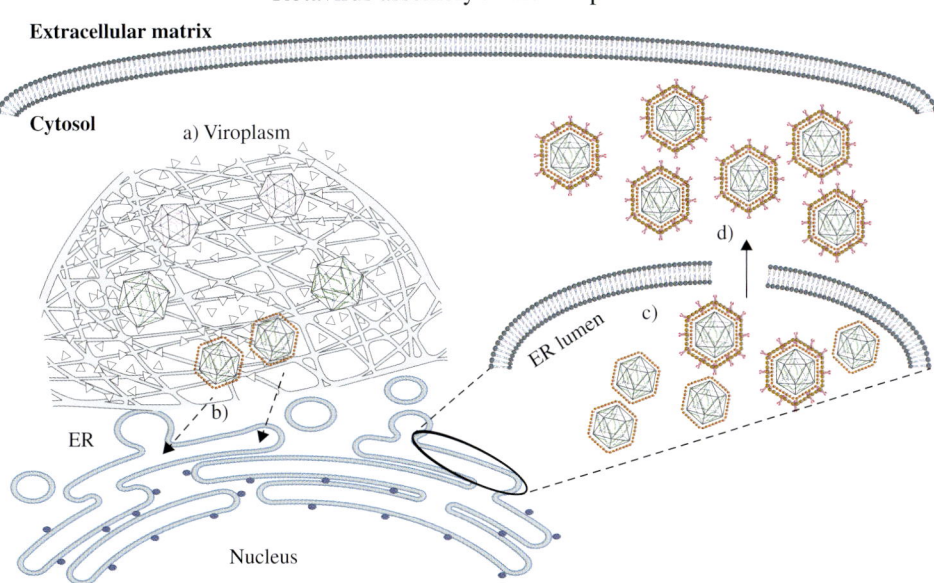

Rotavirus assembly in the viroplasm

FIGURE 5.6   Rotavirus assembly and transport. (a) The assembly of the inner and intermediate shells around 11 ssRNA segments occurs in the viroplasm. (b) DLPs are transported into the lumen of the ER. (c) DLPs acquire the outer shell to produce the TLPs. (d) TLPs are released from the ER lumen into the cytosol. *Credit:* Figure partially created in BioRender.

synthesis. In addition to its role as a PABP surrogate, NSP3 also has higher affinity for eIF-4G and can possibly prevent PABP from binding to eIF-4G, thereby blocking the translation of any cellular mRNAs that manage to be exported into the cytosol. The eIF-4G molecule must dimerize with eIF-4E to regulate cap-dependent translation initiation. Typically, a host cell activates a number of enzymes when it is under stress for many reasons including heat shock, presence of dsRNA, stress on the ER, and amino acid starvation. Once activated, these enzymes phosphorylate eIF-2α to reduce the translation of cellular mRNAs. The eIF-2α molecule must remain unphosphorylated to recruit the methionine-charged initiator transfer RNA (tRNA) to the small ribosomal (40S) subunit. The complex of 40S subunit and methionine-charged tRNA forms the 43S subunit of ribosome. The 43S subunit can bind to mRNAs to initiate translation, but the 40S subunit cannot bind to mRNAs. Somehow, the translation of rotavirus mRNAs is not affected by this phosphorylating activity; however, the mechanism is still not known.

The replicase particle becomes a full virion once the outer capsid is assembled to form a TLP. To allow efficient release of viral particles, rotavirus uses several mechanisms to induce host cell death to maximize viral dissemination and pathogenicity. For instance, NSP4 can regulate virus-mediated apoptosis. This virus can also upregulate the mixed lineage kinase domain-like protein to mediate the necroptotic pathway concomitantly with apoptosis.

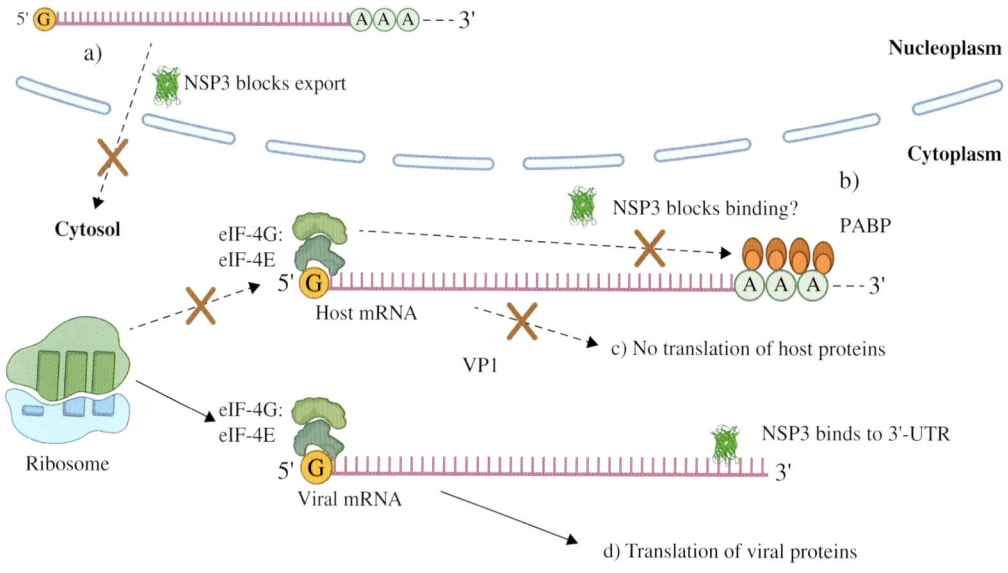

FIGURE 5.7   Downregulation of host protein synthesis. (a) NSP3 localizes host mRNAs in the nucleoplasm, preventing them from being exported into the cytoplasm. (b) NSP3 may act as a competitive inhibitor of PABP for binding to eIF-4G, therefore preventing the recruitment of the ribosomal 43S subunit for translation initiation. (c) Thus, no host protein synthesis occurs. (d) The 43S subunit can still bind to the viral mRNA for translation initiation. Even though the poly(A) tail is absent, the 3'-UTR of the viral capped mRNA can be bound by NSP3 molecules to interact with eIF-4G for translation. *Credit:* Figure partially created in BioRender.

Dissemination in the Host Body

Rotavirus infects and replicates in mature enterocytes. There are two types of epithelial cells lining the small intestine, enterocytes and crypt cells. Crypt cells lack well-defined microvilli and are the progenitor of villus enterocytes. Enterocytes are mature, differentiated cells covering the top of villi. They secrete digestive enzymes as well as regulate absorption of nutrients and water. When these enterocytes die, the viral particles are released into the mucous layer, then make their way through the digestive system and shed in the host's stools.

Significant extraintestinal dissemination of rotavirus in an immunocompetent host is possible but not likely. However, for immunocompromised individuals, this virus had been found to replicate in the liver, the biliary system, and the pancreas. Rotavirus dsRNA, antigen, and sometimes infectious viral particles can be found in serum and other host body sites.

Rotavirus infection had been linked to biliary atresia and pancreatitis. Biliary atresia is a condition in infants where bile ducts are blocked, thus preventing the flow of bile into the duodenum of the small intestine. Bile that builds up in the liver causes liver damage, leading to scarring, loss of liver function, and cirrhosis.

Rotavirus pancreatitis, which damages the pancreas, manifests with vomiting, diarrhea, fever, and sometimes abdominal pain.

Pathogenesis and Clinical Manifestation

Rotavirus is a contagious virus, and its infection is the leading cause of severe, dehydrating gastroenteritis in children younger than 5 years of age. Most children are infected by this virus between 3 months and 3 years old. The earliest infection most likely occurred during the first year of life. Reinfection is common, especially during the first 2 to 3 years.

Repeated infections are associated with the presence of increased serum antibodies and immunoglobulin A (IgA) in the mucous layer. Rotavirus primarily infects enterocytes, causing damage to the epithelial lining of the small intestine at the site of infection. Malabsorption due to major damage to absorptive enterocytes leads to symptoms of vomiting, watery diarrhea, and fever.

The glycoprotein rotavirus NSP4 possesses multiple functions and is mainly responsible for tissue damage. We've already discussed its role in the formation of the DLP as well as acting as a viroporin to elevate the intracellular Ca^{2+} levels required to stabilize the TLP by interacting with the cellular sensor molecule stromal interaction molecule 1.

Additionally, NSP4 can induce cell damage by activating a Ca^{2+}-triggered, kinase-dependent pathway to initiate autophagy, leading to cell lysis. NSP4 can also alter plasma membrane permeability and disrupt tight junctions, along with destabilizing other cell-to-cell junctions, causing paracellular permeability (Figure 5.8). The increased

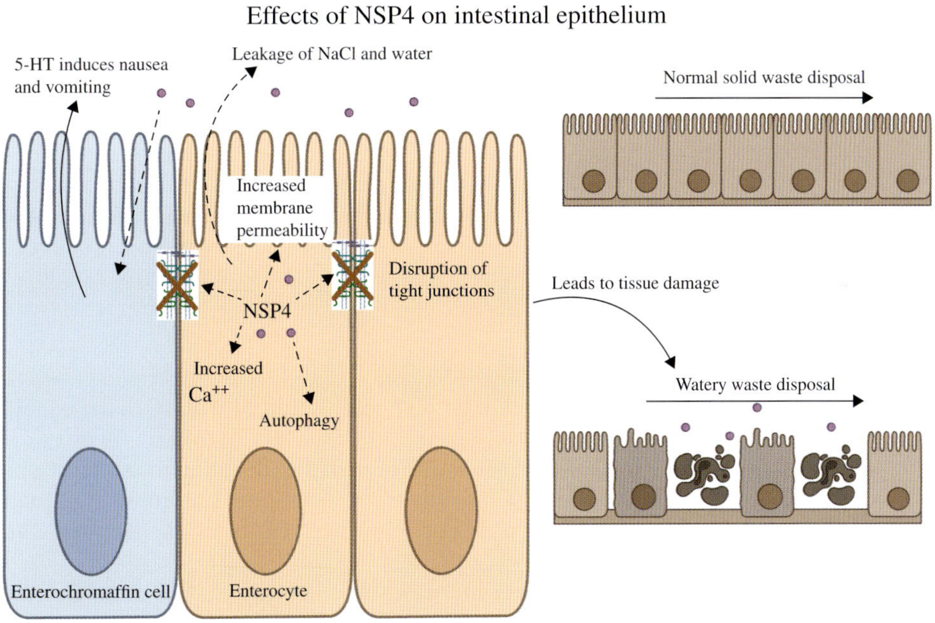

Effects of NSP4 on intestinal epithelium

FIGURE 5.8 Host cell-damaging activities of NSP4. NSP4 induces an increase in cytosolic Ca^{2+} concentration and disruption of tight junctions and destabilizes other cell-to-cell junctions and can cause autophagy to lyse the host cell. NSP4 also causes increased permeability of the plasma membrane, leading to the leakage of NaCl and water from the host cell into the intestinal lumen. Once secreted into the mucous layer of the intestinal tract, NSP4 can induce neighboring enterochromaffin cells to release 5-HT to induce nausea and vomiting. *Credit:* Figure partially created in BioRender.

permeability of the plasma membrane allows NaCl and water to leak into the intestinal lumen. Most critically, NSP4 can be secreted by the infected cell and acts as a viral enterotoxin to exacerbate the problem with diarrhea. Enterotoxin is a substance that can harm the digestive system and cause gastrointestinal symptoms. NSP4 has paracrine effects by triggering the phospholipase C–inositol 1,3,5-triphosphate cascade in uninfected cells to increase the cytosolic Ca^{2+} concentration. When released from the infected enterocyte, NSP4 can stimulate enterochromaffin cells (type of enteroendocrine cell) to release 5-hydroxytryptamine (5-HT), which is a neurotransmitter that regulates gastrointestinal motility and induces nausea and vomiting.

Active viremia of rotavirus commonly occurs in infected children. Viremia can lead to systemic sites of infection; however, this is an infrequent occurrence. Detection of antigenemia (or rotavirus RNA) in children's blood is associated with more severe manifestations of acute gastroenteritis and may suggest extraintestinal involvement in rotavirus pathogenesis. The destruction of the digestive epithelium prevents the body from reabsorbing water, therefore leading to dehydration, which can be fatal. Fatality occurs mostly in individuals with a weak immune system, such as very young children, the elderly, and individuals who have an immunosuppressed or immunocompromised system. And, the cause of death is mainly due to severe dehydration and cardiovascular failure. The pathogenesis of vomiting is connected to the infection of enterochromaffin cells, whose function is to synthesize and secrete the hormone serotonin to activate vagal afferent nerves and stimulate brain stem structures that control vomiting.

Symptoms for rotavirus infection are ultimately characterized by mild to severe vomiting and watery diarrhea. It takes about 2 days after infection for symptoms to develop. The most common symptoms in a child are nausea, vomiting, watery diarrhea lasting up to a week, abdominal pain, fever for 1–2 days, and possibly loss of appetite and dehydration. Signs of dehydration include reduced urination, dry mouth and throat, thirst, dizziness, crying without tears, and eyes that look sunken; for babies, these signs could be unusual fussiness and sunken fontanelle (or baby's soft spot). It is common for an individual to have multiple reinfections of rotavirus throughout the lifetime. The first infection usually causes the most severe symptoms; however, subsequent, repeated infections tend to yield less severe disease symptoms, possibly due to the production of rotavirus-specific IgA.

Diagnosis, Treatment, and Prevention

A major sign of a rotavirus infection is the presentation of watery diarrhea, although there are a number of illnesses that present diarrhea as a symptom. A physical exam and stool sample analysis are performed to confirm the diagnosis. Viral shedding occurs and can remain in the host digestive tract for up to 2 months. The quantity of viral RNA detected in the stool correlates with the severity of the diarrheal symptom and infection.

Techniques like enzyme-linked immunosorbent assay and immunochromatography are most frequently applied to detect antigens from the rotavirus in a routine diagnostic laboratory. For greater sensitivity and suitability to determine the genotype of the virus, such assays are often accompanied by the use of reverse transcription polymerase chain reaction (RT-PCR) to detect viral nucleic acid.

Unfortunately, there is no known specific treatment for rotavirus infection. Nonspecific treatment is focused mainly on rehydration due to the massive loss of body fluid from watery diarrhea and vomiting. The infection usually resolves within 3–7 days.

Additionally, electrolyte and dietary management are important, and using probiotics and antiviral drugs may be necessary.

Proper hygiene and avoiding contact with known infected individuals are good means of preventing rotavirus infection. However, immunization is recommended by the Centers for Disease Control and Prevention and is the most effective means of prevention.

Currently, there are two approved vaccines available: RotaTeq (by Merck Sharp & Dohme Corp.) and Rotarix (GlaxoSmithKline). Both vaccines are a live attenuated and are recommended to be given in multiple doses to babies starting at 2 months old. Clinical trials showed that RotaTeq demonstrated 98% protection, while Rotarix demonstrated 85%–96% protection against severe rotavirus gastroenteritis. However, for reasons that are not well understood, these otherwise very effective vaccines appear to be less effective in low-income countries.

Current Status

Today, for young children the death rate resulting from gastroenteritis is dramatically lower since the introduction of vaccines. Other factors that contributed to this decline include better hygiene habits, proper sewage disposal, and advances in treatment. Unfortunately, rotavirus infection remains the most common cause for diarrheal deaths worldwide, with diarrhea itself the fourth leading cause of death among children under 5 years old and the eighth leading cause of death among all ages. Du et al. (2022) analyzed the age-standardized death rate (ASDR) associated with rotavirus infection from 1990 to 2019. They determined that mortality due to rotavirus infection decreased 64% from 659,053 in 1990 to 235,331 in 2019. Children under 5 years of age and individuals over 70 years old remain the 2 most affected groups.

Even though the global burden of disease decreased dramatically since the introduction of vaccines, rotavirus infection continues to cause a higher burden in African, Oceanian, and South Asian countries. Three countries in Africa have the highest ASDRs. Half of all rotavirus-associated deaths occur in four countries: India, Nigeria, Pakistan, and the Democratic Republic of the Congo. Burden of disease is the measurement of the level of impact on one's life, where the years of healthy life lost from illness and death is quantified as disability-adjusted life years (DALYs). Poor hygiene practice, malnutrition, and inadequate sanitation may contribute to this discrepancy. The high mortality rate appears to be due to the lack of access to medical care for infected individuals. Vaccines have been shown to be the most effective measures to prevent and control this infectious disease.

Scientific Significance and Discoveries

In 2019, it was estimated that more than 45 million DALYs were attributed to diarrheal diseases, despite the mortality rate being the lowest ever. Since these diseases continue to be the major cause of global morbidity and mortality among children under 5 years of age, with rotavirus infection being the most common cause, vaccine production is definitely critical.

Rotarix and RotaTeq are the two oral rotavirus vaccines licensed in 2006 and are currently used globally. Their effectiveness in reducing rotavirus-associated morbidity and mortality is significant. Unfortunately, the impact of these vaccines has lower efficacy, for yet unknown reasons, in low-income countries that have the highest burden of disease; therefore, the development of new and more effective vaccines is warranted.

A number of live-attenuated rotavirus vaccines have been licensed by countries where they are developed and manufactured. Rotavac was derived from a natural isolate in an infected child by Bharat Biotech International Limited in Hyderabad, India. Another live-attenuated vaccine developed in India is Rotasiil by Serum Institute of India Private, Limited, in Pune, India. In the People's Republic of China, Lanzhou Institute of Biological Products Company developed the Lanzhou lamb rotavirus vaccine, which was licensed for use in the private market in the country. The Center for Research and Production of Vaccines and Biologicals in the Socialist Republic of Viet Nam developed Rotavin-M1, which is an oral, live-attenuated human rotavirus vaccine. Other vaccines are being developed by other countries as well. The US National Institutes of Health UK-Compton bovine rotavirus vaccine is a multivalent bovine-human rotavirus reassortment vaccine. The human neonatal rotavirus vaccine is the only VP4 vaccine (vaccine containing VP4 as the immunogenic entity) developed in Australia.

Picobirnavirus

Historical Perspective

During acute gastroenteritis outbreaks in 1988 in Brazil, human fecal samples were collected and subjected to polyacrylamide gel electrophoresis (PAGE) to detect rotavirus genome segments. The two segments of genome belonging to a new family of virus named Picobirnaviridae were identified instead, *Pico* for "small," *bi* for "two," and *rna* for "RNA." This virus was also detected in black-footed pygmy rice rats at around the same time and since has been identified in a number of domestic and captive mammalian and avian species. Some are related to the Birnavirus, which also possess bisegmented genomes. There is a strong presence of picobirnaviral genome in children and adults with immunodeficiencies, including adults infected with human immunodeficiency virus (HIV). The interesting factor is that it is presently unclear whether this virus is a causative pathogen for gastroenteritis since the viral titer in stools of both symptomatic and asymptomatic human patients exhibiting diarrhea is similar. This is a structurally small virus.

Picobirnavirus is assumed to be a zoonotic pathogen since it has such a broad host range, which includes humans, deer, and cattle, although its original host and source are yet to be discovered. Because the pathogen is found in both symptomatic and asymptomatic cases, it is difficult to identify PBV's role as the primary infectious agent causing gastroenteritis. It is possible that the family is a group of secondary opportunistic agents residing in the digestive tract and therefore have no direct involvement in the etiology of this condition.

Classification and Structure of Picobirnavirus

Picobirnavirus (PVB) belongs to the Picobirnaviridae family and is the only known genus of this family. As opposed to the rotavirus, which is physically and genetically big and complex, picorbirnavirus is small and simple. The structure of the virus is non-enveloped with a diameter of 35–41 nm (Figure 5.9). The two segments of dsRNA genome, totaling roughly 4300 base pairs (bp), are encased in an icosahedral capsid. Each dsRNA segment is attached by a VPg protein.

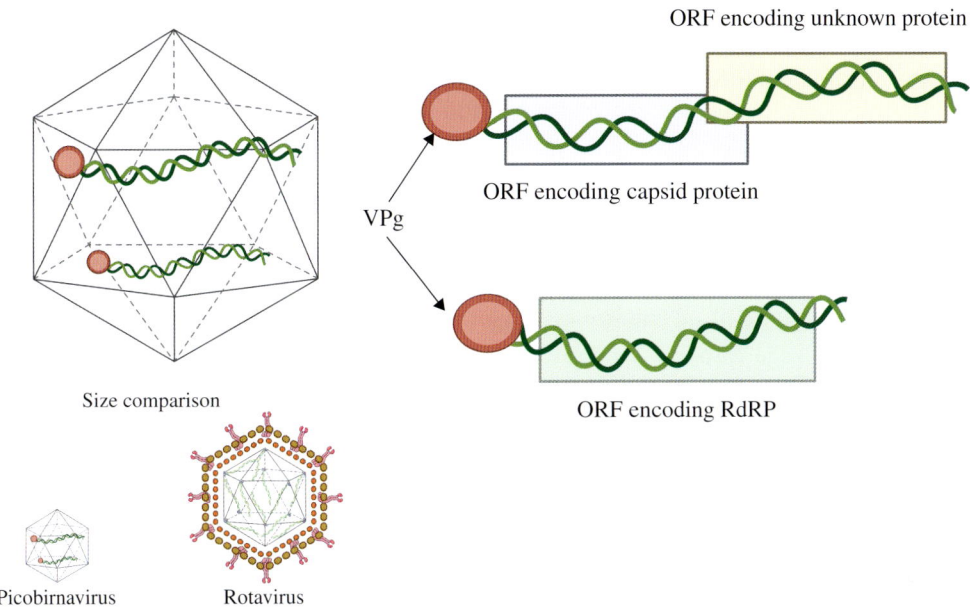

FIGURE 5.9 Picobirnavirus capsid structure and genome. Picorbirnavirus has a smaller and much simpler virion compared to rotavirus. Its 4400-bp genome encodes for only 2 or possibly 3 proteins. *Credit:* Figure partially created in BioRender.

The capsid structure of PBV is significantly different from that of rotavirus. The larger segment 1 contains 2525 bp; this segment houses 2 non-overlapping ORFs, possibly 3, but the presence of this fragment is not definitive. One of these ORFs is known to encode for the capsid protein, while the amino acid sequence encoded by the other ORF shares homology to bacteriophage lysin and the bacteriolytic protein metallopeptidase. The smaller segment 2 contains 1745 bp and houses only 1 ORF that encodes for the viral RdRP.

Infection Cycle of Picobirnavirus

Picobirnavirus has a broad vertebrate host range, infecting at least humans, pigs, rabbits, dogs, rats, snakes, birds, bulls, sea lions, foxes, turkeys, horses, gorillas, marmots, rabbits, roe deer, and chickens. This virus is assumed to be zoonotic because genetically similar genomes were found in the respiratory tracts of both pigs and human patients. Perhaps PBV's simple design and small stature enable this virus genus to dominate over a wide range of hosts, including prokaryotes, fungi, and vertebrates. PBV is a prolific emerging RNA virus whose tropism in vertebrates includes 2 of the most common organ systems (digestive and respiratory), which poses serious public health issues for all their hosts. The natural host for picobirnavirus, however, is still unknown. Their closest known relative when comparing capsid structure and genome organization is the family Partitiviridae, which also contains bisegmented dsRNA genomes. The natural hosts of partitiviruses are fungi and plants.

Picobirnavirus is considered an opportunistic enteric pathogen (infecting the diges-tive tract) associated with clinical disease in humans. It is often isolated with other co-infecting enteric viruses, such as rotavirus, calicivirus, and astrovirus, with other diarrheal causes such as *Escherichia coli* and *Salmonella*, and in immunocompromised individuals such as those infected with HIV. PBV also is likely caused by zoonotic infections since genomes extracted from both human and porcine respiratory tracts are found to be identical.

Picobirnavirus enters the host cell via receptor-mediated endocytosis (Figure 5.10). Once in the endosome, the virion releases a myristoylated peptide that forms pores in the endosomal membrane, which in turn create a passageway for the viral genome to migrate into the cytosol. The VPg attached to each dsRNA segment is cleaved off by the host "unlinkage" protein. Viral mRNA and dsRNA genome synthesis is carried out by its own RdRP in the cytosol. Virion assembly also occurs in the cytosol and becomes vacuolized. Virions are released from the host cell by exocytosis. Virions can also be released via host cell lysis if virus propagation occurs at a high rate.

Pathogenesis and Clinical Manifestation

It is unclear whether PBV members are primary pathogens or simply associated oppor-tunistic gastrointestinal pathogens that cause clinical diseases in humans. PBV infection

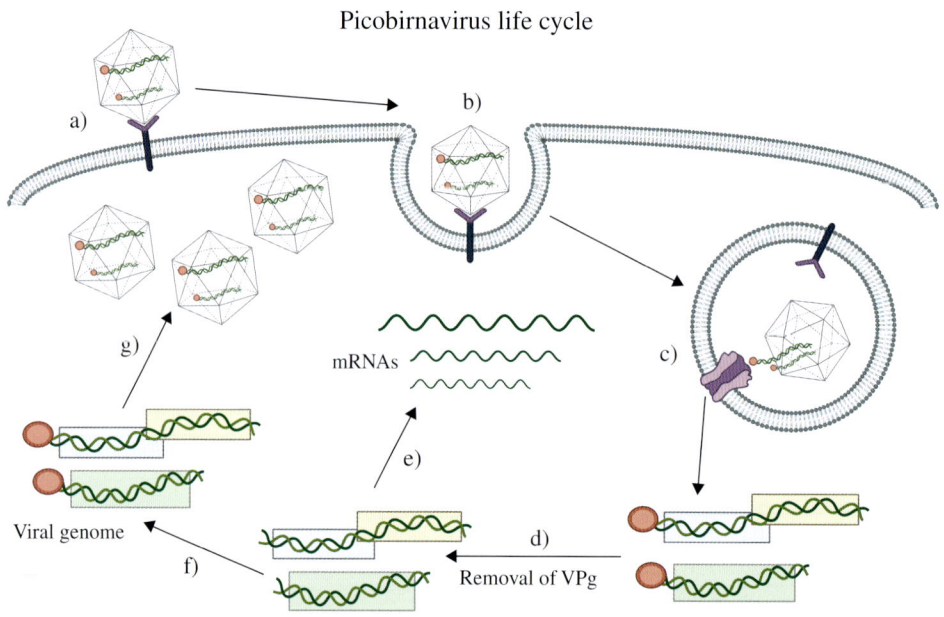

Picobirnavirus life cycle

FIGURE 5.10   Picobirnavirus life cycle in host cell. (a) Attachment of picobirnavirus to the host cell surface. (b) Entry of viral particle via endocytosis. (c) A channel is formed in the endosomal membrane to allow viral genome segments to migrate into the cytosol. (d) VPg at the end of each genome segment is cleaved off by the host "unlinkage" protein. (e) Synthesis of mRNAs for VP synthesis. (f) Synthesis of viral dsRNA genome segments. (g) Virion assembly occurs in the cytosol before exiting the host cell. *Credit:* Figure partially created in BioRender.

has been linked to gastrointestinal disease in individuals with weakened immunity, such as those infected with HIV. Primary infections likely lead to chronic diseases, exhibited by long-term persistence of this virus in the host body, with or without symptoms like diarrhea. However, the role of PBV as the causative agent of acute gastroenteritis is unclear because these virions are found in stools of both symptomatic and asymptomatic patients.

Diarrhea is a common and major symptom of PBV infection in all host species. Of course, there are many factors that can provoke diarrhea, such as nutritional imbalance, poor management, coccidia, chlamydia, and other viral infections (eg, rotavirus). Diarrhea may be caused by PBV alone or by multiple infectious agents acting synergistically, leading to pathogenesis to increase the overall disease burden on the host organism. Infectious agents found to co-infect with PBV include rotavirus, astrovirus, caliciviruses, *E. coli*, and *Salmonella*. The role of PBV as the direct cause of acute gastroenteritis, however, is not yet known. Insufficient cell culture and animal models prevent the cultivation of PBV, which in turn impedes efforts to cultivate and investigate PBV infection.

Diagnosis, Treatment, and Prevention

Picobirnavirus has not yet been isolated; therefore, identification of viral particles relies on the detection of the dsRNA bisegmented genomes. Initially, PAGE, in conjunction with silver staining, was used for direct visualization of the genome segments in samples. RT-PCR is now the main method of detection due to its superior sensitivity over PAGE. The use of RT-PCR to screen metagenomic libraries is gaining acceptance as a legitimate, unbiased method for characterizing viral sequences. Metagenomic analysis is used on bulk samples where there is a mix of nucleic acid from all organisms present.

There is no specific treatment for viral gastroenteritis in general or PBV infection specifically; however, the symptoms can be treated. Getting well hydrated and a lot of rest are of the utmost importance. Nasal decongestants, cough suppressants, and pain or fever medications can relieve symptoms caused by the body's inflammatory responses, especially those of a respiratory infection. Using a room humidifier or taking a hot shower may ease a sore throat and cough. Last, preventive measures involve avoiding activities that may lead to the transmission of PBV.

Prevention of PBV infection includes proper handwashing after using the toilet, helping a child use the toilet, and changing diapers. Rotavirus on fomites can be eliminated by disinfectants. An infected individual should avoid foods that may worsen vomiting and diarrhea, such as apple juice, milk, cheese, and sugary foods and drinks.

Scientific Significance and Discoveries

There is continuing debate about whether PBV is a eukaryotic virus or bacteriophage. PBV has been isolated in the stool of a variety of animals, but researchers are unable to replicate it in cell culture or in animal models. In infected animals, humans included, PBV is always isolated from environments where bacteria also reside. This could mean that PBV replicates and is released from bacterial hosts instead of eukaryotic host cells. Another support for a bacteriophage host hypothesis is that PBV genomes harbor a bacterial ribosomal binding site upstream of their ORFs. Interestingly, some PBV genomes have been found to contain a fungal translational code as well, implying that PBV is a fungal virus.

Further Readings

Amimo J, Raev S, Chepngeno J, et al. Rotavirus interactions with host intestinal epithelial cells. *Front Immunol.* 2021;12:793841.

Arias CF, Silva-Ayala D, Lopez S. Rotavirus entry: A deep journey into the cell with several exits. *J Virol.* 2015;89(2):890–893.

Crawford S, Ramani S, Tate J, et al. Rotavirus infection. *Nat Rev Dis Primers.* 2017;3:17083.

Desselberger U. Rotaviruses. *Virol Res.* 2014;190:75–96.

Du Y, Chen C, Zhang X, et al. Global burden and trends of rotavirus infection-associated deaths from 1990 to 2019: An observational trend study. *Virol J.* 2022; 19(1):166.

Gan T, Wang D. Picobirnaviruses encode proteins that are functional bacterial lysins. *Proc Natl Acad Sci U S A.* 2023;120(37):e2309647120.

Ghosh S, Malik Y. The true host/s of picobirnaviruses. *Front Vet Sci.* 2021;7:2020.

Kashnikov A, Epifanova N, Novikova N. Picobirnaviruses: Prevalence, genetic diversity, detection methods. *Vavilovskii Zhurnal Genet Selektsii (Vavilov Journal of Genetics and Breeding).* 2020;24(6):661–672.

Kashnikov A, Epifanova N, Noviova N. On the nature of picobirnaviruses. *Vavilovskii Zhurnal Genet Selektsii (Vavilov Journal of Genetics and Breeding).* 2023;27(3):264–275.

Kirkwood C, Ma L-F, Carey M, Steele A. The rotavirus vaccine development pipeline. *Vaccine.* 2019;37(50):7328–7335.

Long C, McDonald S. Rotavirus genome replication: some assembly required. *PloS Pathog.* 2017;13(4):e1006242.

Lopez S, Oceguera A, Sandoval-Jaime C. Stress response and translation control in rotavirus infection. *Viruses.* 2016;8(6):162.

Malik Y, Kumar N, Sharma K, et al. Epidemiology, phylogeny, and evolution of emerging enteric picobirnaviruses of animal origin and their relationship to human strains. *Biomed Res Int.* 2014;2014:780752.

McDonald S, Patton J. Assortment and packaging of the segmented rotavirus genome. *Trends Microbiol.* 2011;19(3):136–144.

Medici M, Abelli L, Martinelli M, et al. Clinical and molecular observations of two fatal cases of rotavirus-associated enteritis in children in Italy. *J Clin Microbiol.* 2011;49(7):2733–2739.

Papa G, Borodavka A, Desselberger U. Viroplasms: Assembly and functions of rotavirus replication factories. *Viruses.* 2021;13(7):1349.

Ramig R. Pathogenesis of intestinal and systemic rotavirus infection. *J Virol.* 2004; 78(19):10213–10220.

Reddy M, Gupta V, Nayak A, Tiwari S. Picobirnaviruses in animals: A review. *Mol Biol Rep.* 2023;50(2):1785–1797.

Shah M, Gilchrist J, Forsberg B, et al. Characterization of the rotavirus assembly pathway in situ using cryoelectron tomography. *Cell Host Microbe.* 2023;31:604–615.

Suzuki H. Rotavirus replication: Gaps of knowledge on virus entry and morphogenesis. *Tohoku J Exp Med.* 2019;248(4):285–296.

CHAPTER 6

Single-Stranded DNA (ssDNA)

Erythrovirus B19

Historical Perspective

During an ultrasound at 20 weeks, the fetus of a 32-year-old woman was found to have a number of signs relating to hydrops fetalis, a condition of fluid buildup in the tissues and organs causing extensive swelling or edema. Additionally, indications of reduced blood viscosity due to severe anemia was detected, leading to a risk of cardiac failure. The mother's blood exhibited elevated levels of immunoglobulin (Ig) M and IgG against this particular viral antigen. Amniotic fluid analysis revealed the presence of B19V (also known as parvovirus B19).

Intrauterine blood and platelet transfusions were given to the fetus through the umbilical cord. Fetal hematocrit improved, but was still low, and there were persistent symptoms of hydrops fetalis and an enlarged heart. The female fetus was born at 36 weeks' gestation by emergency cesarean section.

The baby was pale and suffered from respiratory distress. Her abdomen was grossly distended. She was intubated and transferred to a neonatal intensive care unit to receive treatments during the first 3 days of life due to persistent anemia and low reticulocyte (immature red blood cells [RBCs]). Her condition improved quickly, and she was discharged on day 5 with no evidence of congenital abnormalities.

Human B19V (parvovirus B19) was identified by Australian virologist Yvonne Cossart and colleagues in 1975, by chance, while screening a normal blood bank donors' sera for hepatitis B virus. Three B19V genotypes have since been identified, with the most common strain classified as genotype 1. The genomic sequences of genotype 2 and genotype 3 are roughly 11% and 12% divergent from that of genotype 1, respectively. Antibodies against genotype 1 or 2 of human B19V have been shown to be cross-reactive.

Human B19V DNA remains present throughout the life of the infected individual and has even been found in the bones of individuals who had been deceased for 70 years. The persisting virus corresponds to the genotype of the initial infection.

Data show that B19V infection in humans has been occurring for thousands of years. B19V DNA samples were isolated from dental and skeletal samples of over 1000 ancient human remains and subjected to genomic sequencing. These samples were recovered from individuals who lived across Eurasia, Southeast Asia, and Greenland from 500 hundred to 6.9 thousand years. These infected individuals were from different archaeologically defined cultures: Viking age Scandinavians, early Neolithic to early Bronze Age Baikal hunter-gatherers, early Slavs from the Czech Republic, and Tian Shan Hun. Some archaeologic isolates have a 63.9%–99.7% match to present-day B19V genomes.

Genotype 1 appears to have a worldwide distribution. Genotype 2 is mainly present in Northern Europe, while Genotype 3 was present in sub-Saharan and West Africa, South America, and France.

Classification and Structure of Erythrovirus B19

B19V is a member of the *Erythroparvovirus* genus and Parvoviridae family. This family of viruses is divided into three subfamilies, Parvovirinae, Densovirinae, and Hamaparvovirinae. Parvovirinae contains viruses that infect vertebrate host organisms, whereas Densovirinae and Hamaparvovirinae contain viruses that infect arthropod hosts. This is a large family that includes 26 genera and 126 species: *Amdoparvovirus* (5 species; eg, gray fox amdovirus); *Artiparvovirus* (1 species; chiropteran artiparvovirus 1); *Aveparvovirus* (3 species; eg, gruiform aveparvovirus 1); *Bocaparvovirus* (28 species; eg, dromedary camel bocaparvovirus 1); *Copiparvovirus* (7 species; eg, ungulate copiparvovirus 3); *Dependoparvovirus* (11 species; eg, rodent dependoparvovirus 1); *Erythroparvovirus* (7 species; eg, erythrovirus B19); *Loriparvovirus* (1 species; primate loriparvovirus 1); *Protoparvovirus* (15 species; eg, primate protoparvovirus 4); *Tetraparvovirus* (6 species; eg, opossum tetraparvovirus); *Aquambidensovirus* (3 species; eg, decapod aquambidensovirus 1); *Blattambidensovirus* (1 species; blattella germanica densovirus 1); *Diciambidensovirus* (1 species; diciambidensovirus hemipteran2); *Hemiambidensovirus* (2 species; eg, hemipteran hemiambidensovirus 1); *Iteradensovirus* (5 species; eg, Lepidopteran iteradensovirus 1); *Miniambidensovirus* (1 species; Orthopteran miniambidensovirus 1); *Muscodensovirus* (1 species; Pefuambidensovirus); *Pefuambidensovirus* (1 species; Blattodean pefuambidensovirus 1); *Protoambidensovirus* (2 species; eg, lepidopteran protoambidensovirus 1); *Scindoambidensovirus* (3 species; eg, orthopteran scindoambidensovirus 1); *Tetuambidensovirus* (1 species; tetuambidensovirus incertum 2); *Brevihamaparvovirus* (2 species; eg, dipteran brevihamaparvovirus 1); *Chaphamaparvovirus* (16 species; eg, rodent chaphamaparvovirus 1); *Hepanhamaparvovirus* (1 species; decapod hepanhamaparvovirus 1); *Ichthamaparvovirus* (1 species; syngnathid ichthamaparvovirus 1); and *Penstylhamaparvovirus* (1 species; decapod penstylhamaparvovirus 1).

Other related families of virus include Inoviridae (eg, M13 bacteriophage); Microviridae (eg, enterobacteria phage ΦX174); Pleolipoviridae (eg, halogeometricum pleomorphic virus 1); Spiraviridae (eg, *Aeropyrum pernix*); Anelloviridae (eg, chicken anemia virus); Bacilladnaviridae (eg, *Chaetoceros salsugineum* DNA virus 01); Bidnaviridae (eg, *Bombyx mori bidensovirus*); Circoviridae (eg, porcine circovirus); Geminiviridae (eg, maize streak virus); Genomoviridae (eg, human-associated gemyvongvirus 1); Nanoviridae (eg, banana bunchy top virus); and Smacoviridae (eg, babosmacovirus).

The structure of B19V is a non-enveloped virion 23 nm in diameter, which is among the smallest known DNA-containing viruses that infect mammalian cells. A single linear, single-stranded DNA (ssDNA) genome of 5596 bases is enclosed in an icosahedral capsid (Figure 6.1). Unlike ssRNA viruses, the B19V genome can be either a plus or minus ssDNA strand; therefore, a mixture of plus-strand or minus-strand ssDNA molecules can be found in the same virus species and population. Both the 5' and 3' ends of this genome contain inverted terminal repeats (ITRs) that allow the formation

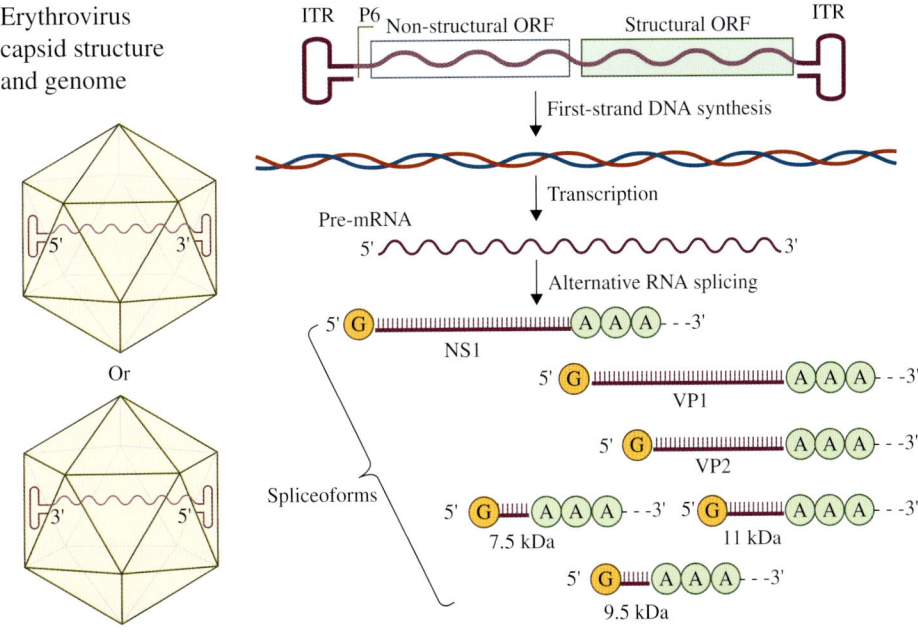

FIGURE 6.1 Erythrovirus B19 capsid structure and genome. The B19 virion is small and simple. The genome can be a single plus-strand DNA or minus-strand DNA, where a hairpin structure is found at each end of the genome. The ssDNA genome is used as the template for the synthesis of its complementary strand, resulting in a dsDNA. The pre-mRNA undergoes alternative splicing to produce spliceoforms (resulting mRNAs) that encode for each of the 6 viral proteins.
Credit: Figure partially created in BioRender.

of base-paired hairpin structures. Since its genome is a linear, ssDNA and the virus doesn't package any nonstructural proteins in its virion, the complementary viral DNA must be synthesized first to yield a double-stranded DNA (dsDNA). The free hydroxyl group at the 3′ hairpin (3′-OH) is used as a self-primer for the synthesis of the complementary strand of DNA using host cellular replication proteins through a process known as "first-strand DNA synthesis." Since the viral genome depends solely on the host cell's machinery for its replication, the only time that the cell has the appropriate machinery to carry out DNA replication of the complementary viral DNA strand is during the synthesis stage of the cell cycle. The complete mechanism for parvovirus genome replication has still not yet been worked out.

The two major open reading frames (ORFs) are flanked by ITR regions, which are 383 bases in length. The function of the ITRs is to regulate self-priming during viral genome replication. The remaining 4830 base pairs contain 2 major ORFs. One ORF encodes for the nonstructural NS1 protein and the other encodes for structural viral protein 1 (VP1) and VP2 (capsid proteins). There are also 3 minor ORFs that encode for 7.5-kDa, 9.5-kDa, and 11-kDa protein. There is only 1 promoter sequence (P6), along with a few upstream enhancer sequences, used to regulate the synthesis of a single precursor (pre-) messenger RNA (pre-mRNA). Mature mRNAs are produced through alternative RNA splicing to be used as the template for the translation of all VPs.

NS1 has been shown to be the only protein essential for B19V DNA replication. NS1 possesses both site-specific endonuclease and DNA helicase activities as well trans-activation domains. The 11-kDa protein involves caspase 10 to activate an apoptotic pathway in the host cell. The 11-kDa protein also affects virion production by regulating VP2 production and its distribution. To date, the function of the 7.5-kDa protein and 9.5-kDa protein are largely unknown.

Host Range, Transmission, Tropism, and Susceptible Host Cell

B19V is the only species in the *Parvovirus* genus known, thus far, to infect humans and is not known to be zoonotic. The host range of this genus, however, includes birds, insects, and other mammals, such as rat, mouse, and mink.

In humans, B19V exhibits high tropism for human erythroid progenitor cells in the bone marrow and fetal liver. B19V causes mild to severe hematological disorders in patients and generally target dividing cells.

In in vitro cultures, B19V replicates in erythroid burst-forming units and erythroid colony-forming units. Both structures are aggregates of different stages of erythroid progenitor cells that generate erythrocytes.

Neither hematopoietic stem cells nor myeloid lineage cells are susceptible to B19V since neither cell type exhibits the surface marker CD36+, which designates permissive cells. Viral replication has a cytotoxic effect on these precursor cells, resulting in decreased erythropoiesis, which in turn leads to erythroid aplasia (abnormal development).

Activation of the erythropoietin pathway and a hypoxic condition, such as the microenvironment in bone marrow, are also required to create an intracellular environment permissive for B19V replication. This is why erythroid progenitor cells are the only permissive cell type for this virus.

B19V can also infect nonerythroid lineage cells, such as myocardial endothelial cells circulating angiogenic cells, endothelial progenitor cells from the bone marrow, and endothelial cells of various tissues (eg, the aorta, umbilical vein, and pulmonary arteries). In addition, a number of cell lines have been used for B19V infection, such as megakaryoblastoid cell lines, erythroid leukemia cell lines, human pulmonary artery cells, circulatory angiogenic cells, umbilical vein endothelial cells, and human aortic endothelial cells.

The principal mode of exit for erythrovirus B19 virus from the host organism is the expelling of respiratory droplets (eg, saliva, sputum, and nasal mucus) containing viral particles while exhaling, coughing, sneezing, speaking, and such. Hence, the mode of transmission is via direct contact with respiratory secretions (droplets) and saliva. In addition to the respiratory route, B19V can be transmitted via the blood-borne route, including passage from the mother to the fetus through the placenta (referred to as transplacental infection) and transfusion of contaminated blood and blood products.

Contamination is problematic because B19V is not only heat stable (eg, stable after incubation at 56° for 60 minutes) but also so small that it passes through filters (which is the characteristic of most known viruses) used to filter blood products. Additionally, these viral particles are stable when exposed to a range of pH, from 3 to 9, and are resistant to lipid-solubilizing solvents such as ether, chloroform, benzene, and acetone.

Infecting the Host Susceptible Cell, Viral Particle Replication, and Tissue Damage

Erythrocyte P antigen (or glycosphingolipid globoside) is the primary receptor for B19V; however, not all cells that express P antigen are permissive to this virus.

Permissive cells must also exhibit co-receptors such as Ku80 and integrin α5β1. When infecting nonerythroid cells, B19V uses an alternative entry route by interacting with surface antibody and entering through complement factor C1q and C1q receptor–mediated endocytosis.

With the exception of placental endothelial cells, however, there's no evidence that B19 virions are produced in nonerythroid cells and are considered nonproductive (ie, no virion production). B19V infection of erythroid progenitor cells can lead to apoptosis, while infection of nonerythropoid cells can lead to a number of inflammatory disorders.

Attachment of the virus to the P antigen causes the viral capsid protein (CP), VP1, to undergo a conformational change, then thus exposing a domain ("VP1 unique region" or VP1u) that in turn interacts with a co-receptor to initiate viral particle internalization into the host cell. Entry of the virus occurs through clathrin-dependent, receptor-mediated endocytosis, but escapes lysosomal degradation (Figure 6.2). The virion is transported out of the endosome before partial capsid disassembly occurs in the cytosol.

The precise mechanism of endosomal externalization, uncoating, and import of the nucleocapsid into the nucleus is still unclear. However, exposure to chelating agents or buffers with chelating properties appears to trigger uncoating in vitro. Data suggest that B19V is small enough to be imported through the nuclear pores (of the nuclear

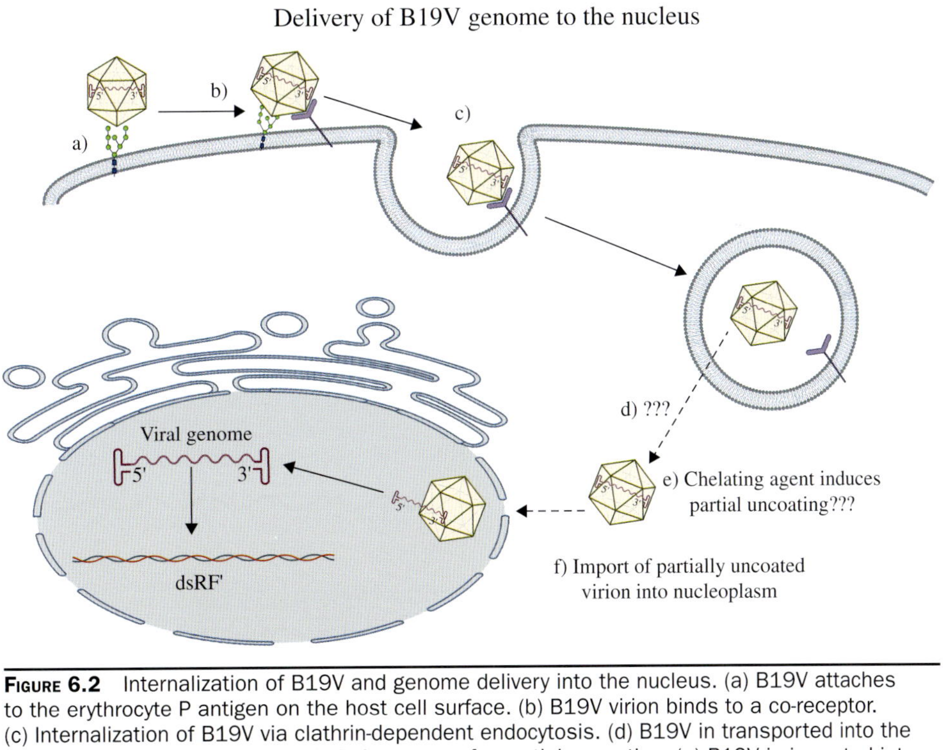

FIGURE 6.2 Internalization of B19V and genome delivery into the nucleus. (a) B19V attaches to the erythrocyte P antigen on the host cell surface. (b) B19V virion binds to a co-receptor. (c) Internalization of B19V via clathrin-dependent endocytosis. (d) B19V in transported into the cytosol, where it interacts with chelating agents for partial uncoating. (e) B19V is imported into the nucleus. (f) B19V genome is extruded into the nucleoplasm to be used as the template for the synthesis of the complementary DNA strand. *Credit:* Figure partially created in BioRender.

membranes) via nuclear localization signals without complete capsid disassembly. The viral genome is then extruded into the nucleoplasm. Subsequent genome replication, transcription, and virion packaging occur in the nucleoplasm.

Once in the nucleoplasm, the ssDNA genome is converted into a double-stranded replicative form of DNA (dsRF DNA) by DNA repair enzymes and is used as the template for both genome replication and gene expression. The host cellular machinery is required for all these activities.

Since the host cellular DNA replication machinery is required for viral genome synthesis, only susceptible cells that are progressing through the synthesis (S) stage are permissive of the cell cycle, even though the viral particle can enter permissive and nonpermissive cells equally well, as long as the cell is susceptible to the virus. The dsRF DNA is produced through a process referred to a "first-strand DNA synthesis."

The host's DNA polymerase uses the hydroxyl group at the 3'-end hairpin (3'-OH) as the primer to extend the original viral ssDNA genome toward the 5' hairpin, producing dsRF DNA (Figure 6.3). Ligation of the 3' end of the newly synthesized complementary strand and the existing 5' end produces a partial circular replicative form DNA molecule (cRF DNA), which is the primary conversion product of the B19V genome. The viral NS1, which possesses both site-specific endonuclease and DNA helicase activities, along with host machinery carries out strand displacement and "rolling hairpin-dependent" DNA replication to produce viral ssDNA genomes to be packaged into the capsid. To date, details of this process have not been worked out.

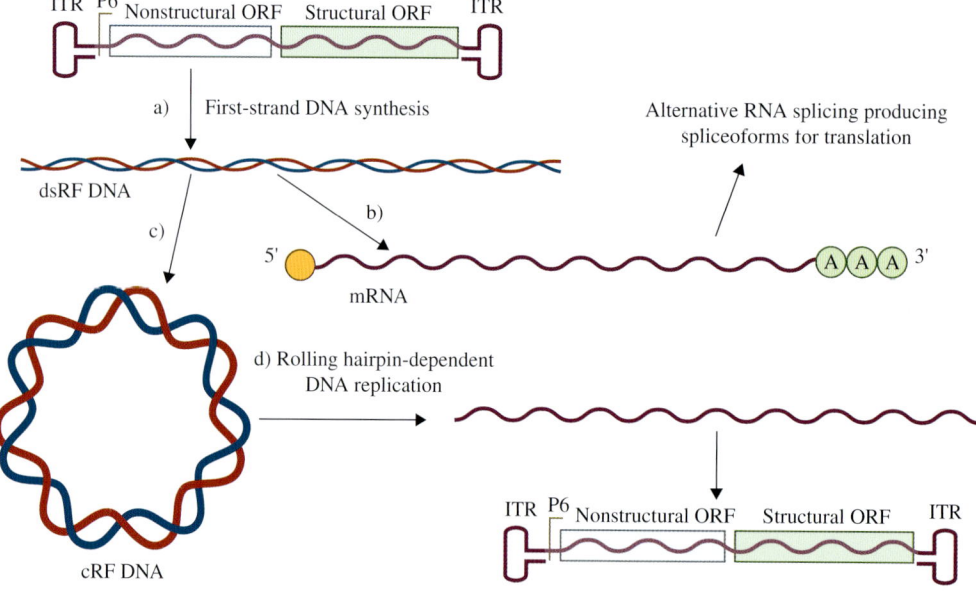

Figure 6.3 B19V genome synthesis. (a) Host DNA repair enzymes carry out first-strand DNA synthesis of dsRF DNA. (b) Host RNA polymerase binds to the dsRF DNA to transcribe mRNAs for viral protein synthesis. (c) Ligation of one of the strands in dsRF DNA produces cRF DNA. (d) NS1 and host machinery carry out "rolling hairpin-dependent" DNA replication to produce viral ssDNA genomes. *Credit:* Figure partially created in BioRender.

As stated, the dsRF DNA is used by the host cell's RNA polymerase to carry out viral mRNA synthesis. Recall that RNA polymerase only recognizes and binds dsDNA. Along with the host cell's transcription factors, NS1 is also required to regulate viral mRNA synthesis. There is only one promoter that is used to regulate the synthesis of a single pre-mRNA, which in turn undergoes alternative RNA splicing to produce specific smaller mRNAs. Each splice product is capped and polyadenylated before being transported into the cytosol to be used as a template for cap-dependent translation using the host cell's machinery. After viral protein synthesis and viral genome replication has occurred, virions are assembled in the nucleus.

As noted in the Infecting the Host Susceptible Cell, Viral Particle Replication, and Tissue Damage section, B19V virions are small enough to be exported into the cytosol through nuclear pores (Figure 6.4). During the early phase of infection, vacuolization around each virion occurs in the cytosol before their release from the cell by exocytosis. However, the bulk of virions are released as the result of the host cell having undergone apoptosis as the viral infection continues.

The nonstructural viral protein, NS1, can induce the host cell to undergo G2/M checkpoint arrest and eventually apoptosis through multiple mechanisms. NS1 can activate the DNA damage response (DDR) pathway in response to DNA double-strand breaks (DSBs), stalled replication forks, and non-hologous end joining (NHEJ) repair of DSBs. These kinases perform important functions in DNA repair, cell cycle control, and preventing genomic instability. NS1 can also activate a pathway that leads to cell cycle

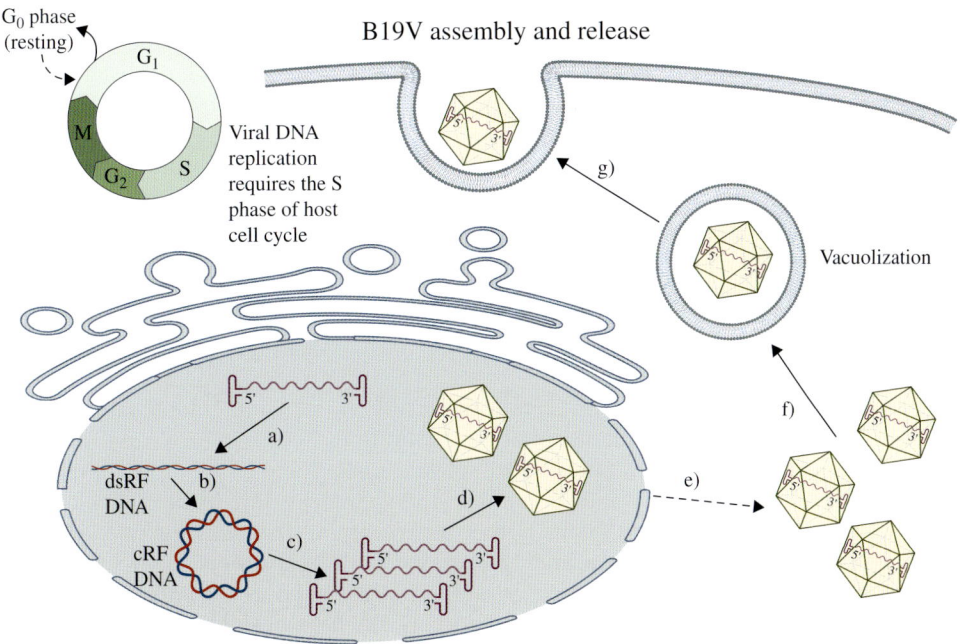

Figure 6.4 Virion assembly and release from host cell. (a) First-strand DNA synthesis producing dsRF DNA. (b) Ligation of one of the DNA strands producing cRF DNA. (c) Genome synthesis via rolling hairpin-dependent DNA replication. (d) Virion assembly in the nucleoplasm. (e) Export of virions through nucleopores. (f) Vacuolization of B19V. (g) Release of B19V via exocytosis. *Credit:* Figure partially created in BioRender.

checkpoint arrest, which can eventually cause apoptosis. CDC is the cell division cycle gene and CDK is the cyclin-dependent kinase. Last, NS1 can induce apoptosis through the activation of caspase 3 and caspase 9. The intrinsic apoptotic pathway is activated once the above-mentioned activities are prolonged. B19V virus causes tissue damage through apoptosis.

Dissemination in the Host Body

Once in the host body, B19V quickly undergoes hematogenous dissemination from either the gastrointestinal or respiratory tract, leading to active viremia. Of course, passive viremia can also occur via blood-borne transmission. However, B19V does little damage in the bloodstream since differentiated, nondividing blood cells are not permissive to this virus. It does its damage in the bone marrow, where erythroid progenitor cells are located.

Pathogenesis and Clinical Manifestation

Human erythrovirus B19 is associated with a wide variety of conditions, from asymptomatic to fatal. The most common condition in children is erythema infectiosum, which is a mild condition, and arthritis and arthralgias in adults. Erythema infectiosum is also known as "fifth disease," "morbus quintus," "slapped cheek syndrome," and "infectious rash."

Slapped cheek rash, which is the appearance of red rash on the face, often occurs in children. Polyarthropathy syndrome is a condition exhibited by pain and swelling of the joints that is more common in adults than children, especially women. Infected adult individuals with healthy immune systems are often asymptomatic but still contagious. Following the primary infection, B19V can establish lifelong persistence and is detectable in a number of human tissues, including kidney, lymph nodes, heart, testes, and synovial tissue.

Unfortunately, there are a number of serious health complications that may be experienced by individuals with existing, underlying conditions, which might affect the nerves, joints, or blood system. These conditions include (1) a congenital or acquired immunodeficiency such as leukemia, organ transplant, and human immunodeficiency virus (HIV) infection; (2) hematological disorders that require increased erythropoiesis, such as chronic hemolytic anemia, sickle cell disease, and thalassemia; and (3) a combination of immunodeficiency and accelerated erythropoiesis during pregnancy.

B19V infection has been known to cause a severe drop in the blood count, causing anemia in some patients with hematological conditions, which in turn can lead to transient or persistent erythroid aplasia and aplastic crisis. Some of the most common disorders resulting from B19V infection are erythema infectiosum, arthropathy, transient aplastic crisis, and hydrops fetalis. Other less common disorders include congenital anemia, chronic red cell aplasia, and "gloves-and-socks" syndrome. B19V infection has been proposed to cause hematological abnormalities such as thrombocytopenia and leukopenia. There are also a number of disorders that have been linked, but not proven, to be due to B19V infection, such as encephalopathy, epilepsy, meningitis, myocarditis, dilated cardiomyopathy, and autoimmune hepatitis.

Erythema infectiosum is the most common disorder with very recognizable symptoms and generally affects children 4–10 years old. After the initial cold-like symptoms, erythema of the cheeks starts to appear, known as the slapped cheek rash. During the next 4 days or so, the blotchy (or maculopapular) rash spreads to exposed extremities and persists up to 6 weeks. This symptom resolves spontaneously without any permanent detectable sequelae.

Arthropathy is another complication of B19V infection. Polyarthralgia is a symptom marked by pain and swelling in the interphalangeal and metacarpophalangeal joints as well as those of the knees, wrists, and ankles. This symptom is exhibited in up to 60% of infected adolescents and adults but in only 8% of children. Women are twice as likely to experience polyarthralgia than men. Polyarthralgia can resolve within 3 weeks; however, in some cases, it can last for months or even years.

Transient aplastic crisis is experienced by individuals with existing hemolytic anemia conditions where there are decreased levels of erythrocytes. Transient aplastic crisis can cause various disorders, including iron deficiency anemia, HIV, sickle cell disease, spherocytosis, or thalassemia. B19V infection blocks the production of an already small population of erythrocytes in these individuals, which can be life threatening. Decreased erythrocyte production leads to a drastic drop in hemoglobin, which may trigger sequelae such as congestive heart failure, a cerebrovascular accident, or acute splenic sequestration. Fortunately, treatment with blood transfusions can allow patients to recover fully within a couple of weeks.

Hydrops fetalis (or fetal death) may occur in pregnant women. Fortunately, infection usually yields mild symptoms for healthy individuals. Symptoms of B19V infection are usually mild and may include fever, headache, cough, and sore throat. These symptoms are mostly associated with the body's inflammatory response.

For the fetus, however, hydrops fetalis causes fetal tissues and organs to retain large amounts of fluid, causing edema. There is a 30% chance that a transplacental B19V infection from the mother to her baby might occur. The fetal liver and heart may also become infected. The mean lifespan of RBCs in a fetus is roughly 54 days, which is 30% shorter than that of an adult (~70 days). With this already shortened RBC lifespan, decreased erythropoiesis causes the baby to develop severe anemia. If the heart is infected, myocarditis can develop. Worse, congestive heart failure and hydrops fetalis may result from the combined effects of severe anemia and myocarditis. Additionally, B19V infection of the fetus may lead to miscarriage and intrauterine death. Several inflammatory disorders may be sequelae to a B19V infection. These disorders include cardiomyopathy, rheumatoid arthritis, hepatitis, vasculitis, and meningoencephalitis. A possible cause of these disorders is suggested by ex vivo cell culture studies, where NS1 from B19V was found to upregulate signals that control inflammatory responses, while downregulating antiviral responses through the activity of the STAT3 (signal transducer and activator of transcription gene 3) signaling pathway.

Diagnosis, Treatment, and Prevention

For erythema infectiosum, the visible slapped cheek rash on the patient's face is a sure sign of this disorder, and a clinical diagnosis can be made without laboratory testing. However, for confirmation, either a B19V-specific antibody test or viral DNA test are diagnostic tools that can be performed to confirm B19V infection.

In immunocompetent patients, blood testing for the presence of IgM, not IgG, against B19V antigens is used to diagnose acute viral infection. The level of IgM antibodies produced can remain high for up to 3 months after the infection.

The amount of IgG is not useful because it's a measure of previous infections, not the most recent one. Neither type of serum testing is useful for patients exhibiting transient aplastic crisis. They are also not used for immunocompromised patients with chronic infection. Patients with either of these two conditions do not test positive for IgM or IgG even when infected. B19V DNA testing using polymerase chain reaction (PCR) assays is preferred over less-sensitive nucleic acid hybridization assays, but is equally sensitive to IgM antibody assays.

B19V infections are usually mild and self-limiting (ie, will go away on their own). Children and adults who are otherwise healthy usually recover completely. Treatment of B19V infections is limited to relieving symptoms like fever, itching, and joint pain and swelling. There is no antiviral medication available. Patients with erythema infectiosum do not require treatment since this disorder can resolve itself within weeks. Given enough time, the bone marrow of patients with transient aplastic crisis can recover. In the meantime, a blood erythrocyte transfusion can be a temporary treatment to compensate for the low level of erythrocyte. In utero erythrocyte transfusion has also been shown to be effective for patients with hydrops fetalis. Polyarthralgia experienced by patients with arthropathy can be treated with nonsteroidal anti-inflammatory drugs. For severe illnesses, intravenous IgM or IgG targeting B19V antigens has been used to reduce symptoms.

To date, no successful vaccine has been developed to prevent B19V infection. Preventive measures are the same as against any other viruses that are transmitted via the respiratory route. Always wash one's hands thoroughly with soap and water. Definitely avoid touching any part of one's face, especially the eyes, nose, and mouth. Avoid close contact with known infected individuals, especially in closed, tight space.

If a pregnant woman is known to be infected with B19V, blood testing and ultrasounds are recommended. Ultrasounds can be used to diagnose hydrops fetalis.

Current Status

B19V infection is very common and is distributed globally, with no race-based, temporal, and geographic clustering trends observed. It is detected in roughly half of the world's population. In the United States, approximately 90% of adults above 60 years old test positive for B19V antigens. For young children between the age of 2 and 5 years old, the seropositivity rates are 5%–10%. By 15 years of age, this rate increases to 50% and to 60% by 30 years of age. The reason for such a high rate of seropositivity for adults is the fact that throughout one's lifetime, an individual continues to experience reinfections, for some annually.

Once infected and having developed the mild disorder erythema infectiosum, an individual acquires immunity that gives protection from reinfection in the future.

Outbreaks of erythema infectiosum often occur during late winter into early spring, often originating in nurseries and schools. This is likely due to people being more often indoors and in close proximity to one another, even an asymptomatic infected person. Outbreaks appear to be cyclic, occurring every 3–4 years. For childcare professionals, B19V infection is an occupational hazard.

B19V-infected individuals of all ages who are otherwise healthy have an extremely low rate of mortality. The most serious complication of B19V infection is hydrops fetalis, which results in 2%–6% intrauterine fetal death rate when the virus is transmitted from a "nonimmune" woman who is infected for the very first time during pregnancy. Fetal B19V infection is not associated with congenital malformation; however, other abnormalities may still occur, such as intrauterine growth retardation, myocarditis, and pleural and pericardial effusions.

Scientific Significance and Discoveries

B19V is the only known virus in the *Parvovirus* genus that infects human hosts. It is also considered unique among DNA viruses in that its substitution (a type of DNA mutation where a complementary base is replaced by a noncomplementary base) rate is unusually high, similar to that of RNA viruses. Currently, there's no specific treatment to eliminate viral particles in the infected individual, or any vaccine to prevent the development of associated conditions. Therefore, there is an urgent need to develop both antiviral medication and protective vaccines against B19V infection.

The first step to developing antiviral compounds is to identify critical steps during the process of virus replication to target. For B19V infection, processes involving NS1 are the most obvious target since it has been shown that NS1 is the only protein essential for B19V DNA replication.

Inhibitors against the endonuclease activity are being screened using an in vitro nicking assay for NS1 that was developed. Competitive peptide analogues of the VP1u region or neutralizing antibodies are being developed to prevent viral entry into the susceptible host cell.

Some drugs approved by the Food and Drug Administration (FDA) used for treatment of other conditions are being tested to target B19V-induced disorders.

One of these drugs is pimozide, which is used to reduce uncontrolled movements and outbursts of sounds caused by Tourette syndrome. Pimozide works by decreasing the activity of dopamine in the brain. It has also been shown to block B19V replication via the antiviral response STAT3/PLAS3 signaling pathway in erythrocyte progenitor cells. Unfortunately, the maximum dose allowed to be used in patients is too low to inhibit B19V replication; therefore, derivatives of pimozide are being developed.

Hydroxyurea is another known and FDA-approved drug that is being tested. It is an antiproliferative drug that is used to treat sickle cell anemia by inhibiting ribonucleotide reductase, an enzyme that mediates deoxyribonucleotide synthesis by converting ribonucleotides to deoxyribonucleotides.

Vaccine development should be a reasonable goal; however, its development and implementation are not included among the priorities of the World Health Organization. In part, this may be due to B19V infection–induced disorders usually being mild and self-limiting. Consequently, there's no need to expend resources to develop a vaccine.

However, there is merit in developing a vaccine to target at-risk populations that have underlying hematological disorders or are immunocompromised. A vaccine for nonimmune women preparing to bear a child could potentially avoid transplacental transmission.

Some other reasons for the slow development of a vaccine are (1) no available animal model for B19V since this virus only infects humans and (2) growing viral particles in cell culture is quite difficult. Therefore, a first step to vaccine development is to develop cell culture techniques that readily and easily support B19V replication.

Porcine Circovirus 2

Historical Perspective

The first outbreak of the postweaning multisystemic wasting syndrome (PMWS), which increases mortality rates and slows down the growth of animals, occurred in January 1991. PMWS is also known as porcine circovirus 2 (PCV2) systemic disease (PCV2-SD) and belongs to a group of conditions called porcine circovirus–associated disease (PCVAD). A farmer at a healthy, farrow-to-finish swine operation in northeastern Saskatchewan, Canada, reported a dramatic increase in nursery mortality (12%–15%). One year earlier, this farm restocked breeding animals from a "high-health" multiplier herd, which was suspected to be the source of transmission. High-health herds are supposed to contain animals that are free of known swine infections and diseases. Clinical symptoms of those affected were jaundice, diarrhea, and respiratory disease; associated ill thrift, icterus, and sudden death. At the time, no known pathogens were identified after diagnostic examination of tissues and blood samples. There was no follow-up on this problem.

In 1994, a similar disease syndrome was observed in an intensive high-health swine operation, also in Saskatchewan. This farm reported a 4-fold increase in postweaning mortality associated with dyspnea, icterus, and weight loss. All known swine pathogens at the time were excluded after diagnostic examination of carcasses, live animals, blood, serum, water, and feed over the course of the outbreak. However, signs that are now known to be associated with PMWS and PCVAD were detected, including the presence of characteristic botryoid basophilic intracytoplasmic inclusion bodies in cells of monocyte/macrophage lineage. This pathogen was later identified as PCV2.

In Germany in 1974, the first porcine circovirus, PCV1, was isolated from a porcine kidney cell culture, PK-15. This strain of PCV is considered nonpathogenic. PCV2 was identified in Canada in the early 1990s. As of December 2021, there were 8 genotypes of PCV2 (PCV2 a–h) identified. PCV3 was identified in 2018 and considered to be nonpathogenic. PCV4 was identified in Hunan Province, China, in 2019.

Classification and Structure of Porcine Circovirus

Porcine circovirus 2 belongs to the Circoviridae family and genus *Circovirus*. The structure of PCV2 is a non-enveloped, icosahedral virion that is 17 nm in diameter, which makes it the smallest known mammalian virus. Its circular, minus-strand DNA genome of 1766–1769 bases contains 11 predicted ORFs (Figure 6.5). The ambi-sense DNA genome is a minus-strand genome that has at least 1 segment of plus-strand reading frame.

ORF1 encodes replication-associated proteins or replicase (Rep and Rep'). ORF2 encodes the viral CP. ORF3 encodes a protein involved in the induction of apoptosis in vitro. ORF4 encodes a protein that upregulates caspase activity and downregulates the activity of $CD4^+$- and $CD8^+$-T lymphocytes, thus causing immunosuppression. The ORF5 gene product has an inhibitory role against host immune surveillance by inhibiting type I interferon expression and thus supports increased PCV2 replication. The ORF6 gene product may be involved in regulating the caspase cascade as well as the expression of cytokines in the infected host cell. The remaining predicted ORFs continue to be under investigation.

Infecting the Host Susceptible Cell, Viral Particle Replication, and Tissue Damage

Even though PCV2 is the major pathogen for PCVD members of the genus *Circovirus* have been detected in birds (canary, pigeon, duck, finch, goose, bull, starling, and swan); arthropods (tick and mosquito); and shellfish and in environmental samples.

In the swine host, PCV2 has strong tropism for lymphoid tissue but can infect other cells, including epithelial and endothelial cells of the liver, lung, and kidney; monocytic

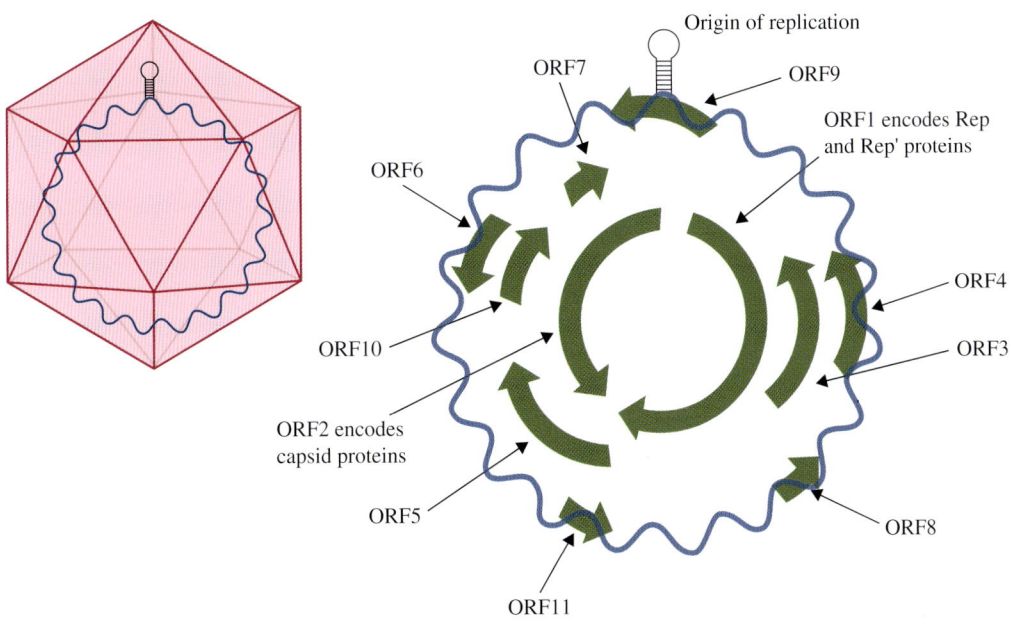

FIGURE 6.5 PCV2 capsid structure and genome. The PCV2 is a simple icosahedral virion without an envelope. Its genome is a single circular, minus-strand DNA molecule. The predicted ORFs are shown on the right of this Figure. There are reading frames on both the plus strand and minus strand.
Credit: Figure partially created in BioRender.

cells including monocytes, macrophages, dendritic cell (DC) precursors, myeloid DCs, and plasmacytoid DCs of lymphoid tissues and organs; and fibrocytes.

Attachment occurs through the interaction between PCV CP and glycosamino-glycans (GAGs) such as heparan sulfate, chondroitin sulfate B, and dermatan sulfate receptors (Figure 6.6). PCV2 internalization into a monocytic cell occurs via clathrin-dependent, receptor-mediated endocytosis; PCV2 entry into host monocytic cells is slow and inefficient. However, internalization of PCV2 by other cell types such as kidney and testicular epithelial cells can occur via actin- and small GTPase (guanosinetriphosphatase)-regulated clathrin- and caveolin-independent pathways. T lymphoblasts (CD4+ and CD8+) use clathrin- and dynamin-2-mediated endocytosis.

Once internalized, uncoating occurs in the endosome, leading to the release of the viral genome into the host cell cytosol. The circular ssDNA is, in turn, imported into the nucleus, where genome replication occurs.

Even though its genome encodes for replication-associated proteins (Rep and Rep'), PBV still requires cellular DNA polymerase for genome replication. This means that permissive host cells must have entered the cell cycle. In addition to the presence of Rep and Rep', viral CP and gene products of ORF3, ORF4, and ORF5 provide support for viral genome replication via the rolling circle DNA replication mechanism.

Assembly is assumed to occur in the cytoplasm, although the exact mechanism of autoassembly is not yet known. PCV2 activates the intrinsic apoptotic pathway to cause lysis to the host cell as a means to exit. PCV2 is transmitted through contact with

Delivery of PCV2 genome to the nucleus

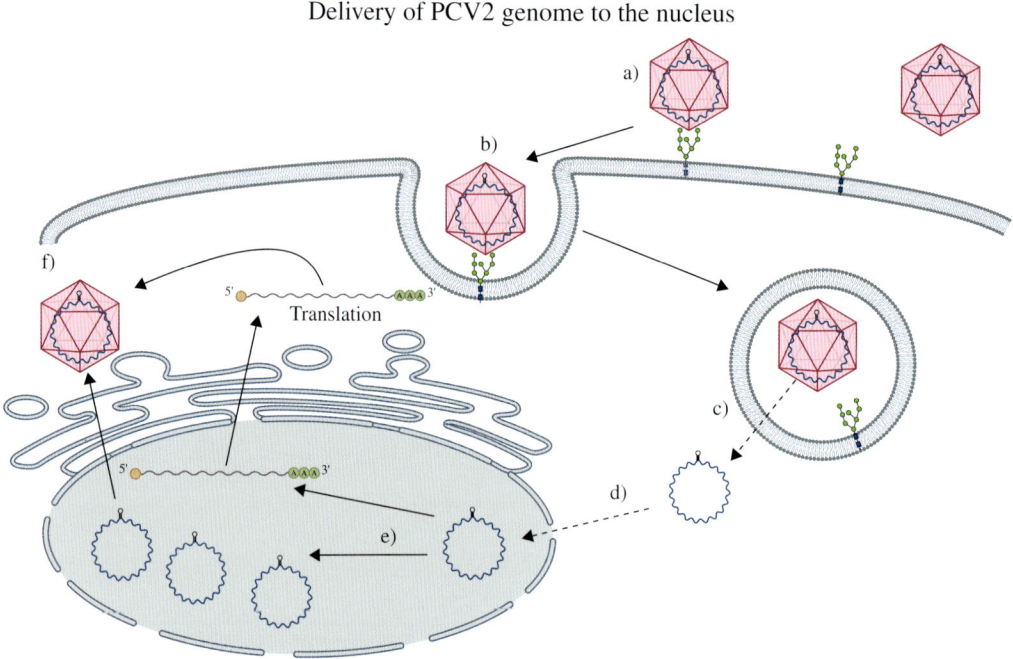

FIGURE 6.6   PCV2 entry into host cell and delivery of genome into nucleus. (a) PCV2 attaches to the host cell's GAG. (b) The virion is internalized via endocytosis. (c) Uncoating occurs in the endosomal lumen, and the viral genome is released into the cytosol. (d) PCV2 genome is then imported into the nucleus. (e) Genome replication and mRNA synthesis occur in the nucleoplasm. (f) Virion assembly occurs in the cytoplasm. *Credit:* Figure partially created in BioRender.

respiratory, oral, urinary, and fecal secretions. Vertical transmission from the sow to piglets may occur but is very rare.

Pathogenesis and Clinical Manifestation

Clinical symptoms exhibited by animals with PMWS caused by PCV2 include poor growth and weight loss, weakness, enlarged lymph nodes, anemia, jaundice, diarrhea, rashes, fever, stomach ulcers, and dyspnea (or difficulty breathing). Sudden death may also occur. PCV2 may also induce prenatal reproductive failure. The infection manifests different symptoms depending on the stage of embryonic and fetal development.

The embryo inside the zona pellucida of the ovarian follicle is resistant to infection (Figure 6.7). However, once ovulation occurs, the embryo loses its defense mechanism and is now susceptible to PCV2 infection. Infection of an embryo in this very early stage leads to death and resorption in utero.

During early fetal development, cardiomyocytes of the heart appear to be the first target of PCV2 infection, followed by hepatocytes, monocytic cells of lymphoid organs, and then lungs. As a result, the fetus undergoes mummification if it dies before the immunocompetent phase (before gestation day 70). Mummification is caused by autolysis and dehydration.

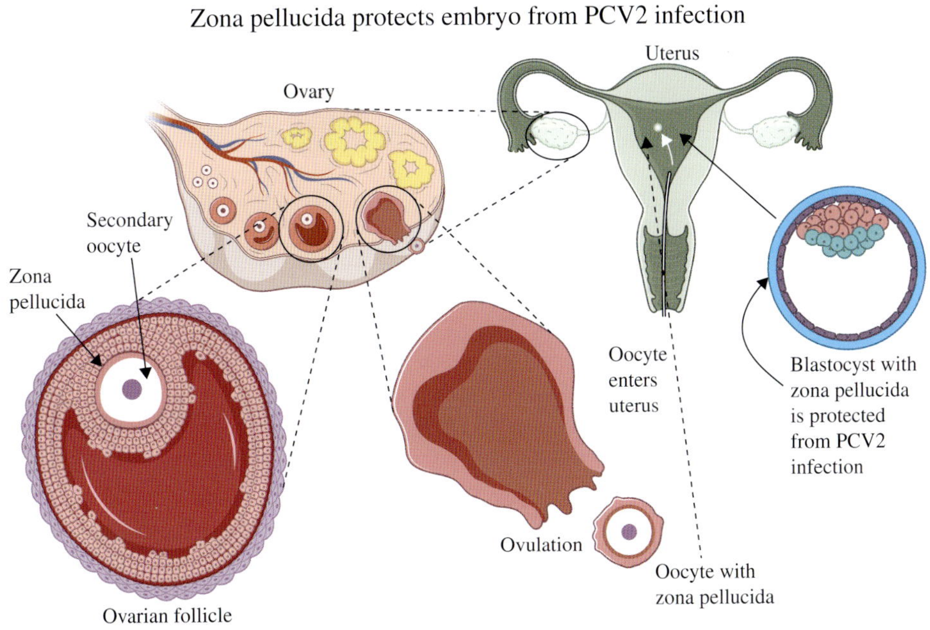

FIGURE 6.7 Protection from PCV2 infection by the zona pellucida. The oocyte in the ovarian follicle is protected by the zona pellucida from PCV2 infection. The zona pellucida continues to protect the oocyte during ovulation and as the oocyte migrates to the uterus. This protection continues after fertilization and during early embryogenesis. The zona pellucida is shed as the blastocyst matures and prepares to implant into the endometrium of the uterus. The blastocyst is now susceptible to PCV2 infection in the absence of the zona pellucida. *Credit:* Figure partially created in BioRender.

Factors that contribute to fetal mummification can be parity, litter size, uterine capacity, nutritional factors, and environmental temperature. If a fetus dies later in gestation, stillbirths or late-term abortions may occur. If the fetus survives, weak-born piglets will result. PCV2 is associated with not only PMWS, but also porcine dermatitis and nephropathy syndrome and porcine respiratory disease complex.

Diagnosis, Treatment, and Prevention

Diagnosis of PVCAD depends on clinical features, histopathology of lymphoid tissues, and PCV-2 detection in damaged tissues (eg, lymph nodes, spleen, tonsils, or Peyer patches). Immunohistochemical tests or in situ hybridization can be used to confirm the presence of PCV2 in tissue samples. PCR can be used to confirm the presence of viral nucleic acid.

While there is no specific treatment, anti-inflammatory agents and antimicrobials are sometimes used to alleviate symptoms. Antibiotics in feed or water can help suppress secondary infections.

PCV2 vaccination is the most cost-effective way to control PVCAD. Good farm management is an effective measure to prevent viral transmission; good management is through good ventilation, thorough cleaning and disinfection of the facility, control of co-infections, single-source of the nursery, increased age at weaning, and lower pig density, for example.

Current Status

PCV2 is main known causative agent of PCVADs, most prominently PMWS. It is found in most swine herds and recognized as an endemic infection of swine in all regions of the world, including European countries, North and South American countries, Asian countries, Oceania, Caribbean, Middle East, and African countries.

Scientific Significance and Discoveries

Even though PCV2s are simple viruses with minimal genome and coding capacity, they do possess a remarkable ability to interfere with host antiviral immune responses, which in turn promote their own propagation. PCV2s affect the expression of interferon α, β, and γ as well as tumor necrosis factor α and interleukins (IL) IL-2, IL-6, IL-10, and IL-12 to disrupt host innate and adaptive immune systems. In addition, PCV2-associated immunosuppression can provoke a more severe disease when co-infecting with another porcine pathogen.

PCV2s have found ways to interfere with host cellular pathways and their downstream functions. More specifically, in the early phase of infection while being internalized by the host cell, ORF2 gene product (CP) activates the phosphoinositide 3-kinase/protein kinase B pathway. This pathway in turn leads to the cleavage and inactivation of poly-ADP ribose polymerase and caspase 3 to maintain an antiapoptotic state of the host cell.

PCV2s do need the host cell to remain alive long enough to support virus propagation. ORF4 gene product also activates nuclear factor kappa B (NFκB) to inhibit pre-apoptotic processes and upregulate CP, Rep, and Rep' expression for virus replication. NFκB also increases the transforming growth factor (TGF) β level. TGF-β in turn stimulates CD4+ T-cell differentiation into Treg cells via the phosphorylation of extracellular signal-regulated kinase. An increase Treg cell production reduces the responsiveness of the host immune system to chronic infections, leading to immunosuppression.

During the late phase of infection, the ORF5 gene product prolongs the cell cycle S phase and induces the host cell to undergo checkpoint arrest to provide more time for virus replication. ORF5 gene product also activates the NFκB pathway to enhance the synthesis of pro-inflammatory cytokines (eg, IL-6, IL-8, and cyclooxygenase 2), leading to tissue damage and thus disease progression. Once viral production is at a high enough level, ORF3 and ORF5 gene products trigger autophagy and apoptosis while downregulating cell survival via the activation of mammalian target of rapamycin.

Further Readings

Caliaro O, Marti A, Ruprecht N, et al. Parvovirus B19 uncoating occurs in the cytoplasm without capsid disassembly and it is facilitated by depletion of capsid-associated divalent cations. *Viruses*. 2019;11(5):430.

Ellis J. Porcine circovirus: A historical perspective. *Vet Pathol*. 2014;51(2): 315–327.

Feher E, Jakab F, Banyai K. Mechanisms of circovirus immunosuppression and pathogenesis with a focus on porcine circovirus 2: A review. *Vet Q*. 2023;43(1):1–18.

Ganaie S, Qiu J. Recent advances in replication and infection of human parovirus B19. *Front Cell Infect Microbiol*. 2018;8:166.

Luo Y, Qiu J. Human parvovirus B19: A mechanistic overview of infection and DNA replication. *Future Virol*. 2015;10(2):155–167.

Mühlemann B, Margryan A, Damgaard P, Jones T. Ancient human parvovirus B19 in Eurasia reveal its long-term association with humans. *Proc Natl Acad Sci U S A*. 2018;115(29):7557–7562.

Nauwynck H, Sanchez R, Meerts P, et al. Cell tropism and entry of porcine circovirus 2. *Virus Res*. 2012;164(1–2):43–45.

Servey J, Reamy B, Hodge J. Clinical presentations of parvovirus B19 infection. *Am Fam Physician*. 2007;7(3):373–376.

Zakrzewska K, Arvia R, Bua G, Margheri F, Gallinella G. Parvovirus B19: Insights and implication for pathogenesis, prevention, and therapy. *Asp Mol Med*. 2023;1:100007.

CHAPTER 7

Double-Stranded DNA (dsDNA)

Herpes Simplex Virus

Historical Perspective

An 11-year-old patient was admitted into a hospital due to ocular redness and decreased visual acuity in the left eye that had been ongoing for about 20 days. Her cerebrospinal fluid tested positive for herpes simplex virus 2 (HSV-2), immunoglobulin (Ig) G, cytomegalovirus (CMV) IgG, and rubella virus IgG, but tested negative for other infections. She was diagnosed with acute retinal necrosis syndrome and viral encephalitis. This patient may have first been exposed to HSV-2 during parturition (or process of giving birth) from the mother. The virus may have remained latent in her body, then reactivated (due to a number of factors) to cause acute retinal necrosis, which may also be associated and complicated with viral encephalitis. She was treated with antiviral, steroid, and prophylactic laser therapy. She was on antiviral therapy for no more than 14 weeks since long-term use of these drugs has been shown to have adverse side effects on children such as nephrotoxicity, gastrointestinal upset, and skin lesions.

In another case, a 66-year-old man was admitted to the hospital for treatment of recurring suprasellar meningiomas, slow-growing tumors that occur near the pituitary gland and optic nerves at the skull base affecting primarily middle-aged individuals. With the exception of reduced vision acuity, preoperative physical examination showed normal health for this patient. However, 1 day after the operation, the patient had a fever of 102°F and appeared lethargic. The surgical site was dry and clean, but a crusted cutaneous lesion was found at the corner of his mouth where his endotracheal tube had been taped. This was assumed to be the result of a bacterial infection, and he underwent antibiotics therapy. Culture of the skin lesion was negative for HSV. His low-grade fever persisted for 2 weeks but increased to 104 °F thereafter. His condition soon deteriorated. Other symptoms include decreased blood pressure, increased pulse rate, abdominal tenderness resulting from bowel ischemia, and bleeding from two intravenous sites. The patient died the following day. Results of liver function tests on the patient before he passed away were consistent with acute hepatic necrosis. Postmortem analysis confirmed that the patient was HSV-1 positive. Fulminant (severe and sudden onset) hepatitis due to HSV primarily occurs in individuals with impaired immunity but can occur in healthy adults.

The Greek word *herpes* means "to creep or crawl" in reference to the spreading nature of these prevalent herpetic skin lesions in ancient Greece. HSV may have existed in human populations as early as 5,000 years ago during the Bronze Age, but infected species ranging from bats to coral may have been found as long ago as millions of years before that time. In the early Greek civilization, Hippocrates described this condition in his writing, and Shakespeare is believed to reference it in *Romeo and Juliet* when he wrote: "O'er ladies lips, who straight on kisses dream, which oft the angry Mab with blisters plagues, because their breaths with sweetmeats tainted are."

It wasn't until 1893 that the French dermatologist Jean Baptiste Émile Vidal recognized that HSV infection is transmissible from one individual to another.

By the 1920s, data showed that HSV can infect the central nervous system (CNS) in addition to infecting skin. The latency characteristic of HSV infection was examined in association with the infection of the CNS. Subsequent members of the Herpesviridae family were identified, such as varicella-zoster virus (VZV) that causes chickenpox and shingles, Cytomegalovirus that causes mononucleosis, and Epstein-Barr virus that increases the chance of Burkitt's lymphoma and Kaposi's sarcoma development.

Classification and Structure of Herpes Simplex Virus

Herpes simplex virus is a member of the Herpesviridae family, which is itself divided into 3 subfamilies: Alpha Herpesviridae, Beta Herpesviridae, and Gamma Herpesviridae. The subfamily Alpha Herpesviridae includes 5 genera: *Iltovirus* (eg, laryngotracheitis virus); *Mardivirus* (eg, Marek's disease virus); *Scutavirus* (eg, chelonid alphaherpesvirus 5), *Varicellovirus* (eg, human herpesvirus 3/varicella zoster virus), and *Simplexvirus*. The *Simplexvirus* genus includes HSV-1 and HSV-2. The Beta Herpesviridae subfamily includes 4 genera: *Cytomegalovirus* (eg, human betaherpesvirus 5), *Muromegalovirus* (eg, muromegalovirus muridbeta 1), *Proboscivirus* (eg, elephantid betaherpesvirus 1), and *Roseolovirus* (eg, herpesvirus 7). Finally, the Gamma Herpesviridae subfamily includes 4 genera: *Lymphocryptovirus* (eg, Epstein–Barr virus), *Macavirus* (eg, macavirus hippotraginegamma 1), *Percavirus* (eg, equid gammaherpesvirus 2), and *Rhadinovirus* (eg, wood mouse herpesvirus).

Other related families of virus include Papillomaviridae (eg, alphapapillomavirus/human papillomavirus [HPV] 16), Polyomaviridae (eg, Merkel cell polyomavirus), Asfarviridae (eg, African swine fever virus), Iridoviridae (eg, lymphocystis disease virus 1), Mimiviridae (eg, acanthamoeba polyphaga mimivirus), Phycodnaviridae (eg, paramecium bursaria chlorella virus 1), Poxviridae (eg, variola virus), Adnoviridae (eg, human mastadenovirus A), Ascoviridae (eg, heliothis virescens ascovirus 3a), Cortico-viridae (eg, pseudoalteromonas virus PM2), Tectiviridae (eg, pseudomonas virus PR4), Alloherpesviridae (eg, ranid herpesvirus 1), Malacoherpesviridae (eg, haliotid herpesvirus 1), Myoviridae (eg, escherichia virus P1), Podoviridae (eg, enterobacteria phage PSA78), Siphoviridae (eg, bacteriophage T5), Lipothrixviridae (eg, acidianus filamentous virus 2), Rudiviridae (eg, acidianus rod-shaped virus 1), Bicaudaviridae (eg, sulfolobus tengchongensis spindle-shaped viruses 1), Ampullaviridae (eg, acidianus bottle-shaped virus), Baculoviridae (eg, nuclear polyhedrosis virus), Fuselloviridae (eg, sulfolobus spindle-shaped virus 1), Globuloviridae (eg, pyrobaculum spherical virus), Guttaviridae (eg, aeropyrum pernix ovoid virus 1), Nimaviridae (eg, white spot syndrome virus), Plasmaviridae (eg, acholeplasma laidlawii virus L2), and Polydnaviridae (eg, bracovirus).

Members of the family Hepadnaviridae (eg, hepatitis B virus) are also double-stranded DNA (dsDNA)viruses; however, they also express a reverse transcriptase and therefore belong to a different Baltimore classification (as defined in Chapter 1). Hepatitis B virus possesses a circular dsDNA genome that is gapped and expresses a reverse transcriptase.

The structure of HSV-1 and HSV-2 is an enveloped, spherical virion with a diameter of 155–240 nm (Figure 7.1). The single linear, dsDNA genome is encased in a capsid with a diameter of 125 nm, which is surrounded by a layer called the tegument, which is sandwiched in between the capsid and the viral envelope. The DNA genome size of HSV-1 is 1.523 kilobase pairs (kb), containing at least 80 protein-coding open reading frames (ORFs). The HSV-2 genome size is 1.547 kb, containing at least 84 unique protein-coding ORFs, possibly more. The 2 genomes share similar overall structure and most of the genes. Each genome contains two unique regions named UL (long unique) and US (short unique). These regions are surrounded by 3 sets of long terminal inverted repeats termed LTRa, LTRb, and LTRb' and LTRc and LTRc'. LTRb (also known as the terminal repeat flanking UL region or TRL) and LTRb' (also known as inverted repeat left flanking UL region or IRL) flank the UL region of the genome. LTRc (or TRS flanking US region) and LTRc' (or IRS) flank the US region of the genome. LTRa sequences (2 flanking the entire genome and 1 separates LTRb' and LTRc') are important for viral genome circularization once in the nucleoplasm. There are 2 origins of replication (Ori) that are used to initiate DNA replication: OriL is located within the UL region, and OriS is located within LTRc.

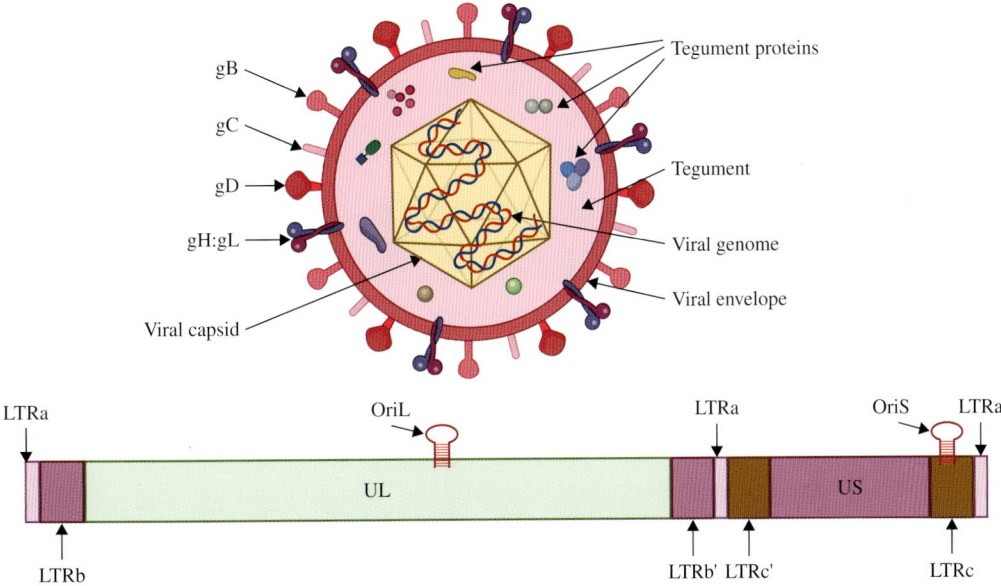

FIGURE 7.1   Herpes simplex virus capsid structure and genome. HSV is a complex structure that contains an envelope exhibiting many matrix proteins on its surface. The single linear dsDNA genome is encapsidated in an icosahedral capsid. The capsid is separated from the envelope by an amorphous structure termed the *tegument*, which contains many tegument proteins. LTRa is located at both ends of the genome. The UL region is surrounded by LTRb and LTRb', while the US region is surrounded by LTRc' and LTRc. *Credit:* Figure partially created in BioRender.

The UL region contains ORFs that encode for 5 of the 6 viral capsid proteins (UL6, UL18, UL35, UL38) and the major capsid protein UL19. The viral envelope contains a number of membrane proteins, many of which are glycoproteins. At least 12 different glycoproteins have been identified: gB, gC, gD, gE, gG, gH, gI, gJ, gK, gL, gM, and gN. Envelope glycoproteins gB, gD, gH, and gL have been shown to be sufficient and necessary for viral entry into the host cell. The HSV genome encodes gene products to regulate its own DNA replication in cooperation with cellular factors. Of these viral gene products, UL9 is an origin binding protein, infected cell protein (ICP) 8 is a single-stranded DNA binding protein, UL30 is the DNA polymerase, UL42 is a processivity factor, and the complex of UL5/UL8/UL52 performs the helicase/primase activity (Table 7.1).

The thick, proteinaceous tegument compartment contains 24 proteins that play crucial roles in many aspects of the HSV life cycle. These roles include the import or translocation of the nucleocapsid into the nucleus, regulation of viral gene expression (especially of the immediate-early genes), viral genome replication, virulence, assembly of the virion, egress from the host cell, and evading antiviral immune responses by modulating host signal transduction pathways and innate immunity.

Tegument proteins that have been identified in the mature virion include UL7, 11, 13, 14, 16, 21, 23, 36 (or VP1/2), 37, 41 (or "virus-host shutoff" [vhs]), 46, 47, 48, 49, 50, 51, 55; US2, 3, 10, 11; RL1/ICP34.5, RL2/ICP0, and RS1/ICP4. These proteins are incorporated into the tegument during different stages. VP (virion proteins) refers to structural proteins that form the viral capsid. ICP (infected cell protein) refers to proteins that play vital roles in evading the host immune system, promoting viral replication, and establishing latency. RL (long terminal repeat region) refers to structures in the viral DNA that are important for genome organization, replication, and regulation of viral gene expression. RS (short terminal repeat region) refers to structures in the viral DNA that are important for genome replication, assembly, and recombination. Vhs (virion host shutoff) protein is an endoribonuclease that degrades both host and viral mRNAs to facilitate efficient replication of viral DNA and gene expression.

For instance, 2 additional tegument proteins, UL31 and UL34, are present in the viral particle in the perinuclear space but are not present in the virion at later stages of assembly or in the mature virion. UL36 (or VP1/2), UL37, UL46, UL47, UL48, and UL49 are only found in the cytoplasmic virion during assembly and are thought to be added after the viral particle has been transported into the cytosol. UL36, UL47, UL48 (also known as VP16), and UL49 are classified as major structurally significant components. Viral protein designated UL are expressed from the UL region of the genome, while US are proteins expressed from the US region.

US3 is another special viral protein that regulates the integrity of the nucleus during nuclear egress of the assembled nucleocapsid (Figure 7.2). US3 is a serine/threonine protein kinase that phosphorylates lamins and the lamin receptor (Emerin).

Host Range, Transmission, Tropism, and Susceptible Host Cell

Despite having the capability of infecting a wide host range, HSV-1 and HSV-2 primarily target humans. They've been known to infect other animal species, albeit quite rarely, such as chimpanzees, bonobos, marmosets, owl monkeys, and domestic rabbits. Their major tropism in humans is the skin epithelium and mucous membrane. HSV-1 has a prevalent tropism for orofacial epithelium, while HSV-2 has a prevalent tropism

A. ORFs within Long terminal repeat LTRb (or TRL) region.

ORF	Gene Product	Function or Possible Function
LTRb (or TRL) ORFs		
ICP0/RL2/α0	ICP0; IE110	E3 ubiquitin ligase activating viral gene transcription; restructuring of chromatin
RL1	RL1; ICP34.5	Neurovirulence factor; evasion of MHC class II responses
LAT	LRP1, LRP2	Latency-associated transcript and protein products

B. ORFs within Unique Long (UL) region.

ORF	Gene Product	Function or Possible Function
UL ORFs		
UL1	Glycoprotein L (gL)	Surface and membrane protein; essential for virion infectivity and cell-cell fusion
UL2	Uracil-NDA glycosylase	Uracil-NDA glycosylase
UL3	UL3	Unknown function
UL4	UL4	Unknown function
UL5	HELI	DNA helicase
UL6	Portal protein UL6	Subunits form capsid portal ring for entrance and exit of DNA
UL7	Cytoplasmic envelopment protein 1	Virion maturation
UL8	DNA helicase/primase complexassociated protein	Facilitate synthesis of RNA primers on DNA template
UL8.5	UL8.5	Shifts *de novo* initiation at origin to rolling circle DNA replication
UL9	Replication origin-binding protein	Replication origin-binding protein; unwind DNA for rolling circle DNA replication
UL9.5	UL9.5	Unknown function
UL10	Glycoprotein M (gM)	Surface and membrane protein; virion maturation and exocytosis
UL11	Cytoplasmic envelopment protein 3	Virion exit and secondary envelopment
UL12	Alkaline nuclease	Alkaline exonuclease required for processing of replication intermediates
UL12.5	UL12.5	Unknown function
UL13	UL13	Tegument protein; serine protein kinase
UL14	UL14	Tegument protein; enhances nuclear localization of UL17, UL26, UL35, and UL33
UL15	TRM3	Processing and packaging of DNA
UL15.5	UL15.5	Unknown function

TABLE 7.1 HSV Proteins and Their Functions

ORF	Gene Product	Function or Possible Function
UL ORFs		
UL16	UL16	Tegument protein
UL17	CVC1	Tegument protein; processing and packaging DNA; localizes preformed capsid to replication compartment
UL18	TRX2; VP23	Capsid protein
UL19	VP5; ICP5	Major capsid protein
UL20	UL20	Membrane protein; prevents fusion of infected cells with adjacent cells
UL20.5	UL20.5	Unknown function
UL21	UL21	Tegument protein; interacts with microtubules for transport of viral genome
UL22	Glycoprotein H (gH)	Surface and membrane protein; enables membrane fusion; induces neutralizing antibodies
UL23	Thymidine kinase	Peripheral to DNA replication; phosphorylates purine and pyrimidine nucleosides
UL24	UL24	Unknown function
UL25	UL25	Processing and packaging DNA
UL26	P40; VP24; VP22A; UL26.5	Capsid scaffolding protein and protease
UL26.5	UL26.5	Unknown function
UL27	Glycoprotein B (gB)	Surface and membrane protein; mediates attachment of virion; component of membrane fusion complex
UL28	ICP18.5	Processing and packaging DNA
UL29	UL29; ICP8	Single stranded DNA binding protein
UL30	DNA polymerase	DNA replication
UL31	UL31	Nuclear matrix protein; directs envelopment at inner nuclear membrane
UL32	UL32	Envelope glycoprotein; processing and packaging DNA
UL33	UL33	Processing and packaging DNA
UL34	UL34	Inner nuclear membrane protein; virion envelopment at inner nuclear membrane
UL35	VP26; NC7	Capsid protein
UL36	UL36; VP1/2	Large tegument protein; essential for egress of virion into cytosol
UL37	UL37	Tegument protein; capsid assembly
UL38	UL38; VP19C	Capsid assembly and DNA maturation
UL39	UL39; RR-1; ICP6	Ribonucleotide reductase (large subunit)
UL40	UL40; RR-2	Ribonucleotide reductase (small subunit)
UL41	UL41; VHS	Tegument protein; virion host shutoff; mediates degradation of RNA early in infection

TABLE 7.1   HSV Proteins and Their Functions (*Continued*)

ORF	Gene Product	Function or Possible Function
		UL ORFs
UL42	UL42	DNA polymerase processivity factor; binds dsDNA
UL43	UL43	Membrane protein
UL43.5	UL43.5	Unknown function
UL44	Glycoprotein C (gC)	Surface and membrane protein; mediates attachment of virion
UL45	UL45	Membrane protein; C-type lectin; aids membrane fusion
UL46	VP11/12	Tegument proteins
UL47	UL47; VP13/14	Tegument protein; import and export of viral RNA
UL48	α-TIF; VP16; ICP25	Trans-inducing factor; virion maturation; activates immediate-early gene expression; virion assembly
UL49	UL49A; VP22	Envelope protein; binds RNA and translocate mRNA from infected to uninfected cells
UL49.5	Glycoprotein N (gN)	Inhibits membrane fusion when complexed with gM
UL50	UL50	dUTP diphosphatase
UL51	UL51	Tegument protein
UL52	UL52	DNA helicase/primase complex protein
UL53	Glycoprotein K (gK)	Surface and membrane protein; prevents infected cells from fusing with adjacent cells
UL54 (α27)	IE63; ICP27	Transcriptional regulation; blocks RNA splicing to enable transport of unspliced RNA into cytosol
UL55	UL55	Unknown function
UL56	UL56	Unknown function

C. ORFs within Unique Short (US) region.

ORF	Gene Product	Function or Possible Function
		US ORFs
US1 (α22)	ICP22; IE68	Viral replication
US1.5		Viral replication
US2	US2	Tegument protein
US3	US3	Serine-threonine protein kinase
US4	Glycoprotein G (gG)	Surface and membrane protein
US5	Glycoprotein J (gJ)	Surface and membrane protein
US6	Glycoprotein D (gD)	Surface and membrane protein; essential for virus entry and membrane fusion; protects infected cell from apoptosis
US7	Glycoprotein I (gI)	Surface and membrane protein; complexes with gE to function as Fc receptor for IgG; gI:gE complex facilitates basolateral spread of progeny virus in polarized cells

TABLE 7.1   HSV Proteins and Their Functions (*Continued*)

ORF	Gene Product	Function or Possible Function
UL ORFs		
US8	Glycoprotein E (gE)	Surface and membrane protein; complexes with gI to function as Fc receptor for IgG; gI:gE complex facilitates basolateral spread of progeny virus in polarized cells
US8.5	US8.5	Unknown function
US9	US9	Tegument protein; involved in anterograde axonal transport of virions
US10	US10	Capsid/Tegument protein
US11	US11; Vmw21	Binds DNA and RNA; packaging of RNA in virions
US12 (α47)	ICP47; IE12	Inhibits MHC class I pathway by blocking transport of antigenic peptides from cytosol into ER lumen to be presented in MHC class I complex

    D.   ORFs within Long terminal repeat LTRc (or TRS) region.

ORF	Gene Product	Function
LTRc (TRS) ORFs		
RS1 (α4)	ICP4; IE175	Major transcriptional activator, essential for progression beyond immediate-early phase of infection
LAT		Latency associated transcript
ORIS RNA		Unknown function
αX and βX RNAS		Unknown function
AL-RNA		Antisense to 5'-sequence of LAT
VZV ORF1		Membrane protein
VZV ORF 13		Thymidylate synthetase

This is a list of identified and studied ORFs found in both HSV-1 and HSV-2. Functions that are described have either been confirmed or projected through conserved peptide sequences.

TABLE 7.1   HSV Proteins and Their Functions (*Continued*)

for genital epithelium. However, they've been known to infect a broad range of cell types, including nerve cells, fibroblasts, and immune cells.

    Heparin sulfate proteoglycan (HSPG), paired immunoglobulin-like receptor α (PILRα), and/or nonmuscle myosin heavy chain (NMHC)-IIA are presented on the surface of susceptible cells. HSPG is expressed by most human cells and is found in abundance in basement membranes. HSPG highly promotes virion adsorption to the host cell but is not essential for membrane fusion, although there is a 100-fold reduction of HSV infection of host cells that lack surface HSPG. PILRα is expressed in immune cells, especially in monocytes, macrophages, and dendritic cells. NMHC-IIA regulates cell division, cell adhesion, cellular movement, cell migration, and contraction and is ubiquitously expressed in human tissues.

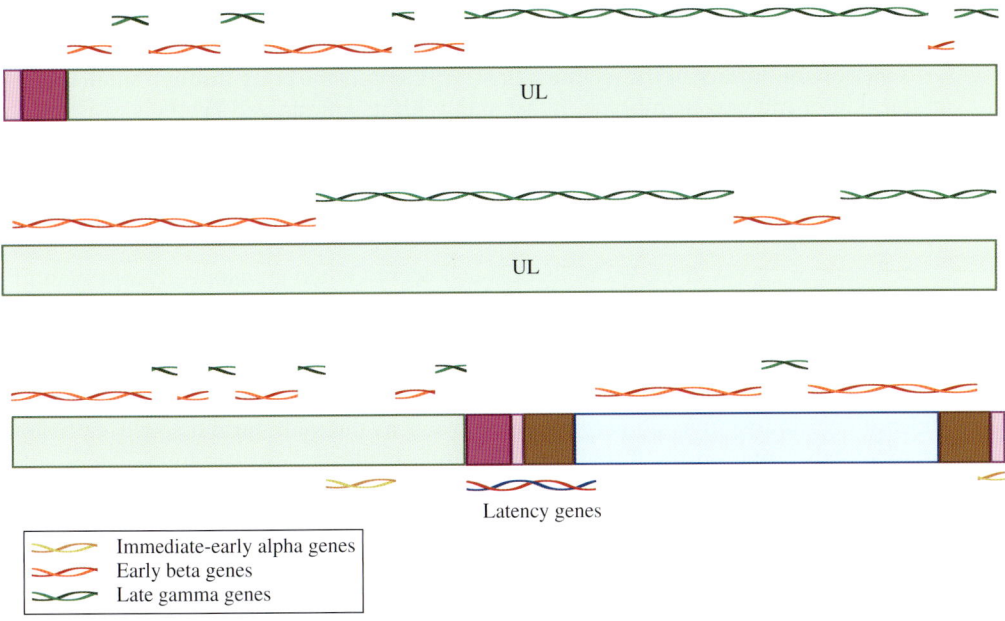

FIGURE 7.2 Location of HSV genes. Immediate-early genes, early genes, and late genes are dispersed throughout the HSV dsDNA. Genes that regulate the latent phase are located within the LTRb' and LTRc' regions. *Credit:* Figure partially created in BioRender.

A number of co-receptors whose function is to regulate membrane fusion are widely distributed among permissive host cells. These co-receptors include herpes virus entry mediator (HVEM), nectin-1, and integrin. HVEM or nectin-1 interact with gD. Integrin interacts with the heterodimer gH:gL. HVEM is expressed in a number of cell types, including epithelial cell, fibroblast, and neuron, as well as immune cells such as T lymphocytes, B lymphocytes, dendritic cells, natural killer (NK) cells, macrophages, and polymorphonuclear leukocytes. Nectin (-1 and -2) is also expressed in a variety of human cells, except T lymphocytes. Integrin is also ubiquitously expressed in human tissues.

The principal exit route for HSV is through infected bodily secretions. These HSV secretions are typically bodily fluids that are produced in either the mouth or genital area. Transmission is through direct contact with contaminated secretions, such as saliva, secretions from lesions in the mouth and lips, secretions from the genital area, or contaminated fomites such as shared eating utensils.

The principal mode of transmission for HSV-1 is through contact with sores, blisters, saliva, and surfaces in the area of the mouth of an infected individual with oral herpes, although transmission may sometimes occur from skin that does not show signs of sores, blisters, or any abnormality since the virus can be shed from the mucosa or skin. Occasionally, HSV-1 can be transmitted to the genital area through oral-genital contact.

The principal mode of transmission for HSV-2 is via contact with genital sores of an infected individual during sexual contact, thereby causing genital herpes. Similar to HSV-1, transmission may occur even when there are no signs or symptoms. HSV-2 can also cause oral infections through oral-genital contact. Both HSV-1 and HSV-2 can be transmitted from mother to newborn during parturition (birthing), causing neonatal herpes.

Once on the surface of the recipient host, HSV enters the body through small breaks in the skin or mucous membrane. Intact skin with its keratinized layers is resistant to viral infection.

Infecting the Host Susceptible Cell, Viral Particle Replication, and Tissue Damage

Attachment of HSV-1 or HSV-2 to its host cell occurs when envelope glycoprotein gB and/or gC interacts with the receptor HSPG (Figure 7.3). Membrane fusion is initiated when glycoprotein gD binds to either HVEM or nectin (depending on host cell type), which in turn leads to interaction between the heterodimer gH:gL and integrin on the host cell's surface. The complex of all these glycoproteins and their cognate receptors is called the "core fusion machinery," where gB acts as the main viral fusogen, or fusion protein. Depending on the target cell, the viral envelope can fuse immediately with the plasma membrane or with the endosomal membrane after clathrin-dependent, receptor-mediated endocytosis.

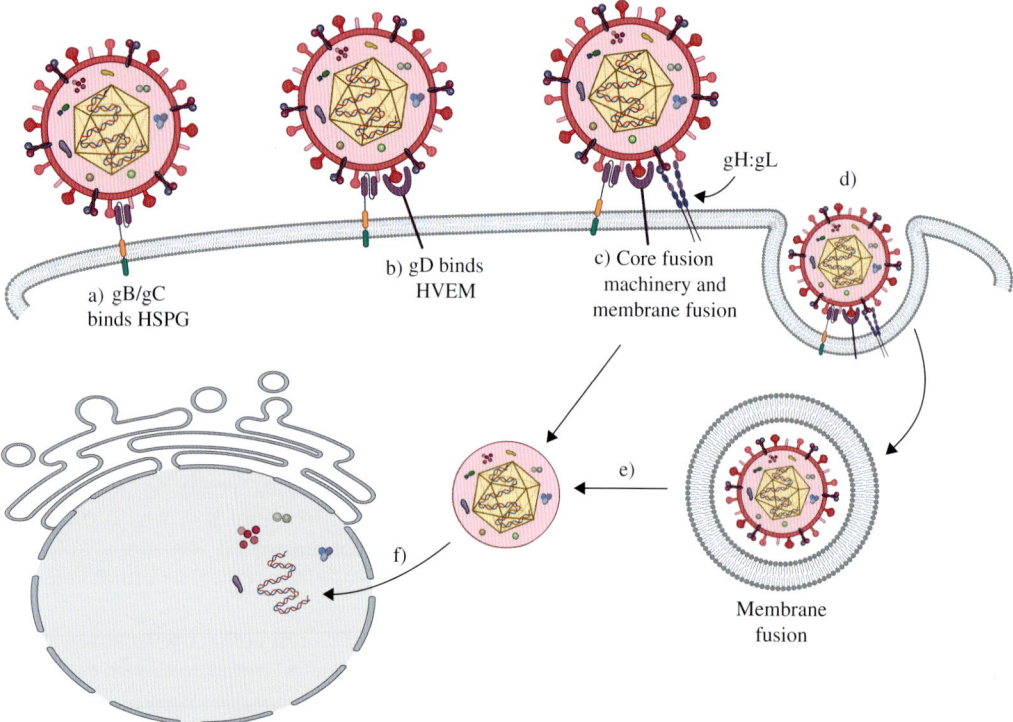

FIGURE 7.3   HSV entry into a host cell and delivery of the genome to the nucleus. (a) HSV attaches to HSPG on host cell surface. (b) Glycoprotein gD binds HVEM to initiate membrane fusion. (c) Heterodimer gH:gL binds to integrin on the host cell surface to form the "core fusion machinery." (d) HSV envelope can fuse directly to the plasma membrane, or endocytosis occurs before fusing with the endosomal membrane. (e) The nucleocapsid surrounded by the tegument is released into the cytosol, then is transported toward the nucleus. (f) The HSV genome, along with a few tegument proteins, is imported into the nucleoplasm. *Credit:* Figure partially created in BioRender.

Once membrane fusion is successful, the nucleocapsid, which is surrounded by tegument proteins, is released into the cytosol. The nucleocapsid travels toward the nucleus, most likely along microtubules. Once arrived at the nucleus, the nucleocapsid docks at a nuclear pore, while the tegument protein VP1/2 anchors it to the outer nuclear membrane. The linear dsDNA genome leaves the virion through the capsid protein UL6, then is injected into the nucleoplasm through a nuclear pore. The empty capsid remains in the cytosol and is eventually targeted for degradation. Some tegument proteins are also imported into the nucleus to perform necessary functions. One of these proteins is UL48/VP16, whose function is to act as a transactivator to promote viral gene transcription. Another is ICP0, whose function is to prevent circularization of the linear dsDNA genome.

Upon its arrival in the nucleoplasm, the HSV dsDNA genome is not associated with histones; therefore, no nucleosomes are detected at this time. Histones eventually bind to the viral DNA, forming nucleosome-like structures where the genome resembles a chromatin (or minichromosome) by 1 hour after infection of the host cell. The amount of bound histones can increase up to 3 hours postinfection. In the meantime, nuclear bodies assemble at the condensing viral DNA, forming a structure called nuclear domain 10 (ND10) (Figure 7.4).

At this point, as discussed further in the chapter, this process can progress to either lytic infection for virus replication or latent infection, the latter of which is

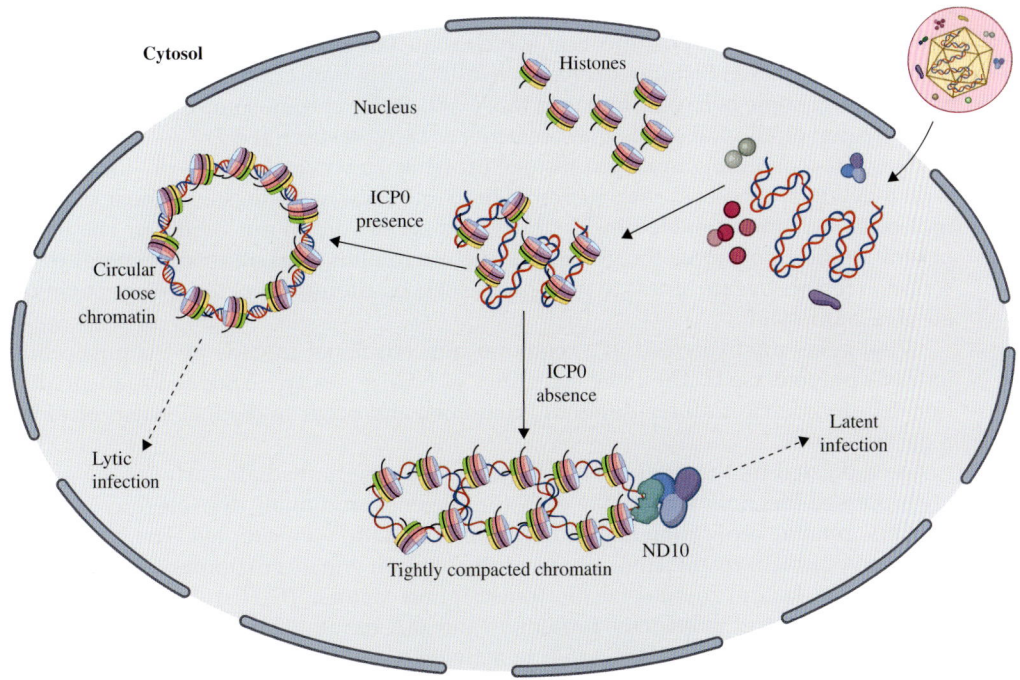

FIGURE 7.4   Preparation of the HSV genome for lytic or latent infection. In the nucleoplasm, the linear dsDNA genome is circularized. Histones are added to the circular dsDNA, forming a "minichromosome." In the presence of ICP0, the minichromosome is a loose chromatin. This structure is accessible to transactivators to initiate gene expression for lytic infection. In the absence of ICP0, the minichromosome is further condensed into a tightly compacted chromatin, which is inaccessible to transactivators, leading to latent infection. This tightly compacted chromatin is localized at ND10 within the nucleoplasm. *Credit:* Figure partially created in BioRender.

more quiescent. The decision to progress toward either path is decided and based on the presence of differential chromatin remodeling mechanism and association with ND10. If the viral DNA structure is compact and localized at ND10, then the virus enters latent infection. The tightly compacted chromatin represses viral transcription.

On the other hand, if the chromatin structure is loose and not associated with ND10, then the virus continues with lytic infection. A loose chromatin structure can be bound by regulatory factors to initiate gene expression to progress toward viral replication.

These opposing conditions depend on the activities of the tegument protein ICP0. ICP0 possesses E3 ubiquitin ligase activity, which performs 2 major functions. First, ICP0 can disrupt the ND10 nuclear structure by targeting its components for degradation at the proteasome. Localization of the viral genome to ND10 restricts viral gene expression. ICP0 can also modify histones, producing a loose chromatin structure that is accessible to transactivators for gene expression. Thus, the presence of ICP0 in the nucleus leads to lytic infection. Without ICP0 in the nucleus, the viral genome would be tightly compacted within 3 hours postinfection and localized at ND10. In this case, the genome circularizes and becomes tightly compacted to prepare to enter latency.

Lytic Infection

Initially, when infecting a nonneuronal cell (eg, epithelial cell), HSV enters the lytic infection process first to amplify the viral particle population. During the period between 3- and 6-hours postinfection, which is the late phase of the lytic infection, the amount of associated histones decreases to allow viral transcription to occur. The primary HSV genome behaves similar to those of cellular chromosomes, where the promoter region of all viral genes that are actively being transcribed are associated with nucleosomes containing modified histone proteins. However, histones do not interact with a newly synthesized viral genome; therefore, newly synthesized linear dsDNA molecules that are assembled inside the capsid are devoid of nucleosomes. The onset of viral replication causes the genomic structure to alter, thus allowing general transcription factors to gain access to previously silenced promoters of the α-genes to initiate gene expression.

Viral genes are expressed in a temporal transcriptional cascade by the host cellular RNA polymerase II. The three categories of HSV genes are expressed sequentially, initiated in order from three different promoters, and occur in distinct stages of virus replication cycle.

As indicated by the term *transcriptional cascade*, the immediate early (or α-) genes are expressed first, followed by the early (or β-) genes, and finally the late (or γ-) genes.

HSV genome replication and transcription regulation are coupled. The viral tegument protein VP16 recruits and forms a transcription enhancer complex with cellular transcription factors (tissue-specific transcription factors, TS-TFs) such as Octamer transcription factor, Specificity protein 1, and GA-binding protein, along with the cellular coactivator Host cell factor-1, to activate α-gene transcription to initiate lytic infection (Figure 7.5). These cellular transcription factors and coactivator regulate the expression of genes that are important for chromatin modification and cell cycle progression.

The α-gene product ICP4 is the major viral transcription factor that activates the transcription of viral β- and γ-genes. ICP4 is required for host cell transcriptional reprogramming to focus on viral gene expression by recruiting other cellular complexes to

HSV gene expression during lytic infection

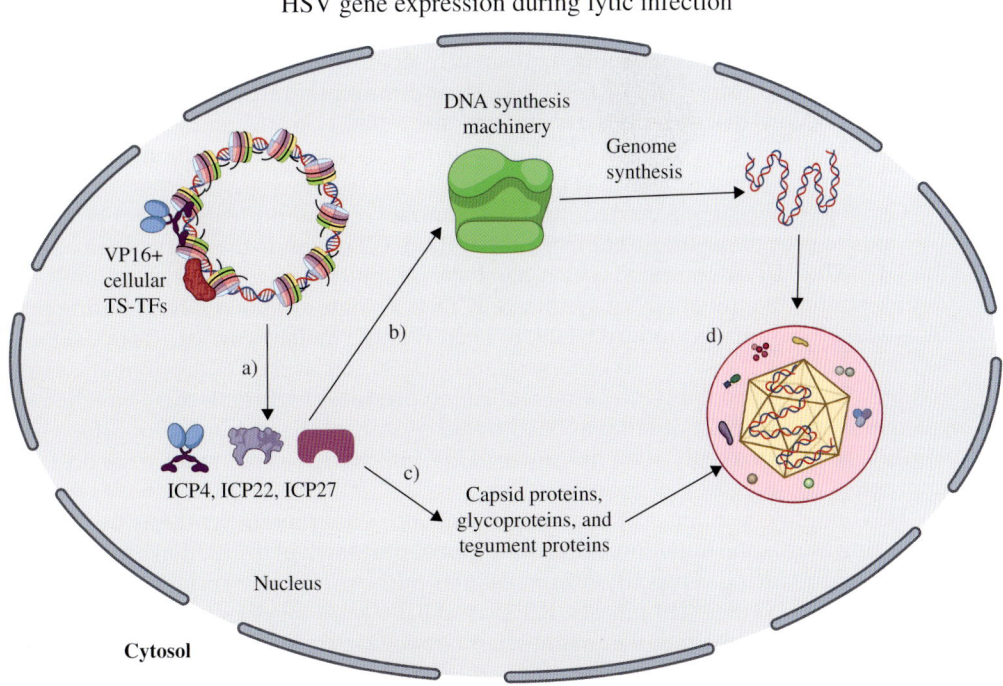

FIGURE 7.5   HSV gene expression. (a) VP16 and cellular TS-TFs regulate immediate-early gene expression, producing ICP4, ICP22, and ICP27. (b) The combination of ICP4, ICP22, and ICP27, along with other cellular TS-TFs regulate early gene expression, producing the DNA synthesis machinery for genome synthesis. (c) The combination of ICP4, ICP22, and ICP27, along with other cellular TS-TFs, also regulate late gene expression, producing capsid proteins, glycoproteins, and tegument proteins. (d) All lead to nucleocapsid assembly, surrounded by the tegument. *Credit:* Figure partially created in BioRender.

aid in transcription initiation, such as cellular TATA-box binding protein (TBP), TFIID, and the mediator complex. TFIID is a general transcription factor and is a member of a multi-subunit complex called the preinitiation complex that plays a crucial role in initiating transcription by RNA polymerase II. Two additional α-gene products, ICP22 and ICP27, are involved in transcription regulation.

The β-genes encode for the viral DNA synthesis machinery: UL9, ICP8, UL30, UL42, and the UL5/UL8/UL52 complex. Once viral DNA replication is initiated, cellular general transcription factors can no longer bind to immediate early promoters; therefore, the initiation of α-gene transcription is attenuated.

As DNA replication continues, β-gene transcription is also attenuated. The onset of genome replication opens up the promoter to allow transcription factors TBP and TAF1, as well as RNA pol II to initiate transcription of γ-genes. This process is also aided by tegument protein ICP4. The γ-genes themselves encode for viral structural proteins (capsid proteins and glycoproteins) and tegument proteins to be assembled into a nucleocapsid surrounded by the tegument. A single infecting viral particle can produce the first infectious progeny within 4–6 hours postinfection and up to 1000 viral particles within 18 hours.

HSV genome replication and transcription regulation are coupled. In fact, genome replication can intensify viral gene expression. There are indications that host cellular factors may provide aid to viral genome synthesis; however, the mechanism has not been elucidated. These cellular factors include the DNA polymerase processivity factor proliferating cell nuclear antigen, mismatch repair proteins, topoisomerases, MRN complex proteins (Mre11 and Rad50, Nbs1), and multiple transcription factors. They regulate functions that are critical to viral genome replication and gene expression, such as DNA repair, DNA recombination, chromatin modification, transcription, and regulation of transcription-coupled RNA processing.

Genome replication is carried out by a complex of viral proteins (or a replisome). U9 binds to each of the origins of DNA replication. U9 unwinds the origin, then allows the UL5/UL8/UL52 complex, which has helicase and primase activities, to melt the double strand and initiate DNA synthesis. ICP8 is a single-stranded DNA binding protein, which is in turn bound by UL29, to keep the two melted strands (from dsDNA) separated during the synthesis of new strands. UL30 acts as the DNA polymerase and has proofreading capability. Genome replication produces high-molecular-weight concatemers, which consist of tandem head-to-tail repeats of the genome. Since this virus brings its own DNA replication machinery, HSV infection does not require the host cell to progress through the cell cycle.

HSV employs a number of strategies to enhance viral protein synthesis by directly enhancing translation initiation, preventing global shutdown of translation by host stress kinases and reducing competition from cellular mRNAs for the translation machinery. Viral proteins US3 (Ser/Thr kinase), ICP27, and ICP6 work together to enhance translation initiation by activating the mammalian taret of rapamycin complex 1 (mTORC1), triggering p38-mnk (mitogen-activated protein kinase-interacting kinase) signaling, and increasing eIF4E (eukaryotic translation initiation factor 4E) and eIF4G interaction, respectively. Cellular enzymes dsRNA-dependent kinase (PKR) and the endoplasmic reticulum (ER) stress kinase (PKR-like endoplasmic reticulum [PER] kinase, PERK) are activated during cellular stress to block a global shutdown of translation initiation. HSV proteins ICP34.5, US11, and gB counter this blockage by reversing the effect of protein kinase activated by dsRNA (PKR) and PERK and inhibit their activation.

Additionally, the tegument protein virion-host shutoff (vhs or UL41) is a viral ribonuclease that is known to degrade both stable host mRNA and viral α-mRNA species preferentially, but not other viral mRNAs. The degradation of the viral α-mRNA might seem counterintuitive but is important in further reducing the amounts of mRNAs that compete for translation initiation of the proteins that are required for virion assembly. To counteract host antiviral, interferon (IFN)-mediated responses within the host cell that aim at detecting viral components and diminishing the progression of viral infection, tegument protein ICP0 inhibits interferon-regulatory factor 3 nuclear accumulation to reduce IFN-β production.

HSV virion assembly occurs in the nucleoplasm. However, histones do not interact with newly synthesized viral genome; therefore, newly synthesized linear dsDNA molecules that are assembled inside the capsid are devoid of nucleosomes (Figure 7.6). Since the nucleocapsid is too large to be exported through the nuclear pore, it must egress from the nucleus across the double nuclear membrane via a two-step process aided by tegument proteins. This two-step process is known as envelopment/de-envelopment.

The primary envelopment step is aided by the nuclear egress protein complex UL31:UL34, as well as UL 13 and US3, where the nucleocapsid buds into the inner nuclear membrane, forming an enveloped virion in the perinuclear space (Figure 7.7).

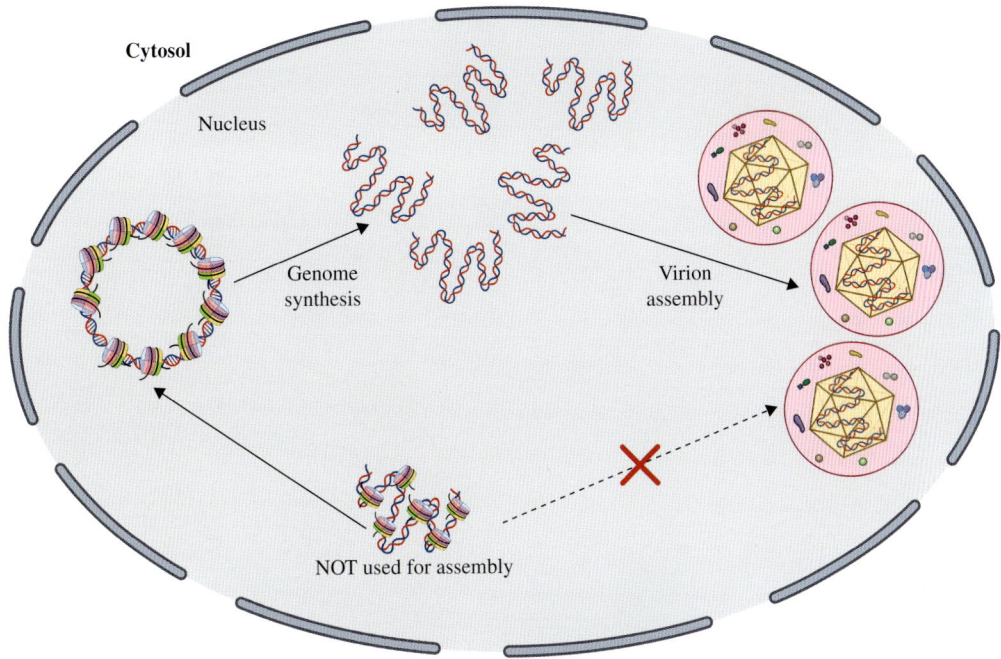

FIGURE 7.6   HSV genome replication. Newly replicated HSV linear dsDNA molecules are not bound by histones. Histone-free dsDNA molecules are assembled into nucleocapsids in the nucleoplasm. Histone-bound dsDNA are not encapsidated. *Credit:* Figure partially created in BioRender.

This enveloped virion in turn undergoes the de-envelopment step where the temporary envelope of the virion fuses with the outer nuclear membrane, releasing a non-enveloped virion into the cytosol. Membrane fusion during the de-envelopment step is regulated by 4 HSV proteins: gB, gH/gL, gK, UL20, and US3. Most of the tegument proteins are recruited onto the nucleocapsid in the cytosol.

The virion undergoes a secondary envelopment step where it again buds into the trans-Golgi apparatus, producing mature enveloped virions in the lumen of the Golgi cisterna. This process involves the interaction between viral envelope proteins that have been delivered to the Golgi membrane and tegument proteins already recruited onto the nucleocapsid. The nucleocapsid UL11 interacts with UL16, the nucleocapsid UL48 interacts with gH, and UL49 interacts with gE.

These interactions, along with UL46 and UL47, regulate the secondary envelopment step. The enveloped virion in the Golgi lumen is delivered to the plasma membrane via the typical secretory pathway and exits the host cell via exocytosis. This process allows the enveloped HSV to exit without causing damage to the primary host cell, which is an epithelial cell.

Despite not causing damage through release by exocytosis, host cells can still be damaged. NK cells and cytotoxic T lymphocytes (CTLs) of the adaptive immune system can identify and destroy viral infected cells through the induction of the extrinsic apoptotic pathway. NK cells identify and destroy cells that do not exhibit self-antigens, referred to as "missing self" cells. CTLs destroy virus-infected cells through the identification of viral peptides presented by the major histocompatibility complex I.

Egress from nucleus and exit from host cell

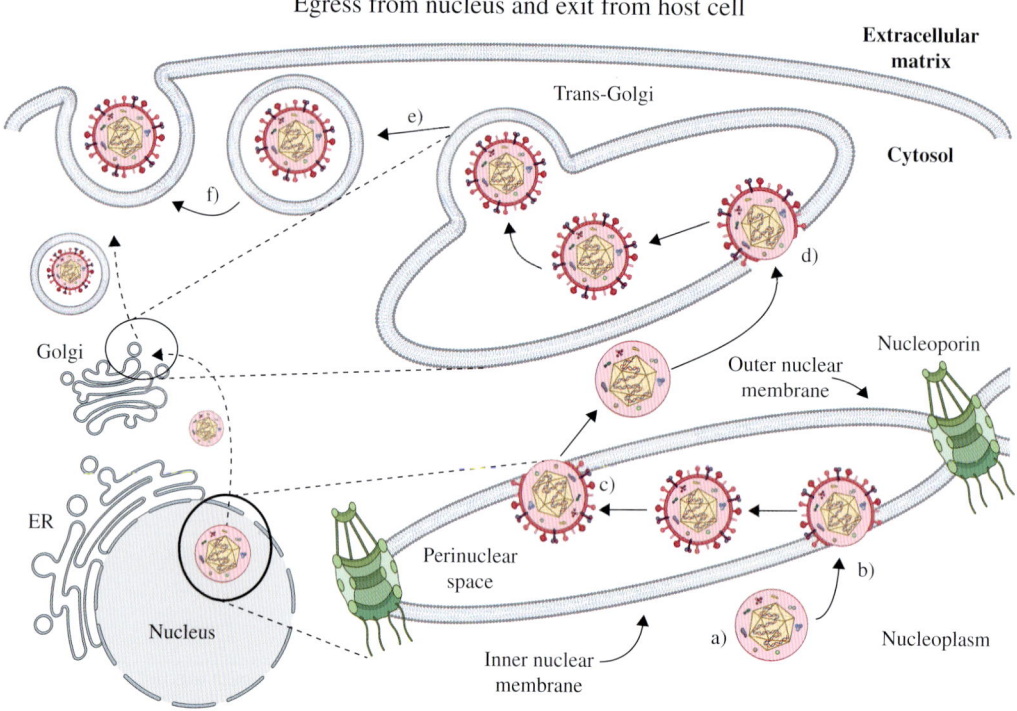

FIGURE 7.7   Egress of viral particles from the nucleus and strategy for exit. (a) HSV nucleocapsid assembly occurs in the nucleoplasm. (b) The nucleocapsid, surrounded by the tegument, buds into the inner nuclear membrane, forming a temporary enveloped virion in the perinuclear space. (c) This enveloped virion egresses via membrane fusion releasing a non-enveloped nucleocapsid, surrounded by the tegument, into the cytosol. (d)This structure undergoes a second envelopment step into the trans-Golgi apparatus to produce a mature, enveloped virion. (e) The enveloped virion is delivered to the plasma membrane via vesicular transport. (f) Release from the host cell is via exocytosis. *Credit:* Figure partially created in BioRender.

Latent Infection

After the initial infection of the epithelial cells, HSV progenies that are released undergo neurogenous dissemination into neurons of sensory ganglia, then enter the latent infection phase. The HSV genome in these neurons can reenter the lytic infection when induced by certain stimuli, such as different types of physical, mental, and environmental stress. Let's start with HSV entry into the sensory neuronal membrane at the end of an axon (Figure 7.8). HSV uses the same mechanism to attach and enter the axon terminal as it employs to enter an epithelial cell. Most importantly, the formation of the heterodimer gK:UL20 is a critical determinant for entry into the neuron. From there, the non-enveloped nucleocapsid is transported by retrograde flow along microtubules to reach the dorsal root ganglion, where the cell body containing the nucleus is located. Uncoating occurs, and the linear dsDNA genome is then injected into the nucleoplasm.

To initiate latent infection, the linear dsDNA genome must first be circularized. The absence of ICP0 is required for the circularization process. Remember that ICP0 is an important regulator of lytic infection. Specifically, the presence of ICP0 in the host cell

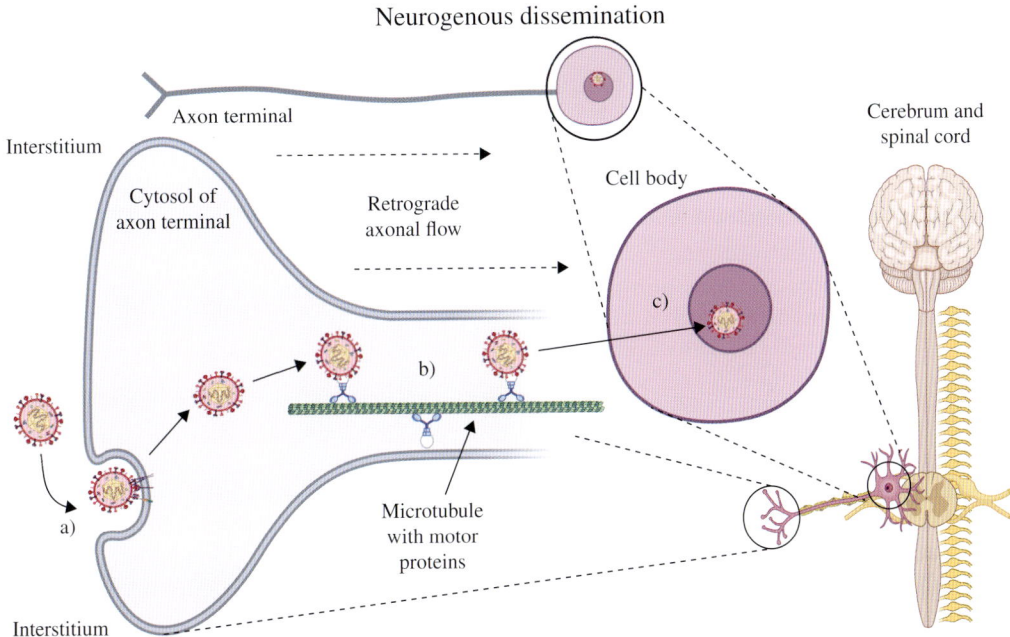

FIGURE 7.8   Dissemination into the dorsal root ganglion. (a) Entry of virion through the axonal terminal of a neuron. (b) Virion is transported by retrograde flow along microtubules. (c) Virion reaches the nucleus. *Credit:* Figure partially created in BioRender.

nucleus leads to lytic infection, while the absence of ICP0 leads to latent infection. Without ICP0 in the nucleus, the viral genome is allowed to circularize, becomes tightly compacted within 3 hours postinfection, localized at ND10, and then is prepared for the latent infection. LTRa sequences located at the ends of the dsDNA are important for viral genome circularization. The compacted chromatin structure is much like a heterochromatin and is referred to as a circular episome. Viral gene expression is repressed to prevent the continuation into the lytic infection.

Latency allows the virus to remain in a noninfectious state for a variable amount of time before reactivation. Of course, not all neuronal infection leads to latency; infection of a neuron may sometimes immediately continue with lytic infection. The advantage of entering the latent infection is to allow the virus (or its genome) to escape host immune surveillance. This latency may have a lifelong persistence in the host organism. Many copies of the circular episome can be maintained in the nucleus of the same host cell, each associated with the ND10. However, HSV must reenter the lytic infection before it can undergo virion replication again and be transmitted to a new host organism to ensure the survival of the virus over time. During the latent infection, a number of activities must be performed:

1. The HSV circular genome must be localized at the ND10 in the host cell nucleus.

2. When a nonneuronal host cell that harbors a latent viral genome undergoes cell division, the viral proteins are expressed to tether the circular genome to host cell chromosomes so that they can be vertically transferred to daughter cells. This activity requires the expression of viral tethering proteins when the

cell divides. Of course, since neurons are nondividing cells, tethering protein synthesis is not necessary.

3. The viral genome must avoid being recognized by the host immune system by limiting the expression of its own genes.

4. Long noncoding viral RNAs, including microRNAs, are expressed to suppress synthesis of proteins that can initiate the lytic infection or regulate cellular gene expression to allow the infected cell to be recognized by the host immune system.

5. Histone modification is modulated to induce the proper viral genome chromatin structure that allows the expression viral genes that can regulate the eventual lytic infection, then infect other host cells and transmit to uninfected individuals.

Not all viral genes are inactive during this period. Viral genes that are canonically active during lytic infection are heterochromatic and inactive during latency. There are regions within the viral genome that are euchromatic and active, but are turned off. The epigenetic organization of the viral genome not only helps maintain latency but also regulates reactivation when required. Most viral genes are inactive (ie, turned off), except for a few select ones; however, none of these RNA transcripts are translated during latency.

The primary viral long, noncoding RNA that is produced is the HSV-1 latency-associated transcript (LAT). LAT performs multiple functions during latency. LAT is important in regulating latent and lytic gene expression for reactivation. LAT has also been implicated in protecting the neuron harboring viral episome from apoptosis. Essentially, LAT represses lytic infection to allow HSV to remain latent in the neuronal cell.

Reactivation

The detailed mechanism for reactivation has not been completely elucidated, but this is what we know. Reactivation of HSV from latency to reenter lytic infection can occur in response to extreme stress induced by external as well as internal signals. Signals include overexposure to ultraviolet radiation from sunlight triggering DNA damage responses; therapeutic radiation; metabolic stress; experience of trauma to the nerves; or symptoms of pathogenic infection such as septic shock, heat shock or fever, menstruation, surgical resection, and hypoxia.

The cellular transcriptional coactivator HCF-1 which is important for the initiation of the primary HSV lytic infection is also required for the reactivation process from latency (Figure 7.9). HCF-1 is also required to mediate chromatin remodeling to promote α-gene transcription initiation and regulate transcription elongation through its association with a complex of transcription elongation components (SEC-P-TEFb). The SEC-P-TEFb is composed of the super elongation complex (SEC) and positive transcription elongation factor b (P-TEFb). This complex is typically required for gene expression during the early stages of embryogenesis. Regulators of SEC-P-TEFb have been shown to drive the reactivation of HSV lytic infection.

Finally, the activity of ICP0 is required once again for reactivation from latency and efficient initiation of lytic infection. Histones are removed from the chromatin, resulting in a loose structure of circular episome, ready for viral gene expression and replication.

Reactivation of HSV infection and release of extracellular vesicles

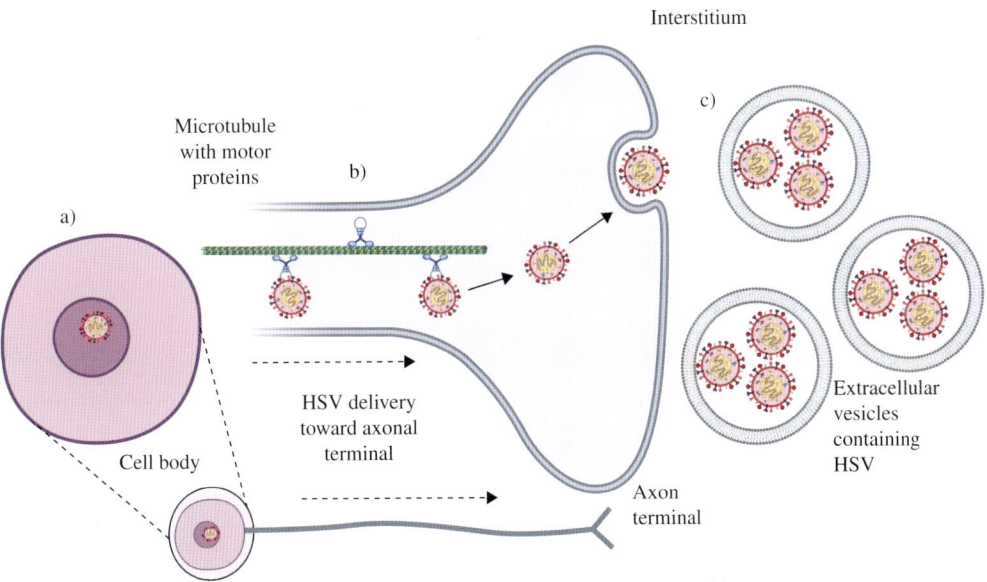

FIGURE 7.9   Reactivation of HSV lytic infection. (a) Reactivation in the karyon (nucleus) of neuronal cell body. (b) Transport of virions along microtubules toward the axonal terminal. (c) Release of virions within extracellular vesicles. *Credit:* Figure partially created in BioRender.

Viral gene expression is regulated like it was during the primary lytic infection. Viral genome replication also uses the same machinery as during the primary lytic infection, except that the genome template is now a circular dsDNA instead of a linear dsDNA. It is thought that this process begins with the "theta mode of replication," which is a mechanism used by prokaryotes, to provide the template for the rolling circle mode of replication. Multimeric concatemers are produced that are, in turn, appropriately cleaved into full viral genome fragments.

Dissemination in the Host Body

The primary HSV infection almost always targets the mucous membrane. Epithelial cells are the major tropism for both HSV-1 and HSV-2, although in different locations in the human body. Even though both serotypes have been detected in orofacial and genital primary infections, HSV-1 is usually associated with orofacial infections in the oral, nasal, or ocular sites, whereas HSV-2 is usually associated with genital infections. HSV replication in nonneuronal cells is prolific.

Once released from these host cells, virions typically disseminate to regional lymphoid nodes that are associated with the submucosa layer, which is internal to the mucosa but external to the muscularis propria. Another round of host cell infection and virus replication occurs. Hematogenous dissemination, resulting in active viremia, may also occur but is extremely rare. The submucosa also contains numerous neurons, including innervating sensory and autonomic neurons. The virus can enter via axon termini, then travel to ganglia, especially trigeminal ganglia (HSV-1) and lumbar-sacral ganglia (HSV-2) of sensory neurons and dorsal root ganglia of autonomic neurons.

Virus reactivated from latently infected neurons migrates back to the site of the initial infection in the mucous membrane (either orofacial epithelium or genital epithelium), producing a second lytic infection. Instead of the budding of individual virions, HSV-infected cells can release extracellular vesicles (EVs), sometimes called exosomes, containing multiple virions into the interstitium of the original primary site of infection (Figure 7.9). Typically, EVs are released by all types of cells to function as a mechanism for intercellular communication. However, HSV-containing EVs can alter its microenvironment and induce immune responses that lead to the formation of cold sores. This mechanism is yet to be deduced.

Pathogenesis and Clinical Manifestation

During the primary HSV infection, infected individuals are often asymptomatic or have only mild symptoms. Symptoms may include fever, body aches, sore throat from oral herpes, headache, and swollen lymph nodes near the site of infection. More severe symptoms of HSV-1 infection include painful lesions called "fever" blisters (or cold sores) in the lips, gums, roof of mouth, tongue, and occasionally the eye. These fluid-filled blisters can break open, leaving open sores called ulcers, which can in turn form a scab. These blisters and ulcers are different from canker sores, which may also appear inside the mouth. Canker sores are caused by an underlying condition called aphthous stomatitis, which may be triggered by food allergies, stress, hormone changes, vitamin deficiencies, hot and spicy foods, and the stress of dental treatment. Blister sores are contagious, whereas canker sores are not contagious and usually go away on their own within 1 to 2 weeks. Severe symptoms of HSV-2 infection often include bumps, blisters, and ulcers around the genitals or anus. Even though HSV-1 is traditionally associated with orofacial diseases and HSV-2 is associated with genital diseases, both serotypes have been found in both locations in the body.

HSV is highly contagious, and transmission occurs through close contact, sometimes as the result of a seemingly innocuous activity, like a simple kiss on the cheek (near the mouth) of a child from an infected, but asymptomatic, family member. HSV-1 is the most likely cause in this case. HSV can also be transmitted from a mother to her baby as the infant passes through the birth canal during delivery, leading to a condition called neonatal herpes, which can be fatal. Rarely, HSV can be transmitted from the pregnant woman to her unborn child. HSV-2 is the most likely cause in this case.

Repeated occurrence (or recurrences) may occur throughout an individual's life. Symptoms of primary infections are accompanied by systemic signs, longer duration of symptoms, and higher rate of complications. Symptoms of recurrences usually do not last as long, or are as severe, as the primary infection. Immunocompromised individual experience more severe, prolonged, and widespread infections and more recurrence. HSV infection can manifest into a number of specific diseases, each with its own characteristics.

Acute herpetic gingivostomatitis is caused by primary HSV-1 infection, typically in children between the age of 6 months and 5 years old; adults can develop this condition as well, but symptoms are less severe. The mode of transmission is direct contact of contaminated saliva of an infected individual. Clinical features include abrupt onset, fever of 102°F to 104°F, anorexia and lethargy, gingivitis, vesicular lesions, tender regional lymphadenopathy, and perioral dermatitis.

Acute herpetic pharyngotonsillitis is caused by an HSV-1 infection in adults more often than gingivostomatitis. Presenting clinical features are fever, malaise, headache, and sore throat. Ulcers on the tonsils and posterior pharynx are formed when virus-filled EVs rupture. Oral and labial lesions are also found in some patients. HSV-2 infection can also cause this condition via orogenital contact and transmission and can occur concurrently with genital herpes.

Herpes labialis (commonly known as cold sores) is the most common disorder of recurrent HSV-1 infection that can occur annually or even monthly. Clinical features include pain, burning sensation, and tingling in the affected site, such as the face or around the lips. Erythematous papules (or red rash) and pustular skin lesions (formed by a collection of leukocytes) may form in the facial area, then eventually lead to ulceration.

Herpetic whitlow is caused by either HSV-1 or HSV-2 infection and manifests as lesions on the end of fingers following direct contact from orofacial or genital infections. Health care professionals, especially dentists, are at risk for this disease. Transmission of the virus may even occur from the infected individual by direct contact. A child who developed herpes gingivostomatitis may, in turn, develop herpetic whitlow as well.

Herpes gladiatorum is caused by either HSV-1 or HSV-2 infection and manifests as lesions on the face, arms, neck, and upper trunk of the torso. This condition is often seen in wrestlers and other contact sport athletes, such as those who play rugby. Transmission can occur during the sporting event or through direct contact with fomites, such as the wrestling mat and rugby ball where sweat drips on them.

Eczema herpeticum typically is caused by an HSV-1 infection and arises from pre-existing skin disease like atopic dermatitis and is more prevalent in children. Clinical presentations include clusters of blisters and ulcers. Bacteremia or bacterial superinfection (a condition caused by *Staphylococcus aureus* infection) can cause this condition to be fatal.

Primary genital herpes is caused by either HSV-1 or HSV-2 infection (but predominantly by HSV-2). The majority of the time (>85%) occurs during oral sex, as opposed to genital-to-genital transmission. Clinical features caused by either HSV-1 or HSV-2 are indistinguishable, which is no surprise since both manifest the same symptoms. In addition to the above-mentioned general mild symptoms, others may include myalgia (muscle aches and pain, which can involve ligaments, tendons and fascia, or the soft tissues that connect muscles, bones, and organs), itching, dysuria, vaginal and urethral discharge, and tender lymphadenopathy. Blisters caused by a first genital infection are usually more painful, last longer, and are more widespread than those caused by a recurrent infection. The rate of recurrent genital herpes is high for both males and females.

Herpes simplex keratitis (or *keratoconjunctivitis*) is caused by HSV-1 infection yielding lesions on the cornea of the eye. Clinical features include a painful sore, tearing, sensitivity to light, and blurred vision. Especially without treatment, the cornea can become cloudy, leading to a significant loss of vision over time.

Herpes encephalitis is a life-threatening illness with a high rate of neurological sequelae after recovering from the infection. This condition is caused mainly by HSV-1 but occasionally by HSV-2 infection through sexual transmission or from mother to newborn. Clinical symptoms begin with confusion, fever, and seizures; ultimately, this illness can be fatal.

Neonatal herpes is caused by HSV-2 infection due to exposure of the infant to the mother during delivery, referred to as perinatal transmission. This condition is rare, occurring in an estimated 0.01% of all births globally. However, it is a serious condition that can lead to lasting neurologic disability and can be fatal. The risk for neonatal herpes is greatest when a mother acquires HSV for the first time in late pregnancy.

Additionally, HSV infection can contribute to a variety of human diseases, including encephalitis, keratitis, neonatal infections and birth defects, Kaposi sarcoma, and nasopharyngeal carcinoma. Infection may progress inside the body, moving through other mucous membranes like those of the esophagus, lungs, or colon. Ulcers in the esophagus cause pain during swallowing, and lung infection causes pneumonia with cough and shortness of breath.

Diagnosis, Treatment, and Prevention

When symptoms arise and an infection is suspected, various diagnostic tests are available. Diagnoses are separated into two approaches: (1) detecting the presence of the virus in lesions or antiviral antibodies in the patient's blood or (2) detection techniques based on both laboratory and point-of-care (POC) devices. Diagnostic technologies to detect HSV infection range from conventional methods to advanced technologies that usually involve detection of viral proteins, viral genetic material, or HSV-specific antibodies in the infected individual's blood. Conventional methods include viral culture, serological tests, biochemical assays, microscopy, and nucleic acid amplification. POC techniques include microfluidics-based tests that enable on-spot testing. Each technique has its limit of detection, sensitivity, and specificity. Laboratory techniques to detect HSV in lesions include microscopy and imaging, detecting viral glycoproteins, and detecting viral genetic material. Techniques to detect viral antibodies in blood (or serological assays) include passive agglutination or hemagglutination assay, Western blot assay, enzyme-linked immunosorbent assay, fluorescence immunoassay, multiplexed flow immunoassay, luciferase immunoprecipitation assay, and microfluidic-based POC devices.

Unfortunately, there is currently no cure for HSV infection, and successful vaccines have not been developed. Treatment includes antiviral drugs, which can reduce the severity of symptoms. These various treatments are most effective when started soon after symptoms are detected or suspected, usually within 48 hours, for example, with the first sign of tingling or discomfort but before the appearance of blisters. Reducing the risk of transmission is an essential part of prevention and treatment for the population. Keeping the infected area clean with gentle washing with soap and water is the simplest and most effective treatment for recurring herpes infection in the lip and genital areas. Cold sores can be soothed and swelling can be reduced by applying ice to the affected area.

Suppressive medications such as acyclovir, valaciclovir, and famciclovir are used for those with recurrent clinical episodes with frequent and painful symptoms. Acyclovir is the most commonly prescribed antiviral agent to relieve discomfort and resolve symptoms during recurrences, including against mild and severe HSV infection. It is a nucleoside analogue that inhibits thymidine kinase (TK) and DNA polymerase activities. Other nucleoside analogues include valacyclovir, penciclovir, famciclovir, trifluridine, idoxuridine, vidarabine, sorivudine, brivudine, ganciclovir, and valganciclovir. The interesting thing about these drugs is that they initially require TK to activate their antiviral activity; however, once TK is inhibited or mutated by these same drugs, these drugs can no longer be activated: a true double-edged sword. To get around this dilemma, nucleotide analogues can circumvent this problem since they are present in a phosphorylated form and therefore do not require activation by TK. Nucleotide analogues include cidofovir, adefovir, and brincidofovir. All the above-mentioned drugs target the elongation step of DNA synthesis.

There are also a number of anti-HSV medications that are neither nucleoside nor nucleotide analogues. Foscarnet is a pyrophosphate analogue that binds and inhibits DNA polymerase activity, with a 100-fold higher affinity for viral DNA polymerase than for human DNA polymerase. Amenamevir (approved for clinical use in Japan) and pritelivir (in clinical trial) are helicase-primase inhibitors that prevent the binding of viral UL5/UL8/UL52 helicase/primase complex. *n*-Docosanol is an over-the-counter, topical medication that prevents the binding and entry of HSV to the susceptible cell by interfering with membrane fusion.

Over-the-counter analgesics such as acetaminophen, naproxen, and ibuprofen can be used to relieve pain related to cold sores. Benzocaine and lidocaine are topical anesthetics that may be used to numb the affected area on the skin or mucous membrane. Trifluridine is an antiviral eye drop used to treat herpes simplex keratitis. Simple therapies like placing an ice pack over the cold sore or drinking cold liquid for oral infection are quite effective in easing symptoms. Sitting in a warm bath for 20 minutes without soap and wearing loose-fitting clothes can help ease the pain of genital infection.

Since HSV is highly contagious, impeding transmission of this virus is probably the most important and effective means of prevention. If infected, avoid oral contact with others, especially when symptoms are apparent. Transmission via fomites is also possible; therefore, avoid sharing objects that touched saliva with others. Individuals with symptoms of genital herpes should abstain from sexual activity while experiencing symptoms. Even though protections can reduce the risk of transmission, areas that are not covered by the condom are still vulnerable. Individuals who have latent HSV infection should avoid known triggers that cause recurrence.

Current Status

The World Health Organization estimates that roughly 67% of the global population, or 3.7 billion people, who are under 50 years old have latent HSV-1 infection. Most of these infections were received during childhood. Another 13% of the population (491 million people) who are between 15 and 49 years old have latent HSV-2 infection. Unfortunately, since sexual transmission from men to women is more efficient, HSV-2 transmission to women is twice as frequent as to men. About 90% of the world's population are infected with either or both HSV serotypes. HSV-2 infection is associated with a 3-fold increased risk of the individual acquiring, then transmitting, human immunodeficiency virus type 1(HIV-1). This is due to the cooperative nature of HSV and HIV infections. HSV-2 infection has been shown to stimulate an increase in dendritic cells and CD4+ T lymphocytes that are susceptible to HIV infection. Additionally, lesions caused by an HSV-2 infection provide an entry portal for HIV. Most individuals are asymptomatic during recurrent viral reactivation. HSV-related clinical manifestations are displayed in only 5%–15% of individuals during reactivation. Adolescents experience that highest incidence of new infections of both serotypes; however, since reinfections often occur, prevalence of infection increases with age.

In the United States, 47.8% of individuals who are 14 to 49 years old are infected with HSV-1, and another 11.9% are infected with HSV-2. Women (of all races and ethnicities), Mexican Americans, and non-Hispanic black individuals experience a higher prevalence of HSV infection.

Scientific Significance and Discoveries

Even though anti-HSV drugs can inhibit viral replication, they have no effect on latent infection; therefore, identification and inhibition of viral protein expressed during latency is critical to thwart reactivation. A number of experimental approaches are ongoing to solve the problem of preventing reactivation, even destroying the latent viral DNA episome. Ongoing experimentation includes the cleavage of viral episomes in latent infected cells by HSV-specific endonucleases and CRISPR (clustered regularly interspaced short palindromic repeats)/Cas technology. Drugs such as the proteasome inhibitor bortezomib can be used to stimulate virus reactivation, followed by anti-HSV treatment with ganciclovir is being investigated. Other drugs, such as supplemental glutamine, have been shown to reduce reactivation of both HSV-1 and HSV-2 in mice and guinea pigs. Inhibitors of chromatin remodeling enzymes may be another means to prevent reactivation by maintaining latent viral DNA episome in the heterochromatin, thus inactive, structure.

MicroRNAs (miRNAs) that are expressed during HSV latent infection are promising targets for the development of therapeutic agents such as miRNA inhibitors like miRNA mimics and anti-miRNAs. The most recent development in HSV treatment research is the use of meganuclease-mediated gene editing, not CRISPR/Cas9, to purge latent HSV episome in vivo, thus eliminating the possibility of viral reactivation and the development of diseases. These lines of research have yielded promising results so far.

Human Polyomavirus

Historical Perspective

Dr. Ludwik Gross published his discovery of the murine or mouse polyomavirus (MPyV) while studying murine leukemia virus. He found that, instead of developing leukemia, the experimental animals developed tumors in their parotid and salivary glands. He described his finding in 1953 on page 954: "It is clear that the leukemic agent is filterable, since typical generalized leukemia could be reproduced with Ak-leukemic extracts that had been filtered through Seitz, or Berkefeld, filters." This filterable agent had the ability to induce salivary gland tumors in experimentally exposed mice. In 1960, the simian vacuolating virus 40 (SV40) that infects rhesus monkeys was identified. This species of virus was found as a contaminant in polio vaccines that were administered between 1955 and 1963. The polio vaccines were prepared in rhesus macaque kidney cells. This resulted in millions of people globally being inadvertently exposed to this virus. Fortunately, even though SV40 is closely related to human polyomavirus (HPyV), as was later found, the virus did not seem to cause discernible disease in these vaccinated individuals. There was no evidence of increased cancer risk. However, SV40 has been found repeatedly in numerous human tumors, such as mesotheliomas, brain tumors, and lymphomas, through the years. Thus far, studies have shown no convincing evidence, neither epidemiological nor molecular, that links this virus to human cancers. This is why the Institute of Medicine's Immunization Safety Review Committee continues to monitor the possible effect of SV40 in humans.

The first two HPyV species, human polyomavirus BK (BKPyV or HPyV1) and JC polyomavirus (JCPyV or HPyV2), were not identified until 1971. BK and JC are initials of the two patients where these PyV strains were isolated. BKPyV was isolated from the urine of a kidney transplant recipient, whose initials were B.K. This virus has been shown to cause kidney damage. JCPyV was detected in a patient named John Cunningham who had a history of Hodgkin lymphoma and progressive multi-focal leukoencephalopathy (PML). More than three decades later, in 2007, three more human viruses were identified: KIPyV, WUPyV, and MCPyV. Karolinska Institute PyV (KIPyV or HPyV3) and Washington University PyV (WUPyV or HPyV4). In 2008, Merkel cell PyV (MCPyV or HPyV5) was isolated from Merkel cell carcinoma (MCC) specimens; this carcinoma is a rare human cancer of neuroendocrine origin. A number of additional HPyV species have been to now, including: HPyV6 and HPyV7 from skin samples in 2010; trichodysplasia spinulosa-associated PyV (TSPyV or HPyV8) in skin lesions from patients with trichodysplasia spinulosa (TS), which is a rare dermatologic condition; HPyV9 from a kidney transplant patient; Malawi PyV (MWPyV or HPyV10) from stool specimens of a 1-year-old Malawi child with gastrointestinal symptoms, mostly diarrhea, in 2012; HPyV10 was detected in condyloma acuminata specimens (or anogenital warts) from a patient with a rare genetic disorder known as warts, hypogammaglobulinemia, infections, and myelokathexis syndrome (WHIM); Saint Louis PyV (STLPyV or HPyV11) from a child from the United States with gastrointestinal symptoms in 2013; Mexico PyV (MXPyV or HPyV12) from stool samples of a Mexican child presenting with diarrhea and from human liver tissue during screening of organs donated by the German Foundation for Organ Transplants (Deutsche Stiftung Organtransplantation, DSO); New Jersey PyV (NJPyV or HPyV13) from a muscle biopsy of a pancreatic transplant recipient suffering from retinal blindness and vasculitic myopathy; and Lyon IARC polyomavirus (LIPyV or HPyV14) in 2017.

Classification and Structure of Human Polyomavirus

Human polyomavirus is a member of the Polyomaviridae family. This dsDNA virus family includes 6 genera and 117 species: *Alphapolyomavirus* (eg, *Alphapolyomavirus tertipanos*), *Betapolyomavirus* (eg, *Betapolyomavirus hominis*), *Gammapolyomavirus* (eg, Anser polyomavirus 1), *Deltapolyomavirus* (eg, HPyV6), *Epsilonpolyomavirus* (eg, *Epsilonpolyomavirus bovis*), and *Zetapolyomavirus* (eg, dolphin polyomavirus 1). To date, 14 species of polyomavirus are known to infect humans.

The structure of HPyV is a non-enveloped virion 40–45 nm in diameter. The single circular, ds DNA genome of 4700–5400 bp is encased in an icosahedral capsid (Figure 7.10). The viral capsid remains stable and fully infectious at up to 75°C for 1 hour, is resistant to ether and an acidic environment, and is resistant to treatment with formalin and organic solvents. Inside the virion, the genome itself is associated with host cell's histones H2A, H2B, H3, and H4 to form a "mini-chromosome" consisting of nucleosomes; however, once in the host cell nucleus, H1 is added.

Viral genome replication is initiated from a unique origin of DNA replication. Different from the herpes simplex virus genome, which is divided into more than 80 ORFs, the HPyV genome is divided into only 3 functional domains: (1) an early transcriptional unit (or coding region) encoding T antigen (Tag) regulatory proteins: large T antigen (LTag) and small T antigen (sTag), and middle T antigen (MTag) for some PyV species; (2) a late transcriptional unit encoding capsid proteins: VP1 (major), VP2, VP3, VP4 (in some HPyVs), and agnoprotein; and (3) a noncoding control region (NCCR) containing the origin of DNA replication and promoter/enhancer elements that regulate the expression of both early and late transcriptional units. Some PyV species also express agnoprotein from the late coding region. MPyV genome expresses an additional

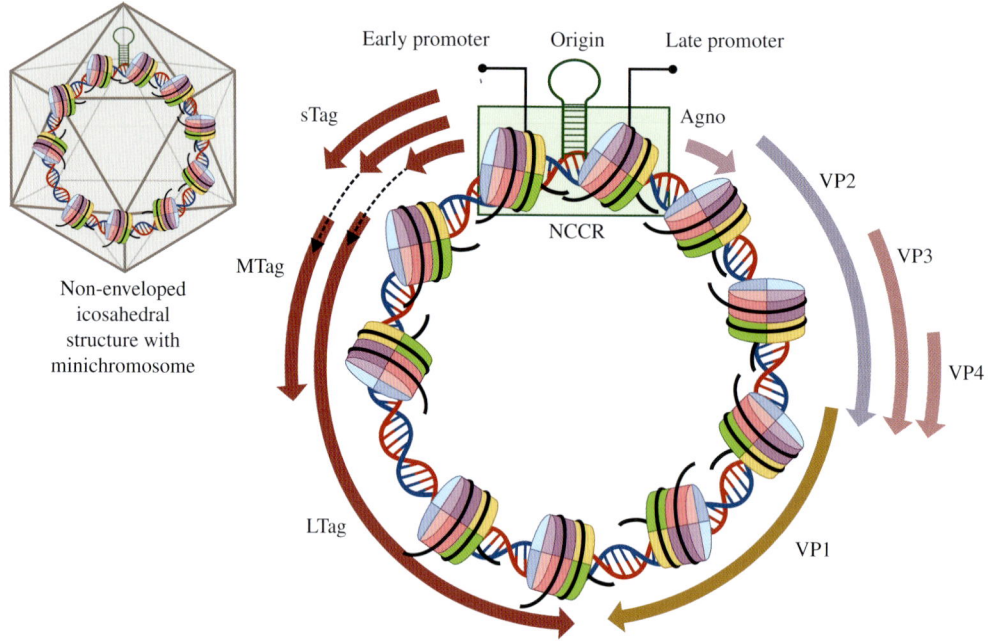

FIGURE 7.10   Human polyomavirus capsid structure and genome. HPyV is a non-enveloped virion with an icosahedral capsid encapsidating a single circular dsDNA genome. Transcription is controlled by 2 promoters that are used to regulate the expression of early and late transcriptional units. VP, viral protein; Agno, agnoprotein. *Credit:* Figure partially created in BioRender.

membrane-bound MTag. The early and late transcriptional units are arranged in opposite directions (ie, on both strands) in the DNA genome, separated by the NCCR. Alternative RNA splicing and alternative overlapping ORFs allow the genome to encode for numerous translational products.

The early coding region encodes for a single transcript, which is then alternatively spliced to produce both LTag and sTag. The synthesis of these proteins occurs early and must be before PyV DNA replication. This is especially true for LTag since it is a nuclear protein that regulates both viral transcription and genome replication. It's been shown to promote transformation of cells in culture and oncogenesis in vivo in model animals.

The late coding region encodes for VP1, VP2, VP3, and VP4 as well as an accessory protein, agnoprotein, that are expressed at later time points. The transition from T antigen expression to structural protein expression may be partially controlled by the expression of miRNAs that are encoded by the late strand transcript itself. Studies show that they may even be able to exert regulatory effects on the host cell. Like the T antigens, these structural capsid proteins are alternative spliced from a single pre-mRNA transcript. VP1 is the major structural protein that is exposed on the surface of the capsid and is involved in receptor binding on susceptible cells. VP2 and VP3 are produced from the same transcript using alternative start codons. These two minor structural proteins are involved in stabilizing the virion when this virus is in the extracellular host environment. They continue to perform this function after VP1 proteins are shed from the capsid during uncoating. Additionally, both capsid proteins are important for viral entry into the host cell nucleus (import signal). VP4 is an additional truncated form of VP2/3. VP4 is believed to be a viroporin that promotes progeny release. Last, the function of agnoprotein may include regulation of viral gene expression, virion assembly, virus maturation and egress, and dysregulation of a variety of cellular processes, such as viral release.

Host Range, Tropism, and Susceptible Host Cell

The host range for polyomavirus is mammals (primates, rodents, and cattle), birds, and fish. The 14 identified HPyV species are not known to be zoonotic. Even though the 14 species of HPyV have a wide range of tropism, each species infects very particular susceptible cells. Persistence or reactivation from latency of each viral species at its designated site may stimulate tissue-specific carcinogenesis. Major sites of persistence or latency for BKPyV and JCPyV are the kidney and urinary tract. BKPyV also remains latent in the lung, eye, and epithelial cells of the ureter and urinary bladder, while JCPyV can remain latent in lymphocytes and the CNS. Both can be reactivated at their sites of latency. In addition, BKV genome sequences have been detected in urogenital tissues and fluids, including those of the cervix, vulva, prostate, semen, and brain tissues. MCPyV, TSPyV, HPyV6, HPyV7, and HPyV9 remain in the latent state in the skin and lymphoid tissue; KIPyV and WUPyV in the respiratory tract; and HPyV10 and STLPyV in the gastrointestinal tract.

JCPyV and BKPyV have been isolated in urine, feces, and saliva from infected individuals. Various human PyV particles have been found in sewage pollution or even in recreational waters. Of fomites or environmental surfaces such as door handles, 75% have been found to be contaminated with PyV DNA. TSPyV DNA has been detected in urine and kidney tissue of patients tested for other conditions.

Transmission for most polyomavirus species is via the respiratory, oral, or fecal-oral route. BKPyV have been associated with upper respiratory illnesses and acute tonsillitis had been reported in children. And, both BKPyV and JCPyV have been reported in urine of immune-competent adults, with JCPyV being more prevalent. Both virus species have been found to persist in the kidney recovered from patients taken at surgery or autopsy. Modes of transmission include urine-oral, transplacental, passive viremia by blood transfusion, direct contact with semen, and organ transplantation.

Infecting the Host Susceptible Cell, Viral Particle Replication, and Tissue Damage

Attachment of the HPyV virion occurs when VP1 interacts with disialic acid motifs on gangliosides (or sialylated glycosphingolipid) expressed on the surface of the susceptible host cell (Figure 7.11). These target gangliosides are widely distributed on multiple cell types throughout the body, which provide this virus a wide range of tropism. However, each species uses a specific ganglioside receptor, and some even require an additional co-receptor for attachment and entry. For example, JCPyV binds to the serotonin receptor 5HT2A (5-hydroxytryptamine receptor 2A) and MCPyV binds to heparan sulfate (a member of sulfated glycosaminoglycans).

Virion entry into the host cell is via either caveola-dependent or clathrin-dependent, receptor-mediated endocytosis. Entry into the host cell is through the caveola-mediated endocytic pathway. Caveola is a cup-shaped invagination (50–80 nm in diameter) that forms a subdomain of lipid rafts in the plasma membrane of many vertebrate cell types. Caveolae are especially abundant in endothelial cells, skeletal muscle cells, fibroblasts

HPyV internalization and uncoating

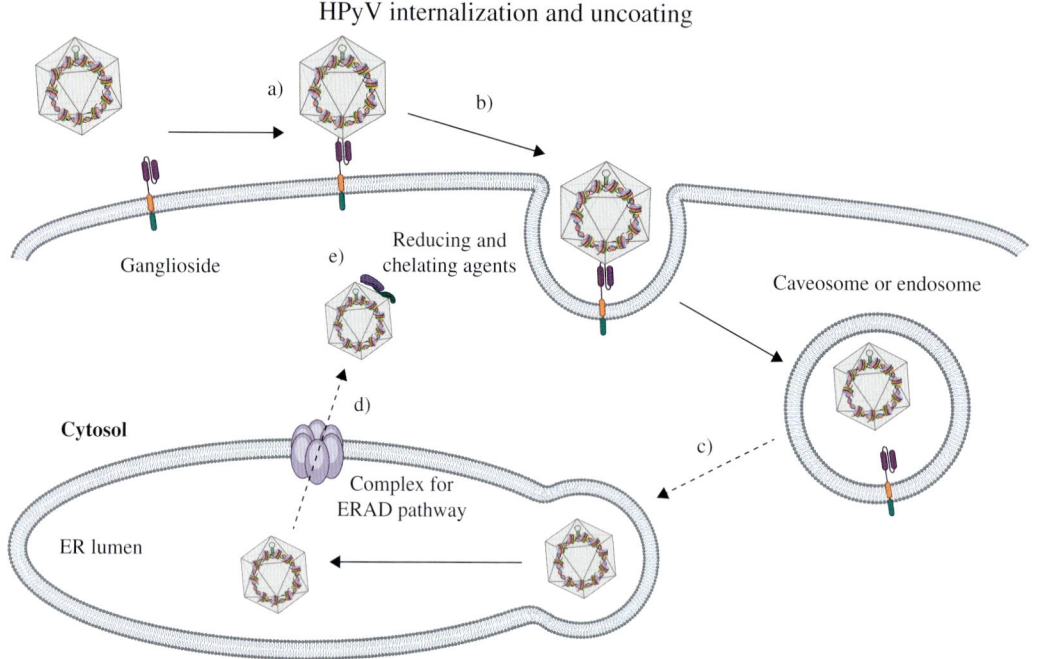

FIGURE 7.11   HPyV entry into host cell. (a) Attachment of HPyV to ganglioside on surface of host cell. (b) Internalization of HPyV by either caveola-dependent or clathrin-dependent, receptor-mediated endocytosis. (c) Delivery of virion to ER via membrane fusion and initial uncoating step. (d) Egress of virion into cytosol using ERAD pathway. (e) Further uncoating by reducing and chelating agents in the cytosol. *Credit:* Figure partially created in BioRender.

adipocytes, and embryonic notochord cells but are undetectable in other cell types. The major protein component of a caveola is caveolin associated with a cavin complex. The function of the caveolin-cavin complex is to regulate caveola-dependent, receptor-mediated endocytosis to internalize the HPyV virion, placing it in a caveosome in the host cell cytosol. A caveosome is an endosomal compartment with neutral pH. Without an acidic environment, uncoating of the HPyV virion cannot occur in the caveosomal lumen. Clathrin-dependent endocytosis, on the other hand, places the virion in an endosomal compartment that has an acidic luminal environment.

In either case, once internalized, both temporary vesicles are trafficked to the ER, where fusion between the vesicular and ER membranes occurs, translocating the virion into the ER lumen. Polyomaviruses target the ER upon internalization inside the cell, which is an extremely interesting and unique feature of HPyV infection, although the purpose is not well understood. Membrane fusion initiates the disassembly of the VP1 layer of the capsid, which in turn exposes VP2 and VP3. The exposed VP2 mimics misfolded protein, leading to the egress of the virion from the ER into the cytosol and requiring the host cell's ER-associated degradation (ERAD) pathways, typically used to target unwanted ER proteins to cytosolic proteasomes.

Further disassembly of the capsid occurs in the cytosol, where the calcium concentration is low, possibly through the activity of reducing agents and chelating agents. The altered viral capsid enables localization to the nuclear pore, and entry of the virion into the nucleoplasm is mediated by VP1, VP2, and VP3. After reaching the nucleus, the capsid is completely disassembled, and the viral genome is exposed and maintained in the episomal form.

Expression of the early coding region is initiated first, producing T antigens (Figure 7.12). Host cellular transcription factors (TS-TFs) and RNA polymerase II are recruited to regulate early gene expression. Cell-specific expression of transcription factors regulating viral gene expression determines if the host cell is permissive to the particular HPyV species. For example, TS-TFs Tst-1, YB-1, c-Jun, NF-1, C/EBPβ, Spi-B, and SRSF1 are important in regulating JCPyV gene expression. Tst-1 plays a role in the development of the nervous system. YB-1 plays a role in DNA repair, transcription, and pre-mRNA splicing. C-Jun is the first oncogenic transcription factor discovered and is involved in a number of cellular activities including promoting cell death, increasing steroidogenic gene expression, regulating the Wnt gene for joint differentiation, and stimulating axon regeneration. NF-1 has multiple roles including nucleosome remodeling, promoting cell differentiation, promoting the survival and maintenance of slow-cycling adult stem cells, and preventing the silencing of telomeric genes. C/EBPβ plays an important role during inflammation, cellular proliferation, and tumorigenesis. Spi-B is involved in B-cell development and function. Serine/arginine rich splicing factor 1 (SRSF1) maintains genome stability, regulates the stability of mRNA, promotes RNA splicing, and regulates the export of processed mRNAs and their translation.

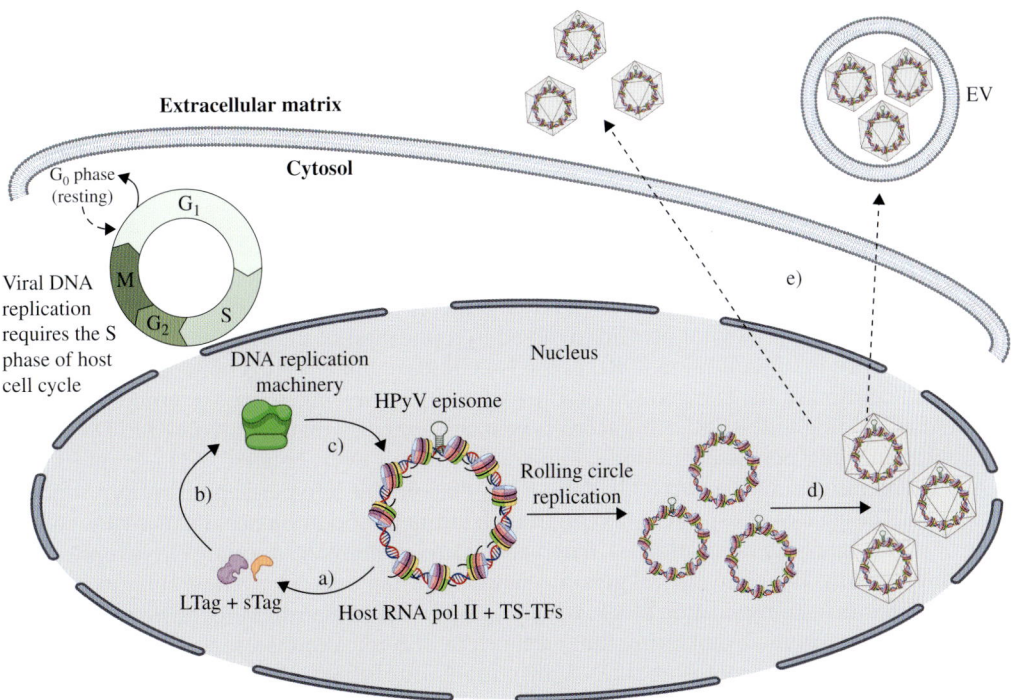

FIGURE 7.12 HPyV genome replication, virion assembly, and release from host cell. (a) Host RNA polymerase II catalyzes the expression of LTag and sTag. (b) Both T antigens stimulate the expression of host cell DNA replication machinery, thus requiring the cell cycle S phase. (c) Host DNA replication machinery carries out genome replication via rolling circle replication. (d) Virion assembly occurs in the nucleoplasm. (e) Release of virions is through either cell lysis or the extracellular vesicular (EV) transport pathway. *Credit:* Figure partially created in BioRender.

Once expressed, both LTag and sTag stimulate the expression of host cell DNA replication machinery and cell cycle progression by regulating central signaling cascades. This is why one of the most important functions of the T antigens is to direct the host cell to enter the synthesis phase of the cell cycle to facilitate viral DNA replication. These cascades are ones that are essential for cell growth, motility, survival, metabolism, and angiogenesis. Both LTag and sTag have been shown to target a number of pathways including: Notch signaling pathway, hedgehog pathway, DNA damage response (DDR) pathway, and ubiquitination-proteasomal degradation pathway, phospholipid signaling pathways, nuclear receptor signaling pathway, and metabolic pathways.

Each T antigen is also associated with a variety of differential cellular pathways. LTag mediates viral genome replication with the help of host epigenetic mechanisms to regulate chromatin remodeling. LTag targets the protein kinase C (PKC) pathway, phosphatidyl-3-kinase/AKT (or protein kinase B)/mammalian target of rapamycin (PI3K/AKT/mTOR) pathway, and interferon signaling pathway. These pathways are important for host cell growth, differentiation, and survival. LTag targets with these pathways to ensure virus replication while inducing host immune evasion. In association with the Wnt pathway, LTag has been demonstrated to upregulate c-Myc and cyclin D1 promoter activities. These promoter activities are essential to permit a quiescent cell to enter the cell cycle. Last, LTag has also been found to block p53 and pRb protein functions to inhibit their growth-suppressive activities, thus permitting cell cycle progression.

While sTag targets the mitogen-activated protein kinase (MAPK) pathway, interacts with protein phosphatase 2A (PP2A) and protein phosphatase 4 (PP4) activities to increase motility and suppressing the canonical NFκB signaling pathway, and activates cell division cycle 42 (Cdc42) and Ras homolog family member A (RhoA). Cdc42 and RhoA are small GTPases that work synergistically to regulate actin cytoskeletal organization, T cell development, cell protrusion, proliferation, and cell survival. PP2A itself is an important regulator that controls cellular metabolism, such as increasing glycolysis, lipid metabolism, and catecholamine synthesis. By inhibiting PP2A, sTag can promote the continuation of cell cycle progression and delay exit from mitosis. This delay gives HPyV additional time to replicate. sTag was also found to inhibit the NFκB signaling pathway and downregulate the expression of NFκB target genes.

LTag is the only viral protein involved in replication of the viral genome. It binds to DNA's origin of replication (ORI) in the NCCR to regulate both genome replication and transcription of the late gene to produce capsid proteins necessary for virion assembly. After binding to the ORI, LTag recruits host cell enzymes that are involved in DNA replication. LTag possesses multiple domains, including DNA helicase, ATPase, and DNA polymerase binding activity. All of these activities enable it to participate in DNA replication. The state of phosphorylation on LTag determines and influences its many functions.

Viral genome replication is carried out bidirectionally via a rolling circle replication mechanism. To enhance viral genome synthesis, LTag overrides a host cell mechanism referred to as re-replication block, which is important in maintaining the integrity of the host chromosome by ensuring that each origin of replication is used only once per cell cycle so that all chromosomal DNA are equally duplicated. Without this re-replication block, the original viral genome can be used as the template for repeated replication. Host cell's histones are recruited to promote the condensation of newly synthesized viral DNA molecules.

Once viral genome amplification begins, LTag performs its additional function, which is to promote expression of the late coding region producing structural capsid proteins. The binding of LTag to the promoter leads to transcription suppression of early transcripts and switches to the expression of late transcripts.

Another viral regulatory protein that is not well studied is agnoprotein, which is expressed in some mammalian PyVs during the late phase. Agnoprotein can bind to LTag, sTag, VP1, and host cellular proteins that regulate cell cycle arrest and DNA repair mechanisms like p53, FEZ1 (Fasciculation and Elongation Protein Zeta-1; involved in axon outgrowth and intracellular transport), PP2A, and Ku70 (subunit of the Ku protein involved in the repair mechanism for double-strand DNA breaks). Therefore, it's been suggested that this viral protein plays a vital role in viral transcription and translation; genome replication; inhibition of the host DNA repair mechanism, virion assembly, and release from the host cell; and is associated with the viral life cycle in general. Binding to p53 allows agnoprotein to induce cell cycle arrest at the G2/M checkpoint, while binding to Ku70 allows this viral protein to block the nonhomologous end joining dsDNA repair mechanism.

Agnoprotein also possesses viroporin activity and has been shown to interact with a fusion attachment protein, which suggests that one of its functions is associated with the exocytotic pathway.

Virion assembly occurs in the nucleoplasm and is released from the host cell, which can cause cell lysis. Virion release through cell lysis occurs when the PHyV infection results in immediate tissue damage and dissemination of the virus. When HPyV doesn't destroy the initial host cell, it's been suggested that virion exit uses secretory extracellular vesicular (EV) transport pathway. Not destroying the host cell might permit the virus to evade the host immune system and allow viral persistence in the host organism. EVs are cell membrane-derived vesicles that are used by most cells to either discard unwanted material into the extracellular space or transmit material to mediate cell-to-cell communication.

Dissemination in the Host Body

Primary HPyV infection usually results in mild symptoms or no observed symptoms at all. Hematogenous dissemination follows the primary infection in most cases. Neural dissemination occurs for species that are associated with neurologic disorders (eg, JCPyV). The mode of dissemination depends on accessibility of susceptible host cells. Viral particles, then, disseminate to specific secondary sites for lifelong persistent infection or latent infection. Latency occurs when the viral genome resides in the host cell nucleus without its genes being expressed. On the other hand, viral gene expression and virion replication are kept at a minimal level in an attempt to avoid detection from the host immune systems in the persistent state. Viral persistent infection is required for virus to stimulate carcinogenesis, while viral latent infection does not result in virion production and the viral genome is only replicated when the host cell divides, thus reactivating virus replication.

Pathogenesis and Clinical Manifestation

Primary HPyV infection is very common in childhood and young adults but causes few to no symptoms, yet they maintain lifelong persistent in the majority of adults. As individuals age, repeated contact may occur, leading to an adult having multiple HPyV species in the latent state. HPyV-stimulated clinical conditions are most common in individuals who are immunocompromised or those with known immunological abnormalities. Although various HPyV species have been detected in organs, tissues, and fluids of immunocompetent individuals, they do not show any sign of disorders. Pathogenesis is always preceded by the compromise of host immune system and increased viral transcription and replication.

Once expressed in high amounts, LTag and sTag oncoproteins can transform cells and induce oncogenesis and tumorigenesis in animals by upregulating rapid host cell cycle progression. In individuals whose immune system is suppressed due to HIV infection, agnoprotein can also bind to HIV-tat protein. HIV-tat is a specialized protein expressed from the genome of the human immunodeficiency virus whose function is described in the Infecting the Host Susceptible Cell, Viral Particle Replication, and Tissue Damage section of chapter 8 (11th paragraph). HIV-tat can regulate both HIV genome and host cellular gene expression, while generating a permissive environment for viral (both HIV and HPyV) replication by suppressing host immune response.

As the permissive host cell proliferates, the circular viral episome is also replicated and transferred vertically to daughter cells. Increased host cell proliferation, directed by both viral oncoproteins, can lead to the increased chance of carcinogenesis. Viral particles can shed into the bloodstream, causing active viremia, and into urine, which causes viruria. Shedding of viral particles is more common in immunocompromised than immune-competent individuals. Currently, 5 of the 14 HPyV species have been linked to specific disorders (Figure 7.13). The others are not strongly associated with any particular clinical condition.

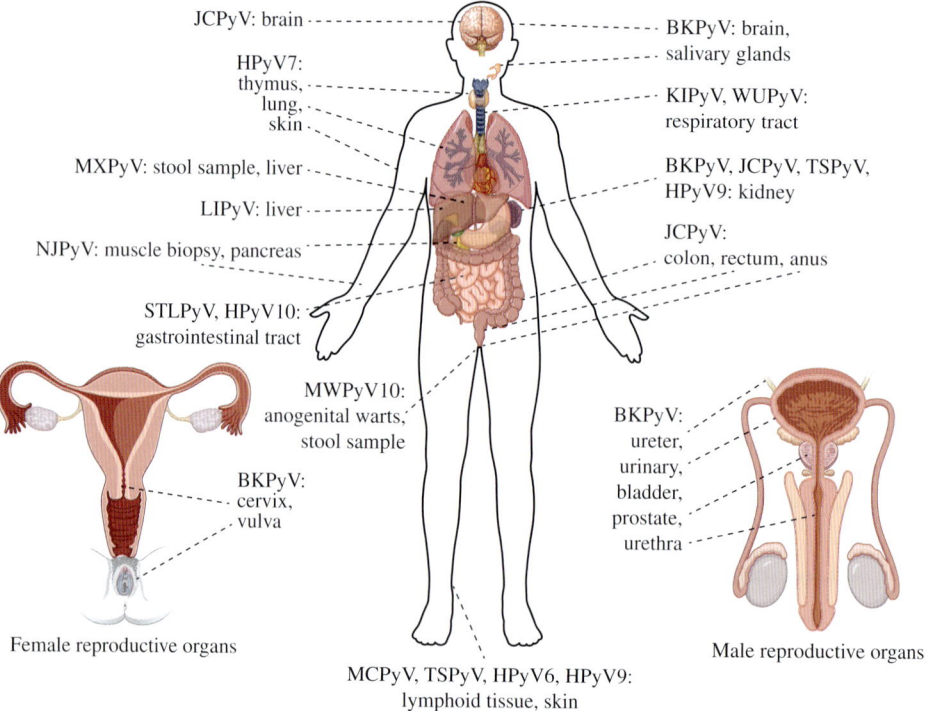

HPyV organ localization causing related disorders

JCPyV: brain

HPyV7: thymus, lung, skin

MXPyV: stool sample, liver

LIPyV: liver

NJPyV: muscle biopsy, pancreas

STLPyV, HPyV10: gastrointestinal tract

BKPyV: brain, salivary glands

KIPyV, WUPyV: respiratory tract

BKPyV, JCPyV, TSPyV, HPyV9: kidney

JCPyV: colon, rectum, anus

MWPyV10: anogenital warts, stool sample

BKPyV: ureter, urinary, bladder, prostate, urethra

BKPyV: cervix, vulva

Female reproductive organs

Male reproductive organs

MCPyV, TSPyV, HPyV6, HPyV9: lymphoid tissue, skin

FIGURE 7.13   HPyV organ localization and related disorders. Human polyomavirus has been isolated in a variety of human tissues and excreted samples. However, only 5 of 14 HPyV species have been linked to specific disorders, including cancer. *Credit:* Figure partially created in BioRender.

BKPyV has been shown to be the causative agent of BKPyV-associated nephropathy (BKVAN) in kidney as well as nonrenal solid organ transplant patients. It has also been linked to ureteric stenosis (narrowing of the ureter) and hemorrhagic cystitis (a diffuse inflammatory condition of the urinary bladder) in hematopoietic stem cell transplantation. Viruria is a likely symptom in these individuals.

JCPyV has been shown to be the causative agent of the rapidly progressive and fatal demyelinating disease called PML. PML is a condition where the myelin layers that insulate neurons in the CNS are not formed due to the destruction of oligodendrocytes that are targeted by JCPyV. This virus also targets astrocytes, whose function is to provide metabolic, structural, homeostatic, regulating the blood-brain barrier, and promoting synapse formation by clearing excess neurotransmitters. The absence of astrocytes would cause the brain to be deregulated, leading to further neurological deficits. Patients who are HIV infected, have hematological malignancies, or are receiving immunomodulatory medication are highly affected by JCPyV. Other neurological disorders that can be caused by JCPyV are granule cell neuronopathy, encephalopathy, and meningitis.

MCPyV is part of the skin microbiota, where 20% to 40% of children under 5 years of age test positive for anti-MCPyV antibodies. By 50 years of age, roughly 80% of individuals test positive for this virus. MCPyV causes asymptomatic infection in the skin throughout one's body. This virus is a major causative agent of MCC, which is a rare but aggressive neuroendocrine skin cancer. Other influencing factors of MCC include immune suppression, exposure to ultraviolet radiation, and age above 50 years.

Merkel cell carcinoma is a rare yet aggressive type of skin cancer. Merkel cells are epithelial neuroendocrine cells located in the stratum basale of the epidermis and oral mucosa. It was discovered that dermal fibroblasts are susceptible and permissive to the MCPyV life cycle. However, transformed cells of MCC exhibit neuroendocrine granules, suggesting that this malignancy likely originated from Merkel cells. No MCPyV DNA or proteins have been detected in tissues adjacent to the tumor.

As with other HPyV species, immunocompromised individuals exhibit a 16-fold higher risk for MCPyV-related MCC than immunocompetent individuals. MCPyV detection is not limited to MCC. It has also been detected in other human tumors.

TSPyV has been shown to be the causative agent for the rare skin disease TS. This clinical condition is more common in individuals who have received kidney as well as nonrenal solid organ transplant. It is also common in patients with lymphocytic leukemia. Symptoms of TS include the development of follicular papules (small rough bumps on the skin) and keratin spines (tiny, skin-colored to yellowish spicules); both are predominantly in the face. Alopecia (hair loss) of the eyebrows and eyelashes may also occur. Inner root sheath cells appear to be susceptible and permissive to TSPyV infection.

Last, HPyV7 has been shown to produce the symptoms of pruritic rash (papular lesions on the skin) and viremia in lung transplant recipients on immunosuppressive therapy. Most recently, HPyV7 was detected in human thymic epithelial tumors.

Diagnosis, Treatment, and Prevention

Diagnosis of polyomavirus is difficult during a primary infection since there are usually few to no symptoms exhibited. The presence of viral components can be detected using antibody assays against specific viruses. Competition assays are often required to distinguish among highly similar HPyV species. Polymerase chain reaction (PCR) to amplify HPyV DNA is a powerful first step in analyses.

Both body fluid and tissue samples can be used for analysis. To detect the presence of the virus in polyomavirus nephropathy, techniques such as urine cytology, quantification of the viral load in both urine and blood, and renal biopsy can be used. Since shedding of infected cells from the kidney into the urinary tract occurs, these cells can be examined with light microscopy to detect nuclear inclusions with the urine cytology technique. Quantitation of the viral load in blood and urine can be done via PCR.

The presence of the virus or viral components in renal biopsies can be tagged with specific antibodies. The sublocalization of the virus in infected cells is also possible with immunohistochemical staining. Blood tests to detect the presence of antibodies against multiple HPyVs have been developed.

Therapeutic treatments for BKPyV-related kidney disease and their mechanism of action include intravenous immunoglobulin for virus neutralization and entry inhibition; levofloxacin to inhibit DNA topoisomerase and helicase; everolimus as an mTOR inhibitor; leflunomide as a tyrosine kinase inhibitor; ciprofloxacin to inhibit DNA topoisomerase and helicase; cidofovir as a nucleoside analogue to inhibit viral polymerase; CMX001 (brincidofovir) as a nucleotide analogue to inhibit viral polymerase; and gallic acid-based small compounds t act as sialic acid receptors binding to prevent entry into host cell. Other treatments include hyperhydration to prevent the urotoxic effect of drug treatments and bladder irrigation, which is an invasive procedure yielding discomfort for the patient.

There are no specific antiviral therapeutic treatments or prophylactic vaccines for JCPyV-related PML. A combination of extended interval dosing of natalizumab and highly active antiretroviral therapy is the most common treatment to reverse the immune suppression that interferes with the normal host response to this virus. Natalizumab, which is a monoclonal antibody used to prevent infected leukocyte infiltration into the brain, has been shown to reduce the incidence and severity of PML.

No vaccine has been developed against MCPyV even though 80% of the general population establishes lifelong viral infection. The reason is mainly because of this huge number of infected individuals, only a very small minority would develop MCPyV-associated MCC. This is an unfortunate cost-benefit issue since developing such a vaccine is not lucrative enough for pharmaceutical and biotechnology companies. Difficulty in identifying a suitable target antigen is another roadblock. CRISPR/Cas9 technology and RNA interference have been shown to be effective in depleting T antigen expression in MCC cell lines, thus impairing cell proliferation and inducing apoptosis. Antiviral drugs such as artesunate and MAL3-101 are being studied.

There is also no known method to prevent exposure to polyomavirus since most individuals are infected during childhood. Viral monitoring in solid organ transplant and hematopoietic cell transplant is an effective tool to prevent transmission.

Current Status

Polyomaviruses are widely spread in adult human populations throughout the world, although primary infections usually occur in childhood or during adolescence and almost always are asymptomatic. A high percentage (60%–70%) of children were seropositive for anti-BKV IgG by 10 years old. BKPyV and JCPyV infections are the two most common, with seropositivity ranging from 72% to 98% depending on the part of the world and in different surveys. In the US population, seroprevalence ranges from 17.6% for HPyV9 to 99.1% for HPyV10. BKPyV maintains lifelong latency in the urinary tissue of immunocompetent individuals but can be reactivated in immunocompromised patients, yielding high levels of virus replication. BKPyV reactivation is observed using serological assays in 45% of kidney transplant patients.

JCPyV can persist in the kidneys and can be reactivated under immunosuppressive conditions, such as lymphoproliferative disorders, HIV/AIDS, and the use of immuno-modulatory drugs. Currently, a number of drugs are being developed or repurposed to inhibit various aspects of the JCPyV life cycle. AY4 is an attachment inhibitor. $5-HT_2R$ inhibitors block virion entry into the host cell. A combination of mefloquine and mirtazapine acts as endosome acidification inhibitors. Sorafenib, U0126, and PD98059 inhibit different steps of the MAPK-ERK (mitogen-activated protein kinase/extracellular signal-regulated kinase) cascade. Typically, ERK regulates the expression of proteins that are involved in cell proliferation, differentiation, survival, and stress response. The dihydroquinazolinones Retro-2^{cycl}, Retro-2.1, and DHQZ36 inhibit intracellular trafficking. Thapsigargin, which is an inhibitor of ER-bound calcium channels, blocks uncoating. Verdinexor inhibits virion transport into the nucleus. AMT580-043 is the most potent inhibitor of Tag. Cytarabine potently inhibits JCPyV replication. Cambinol blocks EV association.

MCPyV DNA can be found on 67% to 90% of healthy skin at various parts of the body, including the face, trunk, upper and lower limbs, hand, and forehead. MCPyV infection occurs early in life, typically transmitted from a relative. There are 37%–42% of children under 6 years of age and 40%–88% of young adults up to 21 years old who have been found to have cutaneous infection. For adults, the range of 63%–84% for 21–40 years old and 72%–92% for 40 years and older. In the United States, 25%–42% of adults are MCPyV seropositive compared to 82% from the Netherlands, 69% from Hungary, and 96% from Italy aged 70–79 years. MCPyV appears to be the major inducer of MCC. MCPyV genome is detected in 80% of the MCC samples from the United States, 70%–85% from Europe, and 90% from Japan, whereas, only 20%–30% is detected in males 85 years of age or older in Australia.

Scientific Significance and Discoveries

Polyomaviruses, especially SV40 and MCPyV, have been used as a model for eukaryotic molecular biology and oncogenesis. In fact, BKPyV and JCPyV served as models for studying cellular pathways in earlier research instead of focusing on human diseases. The knowledge of how p53 and Rb proteins function as tumor suppressors largely came from research using polyomaviruses as models, where protein-protein interaction with the T antigens was the central focus. Through these lines of research, it was also discovered that viral DNA was present in some tumors. Knowledge as common as the differential functions of general transcription factors versus tissue-specific transcription factors came out of studies on polyomaviruses.

Further Readings

Ahmad I, Wilson D. HSV-1 cytoplasmic envelopment and egress. *Int J Mol Sci*. 2020;21(17):5969.

Aubert M, Strongin D, Roychoudhury P, et al. Gene editing and elimination of latent herpes simplex virus in vivo. *Nat Commun*. 2020;11:4148.

Barth H, Solis M. In vitro and in vivo models for the study of human polyomavirus infection. *Viruses*. 2016;8(10):292.

Boothpur R, Brennan DC. Human polyoma viruses and disease with emphasis on clinical BK and JC. *J Clin Virol*. (2010;47(4):306–312.

Buck C, Doorslaer K, Peretti A, et al. The ancient evolutionary history of polyomaviruses. *PLoS Pathog*. 2016;12(4):e1005574.

Cohen J. Herpesvirus latency. *J Clin Invest*. 2020;130(7):3361–3369.

Cook L. Polyomaviruses. *ASM J*. 2016;4(4):2016.

De Gascun C, Carr M. Human polyomavirus reactivation: Disease pathogenesis and treatment approaches. *Clin Dev Immunol*. 2013;2013:373579.

Delbue S, Comar M, Ferrante P. Review on the role of the human polyomavirus JC in the development of tumors. *Infect Agent Cancer*. 2017;12(10):2017.

Deschamps T, Kalamvoki M. Extracellular vesicles released by herpes simplex virus-infected cells block virus replication in recipient cells in a STING-dependent manner. *J Virol*. 2018;92(18):1102–1118.

Dremel S, DeLuca N. Genome replication affects transcription factor binding mediating the cascade of herpes simplex virus transcription. *Proc Natl Acad Sci U S A*. 2019;116(9):3734–3739.

Feltkamp M, Kazem S, Meijden E, Lauber C, Gorbalenya A. From Stockholm to Malawi: Recent developments in studying human polyomaviruses. *J Gen Virol*. 2013;94:482–496.

Giannecchini S. Evidence of the mechanism by which polyomaviruses exploit the extracellular vesicle delivery system during infection. *Viruses*. 2020;12(6):585.

Gross L. Neck tumors, or leukemia, developing in adult C3H mice following inoculation, in early infancy, with filtered (Berkefeld N), or centrifugated (144,000 X g), Ak-leukemic extracts. *Cancer*. 1953;6(5):948–958.

Guo H, Shen S, Wang L, Deng H. Role of tegument proteins in herpesvirus assembly and egress. *Protein Cell*. 2010;1(11):987–998.

Houben R, Celikdemir B, Kervarrec T, Schrama D. Merkel cell polyomavirus: Infection, genome, transcripts and its role in development of Merkel cell carcinoma. *Cancers (Basel)*. 2023;15(2):444.

Jiang M, Abend J, Johnson S, Imperiale M. The role of polyomaviruses in human disease. *Virology*. 2009;384(2):266–273.

Kaiserman J, O'Hara B, Haley S, Atwood W. An elusive target: Inhibitors of JC polyomavirus infection and their development as therapeutics for the treatment of progressive multifocal leukoencephalopathy. *Int J Mol Sci*. 2023;24(10):8580.

Madavaraju K, Koganti R, Volety I, Yadavalli T, Shukla D. Herpes simplex virus cell entry mechanisms: An update. *Front Cell Infect Microbiol*. 2021;10:2020.

Mayberry C, Maginnis M. Taking the scenic route: Polyomaviruses utilize multiple pathways to reach the same destination. *Viruses*. 2020;12(10):1168.

Moens U, Calvignac-Spencer S, Lauber C, et al. ICTV virus taxonomy profile: Polyomaviridae. *J Gen Virol*. 2017;98(6):1159–1160.

Moens U, Macdonald B. Effect of the large and small T-antigens of human polyomaviruses on signaling pathways. *Int J Mol Sci*. 2019;20(16):3914.

Moens U, Passerini S, Falquet M, Sveinbjornsson B, Pietropaolo V. Phosphorylation of human polyomavirus large and small T antigens: An ignored research field *Viruses*. 2023;15(11):2235.

Nath P, Kabir M, Doust S, Ray A. Diagnosis of herpes simplex virus: Laboratory and point-of-care techniques. *Infect Dis Rep*. 2021;13(2):518–539.

Packard J, Dembowski J. HSV-1 DNA replication—Coordinated regulation by viral and cellular factors. *Viruses*. 2021;13(10):2015.

Prado J, Monezi T, Amorim A, Lino V, Paldino A, Boccardo E. Human polyomaviruses and cancer: An overview. *Clinics (Sao Paulo)*. 2018;73(Suppl 1):e558s.

Roller R, Johnson D. Herpesvirus nuclear egress across the outer nuclear membrane. *Viruses*. 2021;13(12):2356.

Sadowski L, Upadhyay R, Greeley Z, Margulies B. Current drugs to treat infections with herpes simplex viruses-1 and -2. *Viruses*. 2021;13(7):1228.

Sandri-Goldin R. Replication of the herpes simplex virus genome: Does it really go around in circles? *Proc Natl Acad Sci U S A*. 2003;100(13):7428–7429.

Silling S, Kreuter A, Gambichler T, Meyer T, Stockfleth E, Wieland U. Epidemiology of Merkel cell polyomavirus infection and Merkel cell carcinoma. *Cancers (Basel)*. 2022;14(24):6176.

Suzich J, Cliffe A. Strength in diversity: Understanding the pathways to herpes simplex virus reactivation. *Virology*. 2018;522:81–91.

Tognarelli E, Palomino T, Corrales N, Bueno S, Kalergis A, Gonzalez P. Herpes simplex virus evasion of early host antiviral responses. *Front Cell Infect Microbiol*. 2019;9:2019.

Yang J, You J. Regulation of polyomavirus transcription by viral and cellular factors. *Viruses*. 2020;12(10):1072.

CHAPTER 8

RNA With Reverse Transcriptase

Human Immunodeficiency Virus

Historical Perspective

The current human immunodeficiency virus 1 (HIV-1) group M is believed to have resulted from multiple cross-species transmissions of simian immunodeficiency viruses that naturally infect African primates. The original transmission into the human population may have occurred sometime between the 1920s and 1940s in southeastern Cameroon, a country in the west Central Africa region, through direct contact with chimpanzee blood during hunting expeditions. Transmission of this original zoonotic human virus appears to have occurred throughout Africa and to other parts of the world. SIV-infected chimpanzees were afflicted with acquired immunodeficiency syndrome (AIDS) long before the appearance of HIV. However, it wasn't until June 5, 1981, that the US Centers for Disease Control and prevention (CDC) reported unusually high rates of a rare lung infection called *Pneumocystis carinii* pneumonia and a rare, but aggressive, cancer called Kaposi sarcoma, among young gay men in New York and California. Both of these rare disorders had already been linked to immunosuppressed individuals. In the ensuing days, more patients diagnosed with opportunistic infections were identified around the United States. On June 15, the first recorded person with what is now confirmed as AIDS was admitted to the clinical center at the National Institutes of Health (NIH); the patient died later that year in October.

As early as 1978, many heterosexual males in Tanzania and Haiti as well as gay men in the United States and Sweden started to exhibit signs of a condition with unknown etiology at the time.

In September 1981, the National Cancer Institute and CDC assembled a conference to discuss this latest epidemic. In December, children at the Albert Einstein Medical College in New York were showing signs of the same disorder that affected early patients, who were gay men. By the end of the 1981, there were 334 individuals who were reported with severe immune deficiency; 16 of these individuals were children, and 130 had died by year's end.

In September 1982, the term *AIDS* was used publicly by the CDC.

In May 1983, the realization that a retrovirus was the causative agent of AIDS was reported by Dr. Françoise Barré-Sinoussi, who was later awarded the Nobel Prize in

177

Medicine in 2008 for this discovery. In November 1983, the World Health Organization (WHO) held a 4-day conference to discuss the global AIDS pandemic and initiate international surveillance.

In April 1984, Dr. Robert Gallo and his colleagues reported that the retrovirus they referred to as strain III of the human T-lymphotropic virus (HTLV-III) was the causative agent for AIDS. HTLV-III is now identified as the human immunodeficiency virus (HIV). Research on the development of a vaccine against this condition quickly ensued.

It was reported that, in 1985, more people were diagnosed with AIDS than in any year before that and had increased by 89% compared to 1984. Of the total number of individuals afflicted by this disorder and died from 1981 through 1985, on average 15 months after diagnosis, 51% were adults and 59% were children.

In May 1986, the International Committee on the Taxonomy of Viruses officially designated the name of this virus as human immunnodeficiency virus (HIV). In October, the CDC reported that AIDS affected African Americans and Latinos disproportionately, raising the question of equity in early diagnosis, delivery of treatment, and education on prevention for different populations.

On March 19, 1987, the US Food and Drug Administration (FDA) approved the first antiretroviral drug, zidovudine (AZT), to treat AIDS patients. This was initially developed as an anticancer drug. In August, the first testing of a candidate vaccine against HIV on humans was sanctioned by the FDA.

By 1988, the CDC reported that the total number of AIDS cases was 82,406, with 46,134 deaths. By 1989, these numbers rose to 100,000 reported cases of AIDS and 59,000 AIDS-related deaths in the United States alone. World governments reported that there was a total of 132,977 cases of AIDS by early this 1989. However, WHO officials believed that these statistics are greatly underestimated. They estimated that at least 300,000 to 350,000 individuals had AIDS, while this range may have been as high as 630,000–1,380,000 cases. An estimated 5 to 10 million people were infected with HIV by this time.

In 1990, the FDA approved AZT to be used to treat pediatric AIDS patients.

AIDS became the most common cause of death for men between the age of 25 and 44 in the United States in 1992. And, in 1993, the FDA approved the female condom to provide women the same protection against HIV as male using condoms during sexual activities. The CDC expanded the list of clinical indicators of AIDS to include pulmonary tuberculosis, recurrent pneumonia, and invasive cervical cancer.

By 1994, AIDS surpassed all other diseases and became the leading cause of death among individuals 25 to 44 years of age. In August, the US Public Health Service recommended AZT to be used by pregnant women to reduce the risk of HIV transmission to her fetus. In December, the FDA approved the first oral, non–blood-based antibody test for HIV.

In June 1995, the FDA approved the use of the protease (PR) inhibitor saquinavir (Invirase®) to be included in treatment regimens through a novel therapeutic approach known as highly active antiretroviral therapy (HAART). By October 31, 1995, the CDC reported a total of 501,310 persons with AIDS and 311,381 (62%) deaths in the United States alone. The Joint United Nations Program on HIV/AIDS (UNAIDS) reported that 3.2–4.3 million individuals were newly infected with HIV, which brought the total of cases to 6 million globally for the year. This was the peak of the pandemic through the almost 2 decades the accumulated number of HIV-positive or AIDS cases rose to

22,000,000. The mortality rate had since greatly declined due to HAART and decreasing HIV infection incidence.

In 1996, new drugs that were approved by the FDA include nevirapine (Viramune®), a nonnucleoside reverse transcriptase (RT) inhibitor, and ritonavir (Norvir®) and indinavir (Crixivan®), which are PR inhibitors. Studies by Dr. David Ho discovered that Kaposi sarcoma experienced by AIDS patients is caused by a co-infection with a herpes simplex virus.

By 1997, the estimated number of HIV-positive individuals was increased to 30,000,000 globally. However, AIDS-related deaths in the United States declined by 47% compared to 1996 due largely to treatment with HAART.

As of 2022, of the world population who are HIV infected, 86% knew their status and 89% of those had access to treatment. Among those who gained treatment, 93% were reported to be virus free. Funding for HIV programs remains strong, roughly 58% from the US government; however, the level of funding has been declining.

In 2023, the UNAIDS indicated 36.1–44.6 million people in the world were living with HIV. Of these, 1–1.7 million individuals were newly infected, and 500,000–820,000 people died from AIDS-related illnesses. Since the start of the pandemic in 1978, a total of 71.3–112.8 million people were infected with HIV, with 35.7–51.1 million deaths from AIDS-related illnesses.

Classification and Structure of Human Immunodeficiency Virus

Human immunodeficiency virus is a member of the Retrovidiridae family. This family is further divided into 2 subfamilies: Orthoretrovirinae and Supumaretroviriae. Orthoretrovirinae consists of 6 genera: *Alpharetrovirus* (eg, Rous Sarcoma virus), *Betaretrovirus* (eg, mouse mammary tumor virus), *Deltaretrovirus* (eg, primate T-lymphotropic virus), *Epsilonretrovirus* (eg, walleye dermal sarcoma virus), *Gammaretrovirus* (eg, gibbon ape leukemia virus), and *Lentivirus* (eg, HIV type 1 and type 2). Supumaretroviriae consists of 5 genera: *Bovispumavirus* (eg, bovine foamy virus), *Equispumavirus* (eg, equine foamy virus), *Felispumavirus* (eg, feline foamy virus), *Prosimiispumavirus* (eg, simian foamy virus), and *Simiispumavirus* (eg, simian foamy virus).

Other related families of virus include Metaviridae (eg, *Drosophila melanogaster* gypsy virus) and Pseudoviridae (eg, *Thalassiosira pseudonana* CoDi5.6 virus).

The structure of HIV is an enveloped virion with an overall diameter of 145 nm (Figure 8.1). Two identical molecules of linear, plus-strand RNA genome of 9749 bases are enclosed in an asymmetric cone-shaped capsid (CA) that is 100 nm long and 50 nm wide. The 5′ end of each RNA molecule is attached to an uncharged lysine transfer RNA (tRNA), while the 3′ end is polyadenylated. The tRNA functions as the primer for the RT, which possesses both RNA-directed DNA polymerase activity and DNA-directed DNA polymerase activity. The 2 strands are noncovalently linked and associated with a number of ribonucleoproteins (RNPs), including the nucleocapsid (NC) protein MA (matrix), p6, RT, Vpr (virus protein r), PR, and IN (integrase). The presence of the P7 protein in the CA protects the viral genome from digestion by exonucleases.

The HIV genome possesses 3 standard open reading frames (ORFs; regions used as the template for trancription) found in all species of *Lentivirus* and are designated Gag (group-specific antigen), Pol (polymerase), and Env (envelope) (Figure 8.2). The Gag encodes the structural proteins' outer core matrix protein (MA or p17), capsid protein (CA or p24), nucleocapsid protein (NC or p7), nucleic acid-stabilizing late

HIV virion structure and genome

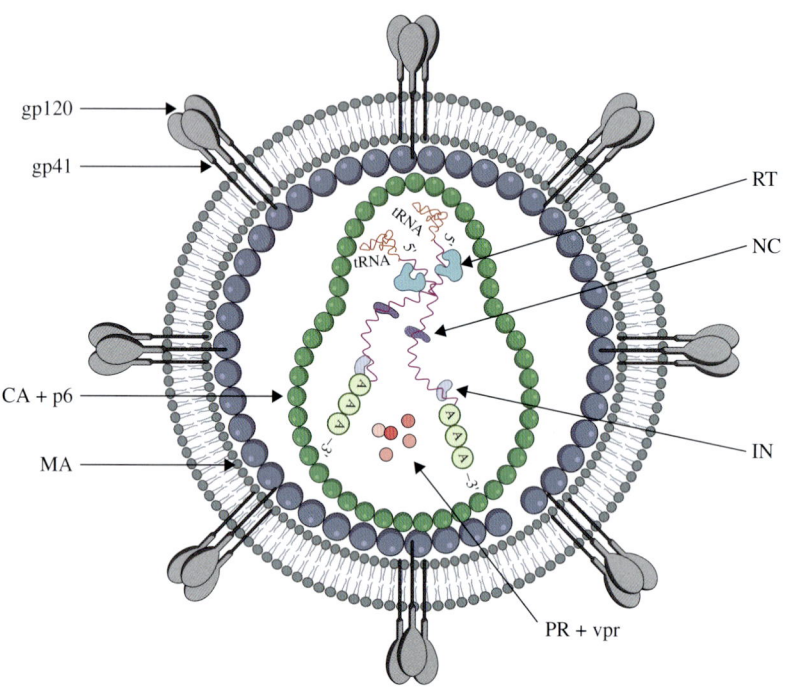

FIGURE 8.1 HIV virion structure. The 2 identical molecules of HIV RNA genome are encased in a conical CA. Each RNA molecule is bound by an RT, NC protein, and IN. Vpr and PRs are also packed inside the CA. This structure is surrounded by an enveloped integrated with gp41 and gp120. The viral envelope is lined with a layer of outer core membrane proteins on the internal face. *Credit:* Figure partially created in BioRender.

assembly protein (LI or p6), spacer peptide 1 (p1), and p2. The Pol gene encodes areverse transcriptase (RT or p51), ribonuclease H (RNase H or p15), protease (PR or p12), and integrase (IN or p32). RT and RNase H interact with each other to form a complex termed p66. The Env gene encodes the envelope glycoprotein gp160, which ultimately is cleaved into surface envelope glycoprotein gp120 (or SU) and transmembrane protein gp41 (or TM) during the maturation process carried out by viral particles that are released from the host cell. A 5′-LTR and 3′-LTR are also found at the end of each RNA molecule.

The HIV genome contains 6 additional ORFs encoding regulatory proteins, which is more complex than other lentiviral genomes. Regulatory proteins transactivator protein (Tat) and RNA splicing-regulator (Rev) are necessary for the initiation of HIV replication. Accessory proteinsnegative regulating factor (Nef), viral infectivity factor (Vif), Vpr, and virus protein unique (Vpu; exclusively in HIV-1) or virus protein x (Vpx; exclusively in HIV-2) have important functions during viral replication, virus budding, and pathogenesis.

Once in the host cell, unlike poliovirus, which also possesses a plus-strand RNA genome, HIV plus-strand RNA genome cannot be used immediately as the template for translation. The viral genome must be imported into the nucleoplasm, where the RT initially uses 1 of the RNA molecules as the template to synthesize a double-stranded DNA (dsDNA) version called a proviral DNA. The proviral DNA must be integrated

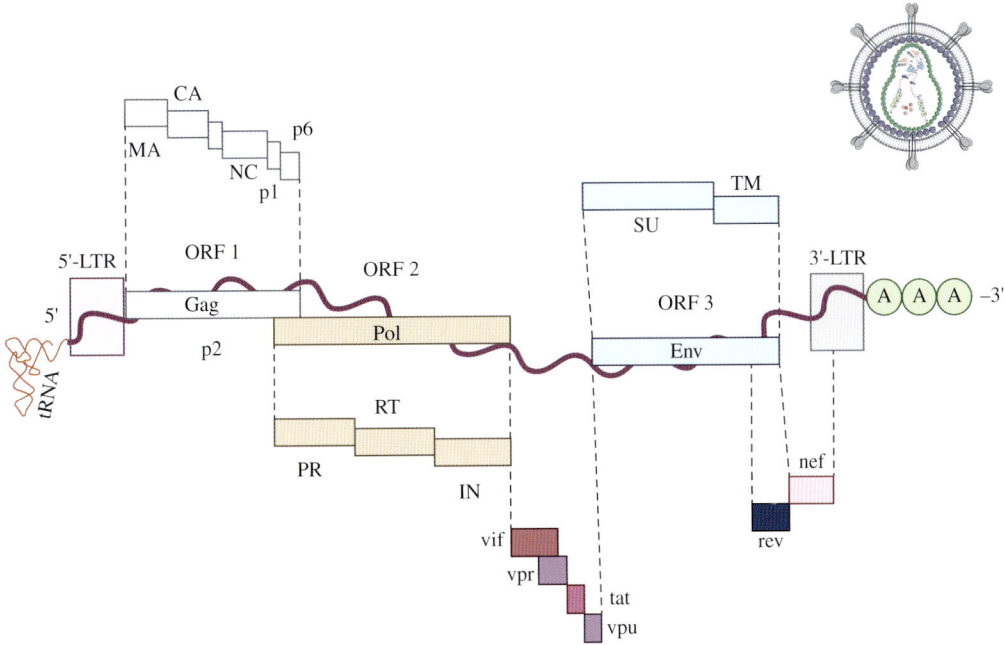

FIGURE 8.2   HIV ORFs and gene products. The HIV genome contains the 3 standard ORFs found in all species of lentiviruses: Gag, Pol, and Env. Each of these ORFs encodes for various viral proteins. HIV contains 6 additional ORFs that encode for regulatory proteins: Vif, Vpr, Tat, Vpu, Rev, and Nef. *Credit:* Figure partially created in BioRender.

into the host chromosome, aided by the IN, before viral messenger RNA (mRNA) synthesis can occur. The proviral DNA is integrated at relatively random sites in the cellular chromosome, although there seems to be certain hot spots. Once integrated into the cellular chromosome, the proviral DNA is termed a provirus.

Host Range, Tropism, and Susceptible Host Cell

Humans are the only natural host for HIV-1 and HIV-2. Susceptible host cells exhibit cluster of differentiation 4-positive (CD4$^+$) receptor on their surface to regulate virion attachment. CD4 (cluster of differentiation 4) is a glycoprotein found on the surface of T helper cells, monocytes, macrophages, and dendritic cells whose function is to coordinate immune responses. However, an additional co-receptor must also be present whose interaction with HIV gp120 is required for membrane fusion and entry into the host cell.

HIV tropism is determined by the presence of CD4 receptor and either C-C chemokine receptor type 5 (CCR5 or R5) or C-X-C chemokine receptor type 4 (CXCR4 or X4). CD4 is a glycoprotein found on the surface of T lymphocytes, monocytes, macrophages, and dendritic cells. CD4 performs important functions during immune responses. Susceptible CD4$^+$ hematopoietic cells include not only predominantly CD4 T lymphocytes but also monocytes, macrophages, and dendritic cells (DCs); however, only CD4 T lymphocyte and macrophages are permissive to HIV. Even though DCs exhibit the surface CD4 receptor, they do not exhibit an identifiable co-receptor.

Moreover, HIV can also interact with DC-SIGN (dendritic cell-specific ICAM-3-grabbing non-integrin) receptor on the surface of the DCs and macrophages; however,

DCs do not exhibit a co-receptor. DC-SIGN is a lectin receptor that helps capture, destroy, and present pathogens to the immune system. The CD4 T lymphocyte is a CXCR4 tropism, and the monocyte and macrophage are a CCR5 tropism. DCs can deliver the viral particle attached to its surface into the lymph node and expose other uninfected CD^{4+} cells to the HIV.

Transmission of HIV is via direct contact with certain contaminated body fluids from an HIV-infected individual. These fluids include saliva, blood, plasma or serum, semen and preseminal fluid, rectal fluids, vaginal fluids, and breast milk. Transmission can also occur through contaminated transplanted organs such as kidney, bone, bone marrow, and cornea.

The mode of transmission can be through vaginal or anal sex, through sharing contaminated needles used for injecting drugs or tattooing, contacting the mucous membrane with fomites like shared sex toys, bite injuries, transplacental passage, and breastfeeding. HIV cannot be transmitted by merely coming into contact with contaminated fluids. The virus must end up in the bloodstream via the mucous membrane of the oral or urogenital organs, through open cuts or sores, or by direct injection causing passive viremia.

Infecting the Host Susceptible Cell, Viral Particle Replication, and Tissue Damage

Attachment occurs when HIV envelope protein gp120 interacts with the CD4 on the surface of the susceptible host cell (Figure 8.3). This interaction causes gp120 to undergo a conformation change, thus enabling it now to interact with a nearby co-receptor, which in turn exposes gp41 to bind to CD4 to mediate membrane fusion between the viral envelope and plasma membrane, thus also creating a membrane pore. Membrane fusion triggers partial uncoating of the conical CA, allowing the RNP complex to be released into the host cell cytosol.

The RNP complex (which includes the viral genome, RT, MA, p6, IN, Vpr, and PR) is, in turn, imported into the nucleoplasm. MA contains a nuclear import signal, which is recognized by the cellular importin molecule. The partially uncoated NC may also be imported into the nucleus before the release of the RNP complex directly into the nucleoplasm. In either case, the very first viral activity in the nucleus is the synthesis of the proviral DNA, which is the dsDNA produced by RT using 1 of the original viral RNA genomic molecules as the template. Proviral DNA synthesis occurs in the nucleus, not in the cytosol, as previously thought.

RT is an RNA-directed RNA polymerase that regulates DNA synthesis using the plus-strand RNA genome as the template to produce a minus-strand complementary DNA (cDNA). The uncharged, lysine tRNA functions as the primer for RT activity. The original plus-strand RNA temporarily forms a hybrid with the minus-strand cDNA. RT is complexed with p15, which possesses RNase H activity that degrades the original RNA strand of the newly formed RNA-DNA hybrid. The minus-sense, single-stranded cDNA is in turn used as the template for plus-strand DNA synthesis, forming a dsDNA molecule. The resulting dsDNA product is referred to as a proviral DNA.

This proviral DNA interacts with other viral proteins, which include MA, Vpr, and IN, to form the pre-integration complex (PIC). The proviral DNA is flanked with LTR sequences at both ends, designated 5' LTR for the upstream region and 3' LTR for the downstream region. The 5' LTR contains enhancer promoter sequences for viral transcription, while the 3' LTR encodes a polyadenylation signal. At this point, pre-integration

Early phase of HIV infection

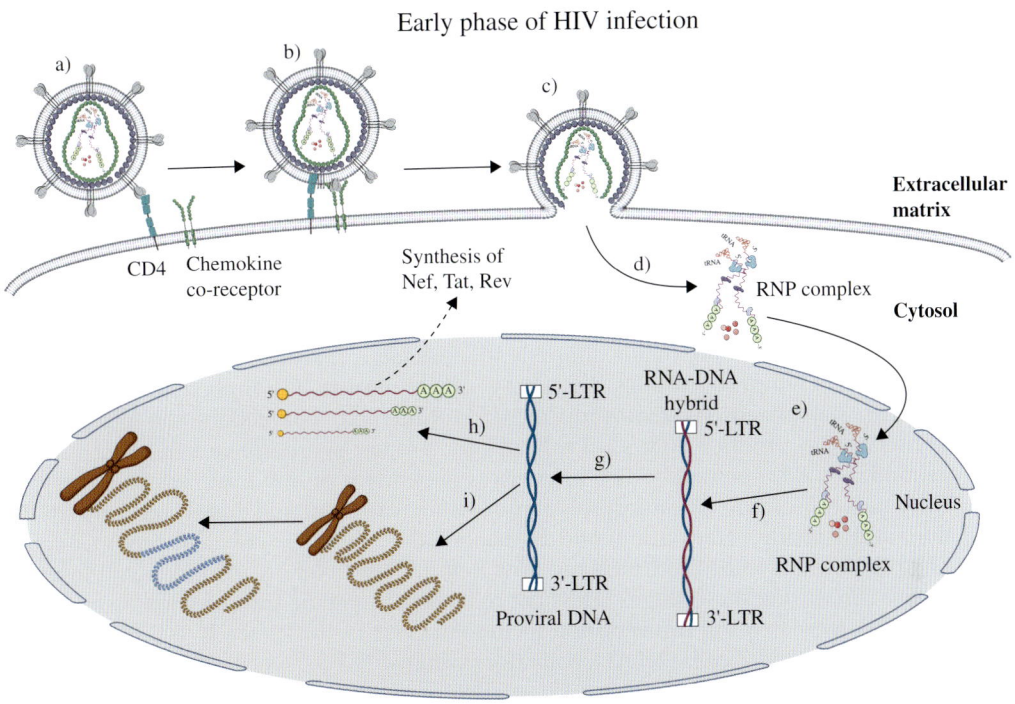

FIGURE 8.3   HIV entry into host cell and early phase of infection. (a) HIV attaches to CD4 on the host cell surface. (b) Conformational change allows gp120 to interact with the chemokine co-receptor, while gp41 interacts with CD4. (c) Gp41 mediates membrane fusion and triggers partial uncoating of the conical CA. (d) The RNP complex is released into the host cytosol. (e) RNP complex is imported into the nucleoplasm. (f) RT regulates minus-strand cDNA synthesis, producing an RNA-DNA hybrid. (g) RT regulates plus-strand DNA synthesis to produce a proviral DNA. (h) Proviral DNA is used for pre-integration expression of Nef, Tat, and Rev. (i) Proviral DNA is inserted into the host chromosome. *Credit:* Figure partially created in BioRender.

transcription occurs, which is a process where the unintegrated proviral DNA is used by host machinery to express a small amount of the regulatory proteins Rev, Tat, and Nef. These proteins are required for subsequent stages of viral replication.

Once pre-integration transcription is accomplished, the proviral DNA is integrated into the cellular chromosome, resulting in a provirus. Integration is regulated by viral IN along with cellular factors, such as LEDGF/p75, where the 5' LTR and 3' LTR are ligated to the host DNA. Most sites in the chromosomal DNA can serve as integration acceptor sites; however, integration usually occurs in chromosomal regions that are actively transcribed or sites with chromatin-associated factors that can interact with IN. The integrated proviral DNA is referred to as a provirus. At this point, HIV infection is divided into early phase and late phase of the viral cycle.

The early phase includes processes that were initiated with attachment, then entry, uncoating, proviral DNA synthesis, and finally insertion of the proviral DNA into the host chromosome referred to as a provirus. This is where we are right now in this chapter. Late phase can either be active infection or latent infection.

Active infection of the late phase includes processes starting with viral mRNA and protein synthesis, then virion assembly, and release of virion by budding. Latent infection is directed by the interaction between Rev and Tat. Tat is required for viral gene transcription so that the virus can continue with active infection. When bound to Tat, Rev regulates the delivery of cytoplasmic Tat to the proteasome for degradation. When the concentration of Tat falls below a critical threshold, the virus enters latent infection.

During the latent infection of the late phase, the proviral DNA remains inactive or dormant; that is, no viral RNA or protein synthesis in the cell until reactivation occurs later. Once integrated, the provirus is replicated along with cellular DNA during cycles of cell division, as with any cellular gene. Reactivation enables the virus to reenter active infection.

Active infection requires the host cell's transcription machinery and can only occur after host cell activation, that is, the activation of naïve, mature T lymphocytes or macrophages in lymphoid tissues. In resting naïve, mature T lymphocytes (or macrophages), the provirus remains dormant. During active infection, the provirus is transcribed to produce both viral mRNAs for protein synthesis and full-length viral genomic RNA for assembly. The provirus is dependent on the host cell machinery to carry out transcription; transcriptional processing (capping, polyadenylation, methylation, and splicing); translation; and translational processing (covalent modification). The previously synthesized viral proteins Rev, Tat, and Nef are also involved in viral replication during these stages. The host cell's tissue-specific transcription factors (TS-TFs) that regulate viral transcription include nuclear factor kappa B (NF-κB), nuclear factor of activated T cells (NFAT), activator protein-1 (AP1), specificity protein 1 (Sp1), lymphoid enhancer binding factor 1 (Lef), E26 transformation -specific (Ets), upstream stimulatory factor (USF). NFAT is involved in T lymphocyte activation and differentiation as well as self-tolerance. AP1 plays a role in cell proliferation, differentiation, migration and apoptosis. Sp1 also plays an important role in cell proliferation and differentiation and apoptosis. Lef induces gene expression of proteins that are involved in embryonic development and stem cell function. Ets is plays a role in cell proliferation and differentiation and development of immune cells. Lastly, USF regulates gene expression in response to stress and immune responses. Note that these are these same TS-TFs that are upregulated as a result of the activation of CD4$^+$ cells.

After transcription, the primary viral RNA transcript undergoes alternative RNA splicing to produce 3 different mRNAs covering 3 different reading frames: doubly spliced viral mRNA (2 kb), single spliced viral mRNA (4 kb), and unspliced, full-length viral mRNA (9 kb). Each mRNA is capped and polyadenylated. The 2-kb mRNA is produced first, then exported into the cytosol for translation, resulting in the production of Tat, Rev, and Nef. Tat and Rev are then imported back into the nucleus (Figure 8.4).

In the nucleus, Tat binds to the Tat-response element sequence in the 5' LTR on both the 9-kb and 4-kb mRNAs to increase the rate of expression of both molecules. Rev binds to RRE (Rev-response element) sequence on both 9-kb and 4-kb mRNAs to direct their export into the cytosol. Note that the RRE is not required for 2-kb RNA export.

The 4-kb mRNA encodes for gp160, which is later cleaved by the host PR, furin, found in the endoplasmic reticulum (ER) to form gp120 and gp41. Gp160 is co-translated with cellular CD4, and both proteins are inserted into the ER membrane. Here, CD4 interacts with gp160 as this complex is delivered to the plasma membrane via vesicular transport. This interaction with CD4 blocks gp160 cleavage by furin, which is required for virion maturation later and the production of gp120 and gp41 as separate subunits. Posttranslational modification converts gp160 into a glycoprotein.

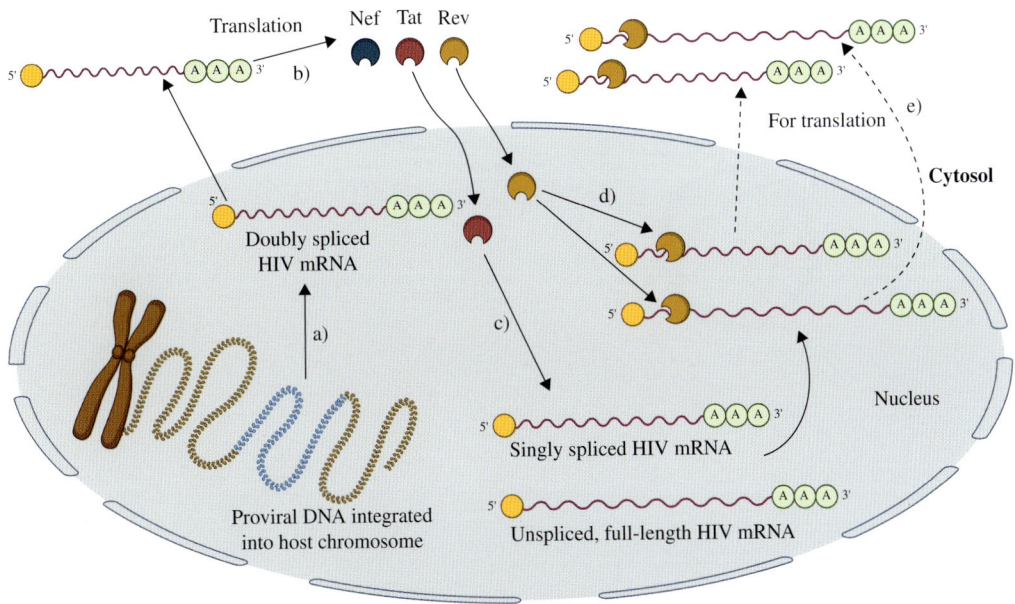

Active HIV infection

Figure 8.4 Expression of Rev, Tat, and Nef and their functions. (a) Doubly spliced HIV mRNAs are transcribed in the nucleoplasm. (b) Doubly spliced mRNAs are exported to the cytosol for translation of Nef, Tat, and Rev. (c) Tat is imported back into the nucleoplasm to increase transcription of singly spliced and unspliced mRNAs. (d) Rev is imported back into the nucleoplasm to direct the export of singly spliced and unspliced mRNAs into the cytosol. (e) Viral protein synthesis leading to virion assembly. *Credit:* Figure partially created in BioRender.

The 9-kb mRNA is synthesized last and encodes for Gag polyprotein and Gag-Pol polyprotein. The Gag polyprotein is cleaved by PR to form the structural proteins MA, CA, and NC. The cleavage of the Gag polyprotein occurs during the maturation process, after budding. The Gag-Pol polyprotein is also cleaved by PR to produce multiple viral enzymes, including RT and RNase H, IN, and PR itself.

The Gag polyprotein, Gag-Pol polyprotein, and the full-length, unspliced transcript form a complex that regulates the linkage of an uncharged tRNA to the 5′ end of the viral genomic RNA. An uncharged lysine tRNA is used for HIV assembly, while other types of uncharged tRNA are used for the assembly of other retroviruses. The viral proteins and RNA genome are assembled into virions at the host plasma membrane (Figure 8.5). Viral proteins cleaved from Gag and Gag-Pol polyproteins bind to the viral RNA genome to form the NC core.

Release of virions is via budding into the extracellular MA. The viral envelope is acquired from the host cell plasma membrane; therefore, it possesses a number of host markers such as major histocompatibility complex (MHC) class I and MHC class II. At this point, the virion is an immature viral particle that is avirulent. Proteolytic cleavage of polyproteins and a maturation process mediated by PR must occur in the immature virion to form a mature, virulent HIV virus.

Assembly and viral particle maturation

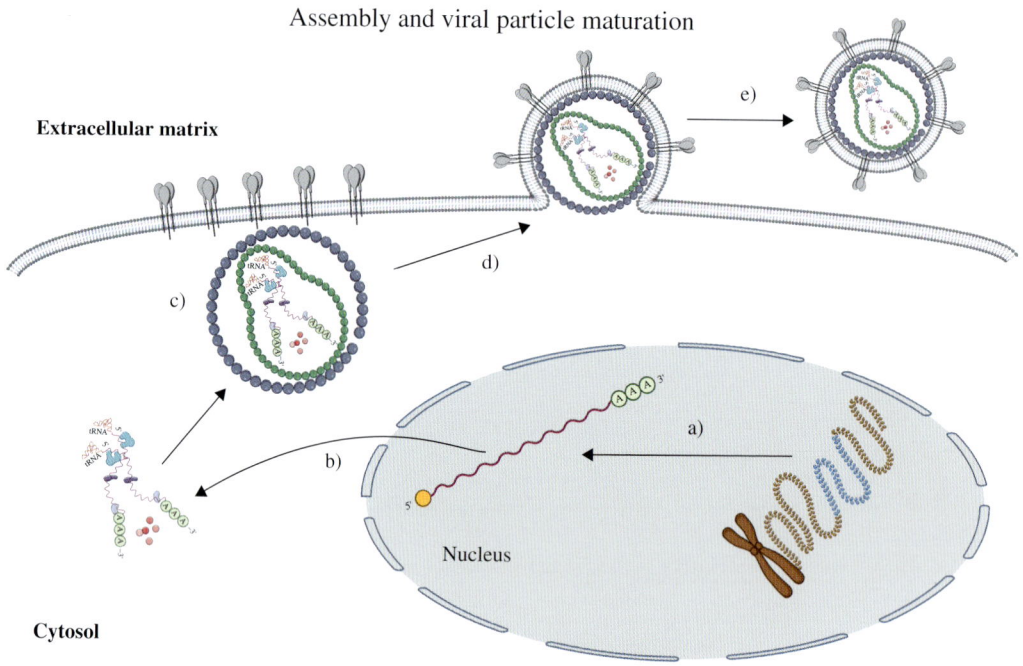

Figure 8.5 Virion assembly and release from host cell. (a) HSV genome replication and gene expression. (b) Viral mRNAs are exported into the cytosol for protein synthesis and virion assembly. (c) Virion assembly occurs at the cytosolic side of the host plasma membrane. (d) Virion exit the host cell via budding. (e) Viral particle maturation occurs when gp160 is proteolytically cleaved into gp120 and gp41. *Credit:* Figure partially created in BioRender.

Host cell infection by HIV and the rate of virus replication are quite slow in contrast to poliovirus (another plus-strand RNA virus), where up to 1 million polioviral particles can be produced within 4 hours of host cell infection. Attachment of HIV to the susceptible cell alone takes up to 2 hours to complete. The synthesis of the proviral DNA isn't completed until roughly 6 hours later, and integration into the cellular chromosome takes an additional 6 hours. The first virion is released 24 hours after infection. Suffice it to say, HIV replication is quite a complex and laborious process.

To increase HIV replication and improve their survival, several accessory proteins are needed to decrease host immune responses and increase the efficiency of viral genome replication. These proteins are Nef, Vpr, Vif, and Vpu/Vpx and are only expressed in HIV-1 and HIV-2, not found in other lentiviruses.

Nef (negative factor) plays important roles in activating host cell apoptosis and increasing virus infectivity (Figure 8.6). Remember that CD4 blocks the digestion of gp160 into gp120 and gp41, thus reducing the level of plasma membrane–bound CD4 necessary to allow gp160 digestion. Nef binds to CD4 to induce clathrin-dependent endocytosis, which in turn causes a reduction in the number of surface CD4 molecules. Nef also reduces the delivery of surface MHC class I to its final destination by preventing MHC class I trafficking from the Golgi apparatus to the host cell plasma membrane. By exhibiting a reduced number of MHC class I on the host cell surface, cytotoxic T lymphocytes (CTLs) are less likely to be able to identify and destroy the HIV-infected cell.

Role of Nef and Vpu on HIV infectivity

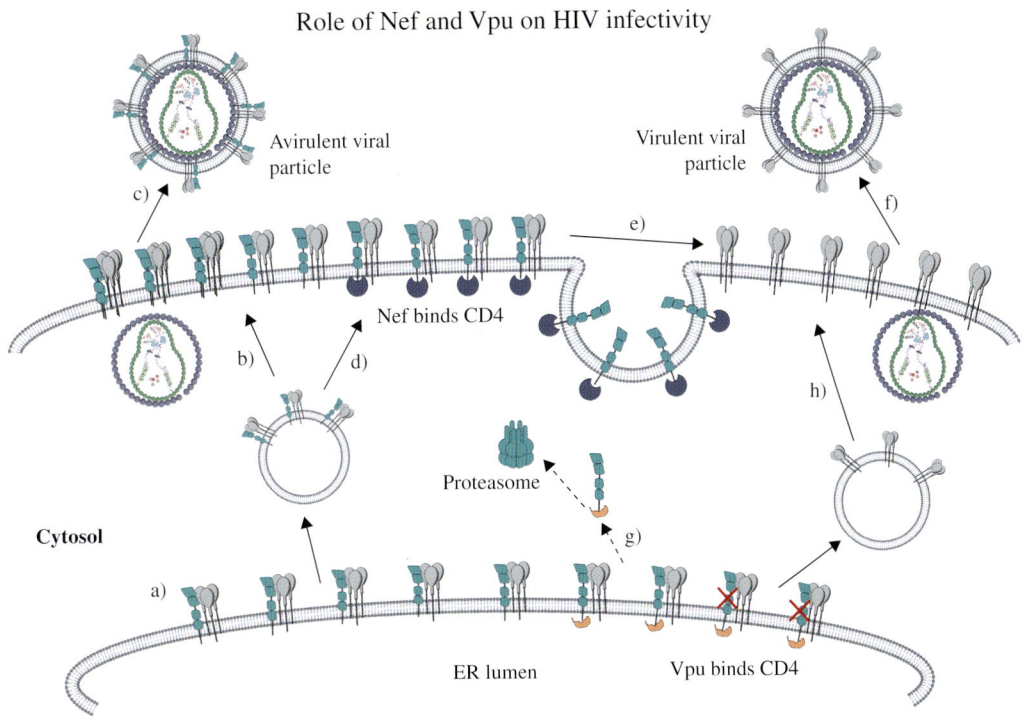

FIGURE 8.6   Role of Nef and Vpu for increasing HIV infectivity. (a) Both gp160 and CD4 are embedded into the ER membrane, where CD4 binds to gp160. (b) This complex is delivered to the plasma membrane via vesicular transport. (c) The viral particle released at this point is avirulent since gp160 cannot be cleaved. (d) The same complex can also be delivered to the plasma membrane where CD4 is bound by Nef. (e) Nef induces the endocytosis of CD4, leaving gp160 unbound. (f) The viral particle released is virulent since gp160 can now be cleaved into gp120 and gp41. (g) At the ER membrane, CD4 can also be bound by Vpu to deliver CD4 to the proteasome for degradation. (h) Gp160, unbound by CD4, is delivered to the plasma membrane and used to form virulent viral particles. *Credit:* Figure partially created in BioRender.

However, these cells may still be identified by natural killer (NK) cells for destruction. NK cells search for cells that lack surface MHC class I complexes to destroy them. Cells that lack MHC class I complexes are referred to as "missing self" cells. Last, Nef can also activate the T lymphocyte that is infected by HIV. Nef-activated T lymphocytes cannot mount an immune response.

Vpu (virus protein u) is specific to HIV-1, and Vpx is specific to HIV-2. These proteins bind to CD4 molecules that are inserted into the ER membrane and target them for proteasomal degradation in the cytosol. With no or a low amount of CD4 available, gp160 can now be cleaved by PR in the immature HIV virion when it is not bound by CD4.

Vpr (lentivirus protein r) plays an important role in the nuclear import of the PIC. Vpr can also induce the infected host cell to undergo G2/M checkpoint arrest by activating the ataxia telangiectasia-mutated (ATM) and Rad3-related (ATR) proteins (Figure 8.7). Normally, ATM and ATR are activated when the cell detects a threshold level of DNA damages. During cell cycle checkpoint arrest, host DNA repair activities increase. Some of these DNA repair mechanisms involve single-stranded DNA (ssDNA) and dsDNA breakage, which results in an increase in the efficiency of proviral DNA integration into the host chromosome. Inducing checkpoint arrest also enables sufficient amount of viral

Role of Vpr in HIV replication

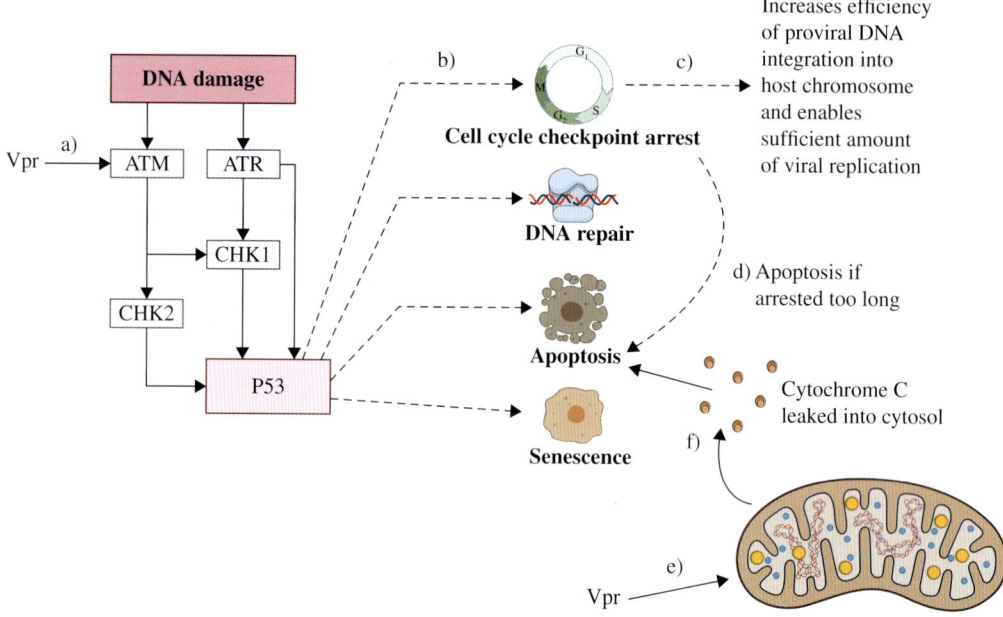

FIGURE 8.7   Role of Vpr for increasing HIV replication. (a) Vpr can activate ATM and ATR to induce the activation of p53. (b) P53 induces the cell to enter cell cycle checkpoint arrest. (c) Checkpoint arrest provides the time and materials for increased efficiency of proviral DNA integration into the host chromosome and a sufficient amount of viral replication. (d) The apoptotic pathway would be triggered if the cell remains in checkpoint arrest for too long. (e) Vpr can also cause leakage of cytochrome C from mitochondria into the cytosol. (f) Cytosolic cytochrome C activates the apoptotic pathway. *Credit:* Figure partially created in BioRender.

replication to occur before the cell completes its cell cycle and divides. The apoptotic pathway can be activated if the cell is halted in G2/M checkpoint arrest for too long. Additionally, Vpr can even trigger apoptosis directly by causing leakage of cytochrome C from mitochondria. The eventual death of the host cell releases all the virions that were produced, ready to disseminate throughout the body. Since HIV likely infect cells of the vascular system and lymphoid tissues, dissemination is easy and fast. Last, Vpr binds to uracil DNA glycosylase, which is a DNA repair enzyme that initiate the base excision repair pathway to prevent mutagenesis. This activity decreases uracil-DNA glycosylase (UNG) activity but increases viral genome uracilation, thereby increasing viral genome synthesis.

Vif (viral infectivity factor) plays an important role in the infectivity of HIV viral particles by preventing cellular proteins from entering the virion during the budding process. Otherwise, these cellular proteins might target viral proteins for proteasomal degradation.

Dissemination in the Host Body

Primary infection of HIV occurs mostly in lymphoid tissues and/or organs and the bloodstream. The replication and release of viral particles into circulation quickly disseminate into lymph nodes that are located throughout the body. DC activity can also exacerbate the problem (Figure 8.8). Even though DCs are susceptible but not permissive to HIV infection, they can still deliver the virus on their surface into lymph nodes. Once in the lymph node, the HIV-bound DC can present the virus to a mature T lymphocyte by forming an infectious synapse, similar to the immunologic synapse formed between MHC on the presenting cell and T-cell receptor (TCR) on the T lymphocyte to be activated.

Swollen lymph nodes can be a sign of either the acute or chronic stage of HIV infection. Infection usually ends in the death of the host cell. The majority of these cells are CD4 T lymphocytes in the lymphoid tissue/organs. To make up for the loss of this cell type, the body overcompensates by increasing the production of total T lymphocytes, thus causing some lymph nodes to swell. Unfortunately, in addition to the destruction of some lymph nodes, most of these replenishing lymphocytes are not able to be activated and differentiate into CD4 T lymphocytes before further tissue destruction, causing immunosuppression of the adaptive immune system, leading to susceptibility to opportunistic infections.

HIV can remain persistent throughout many organs and tissues in the body, including the central nervous system; lymphoid tissues (spleen, thymus, lymph nodes,

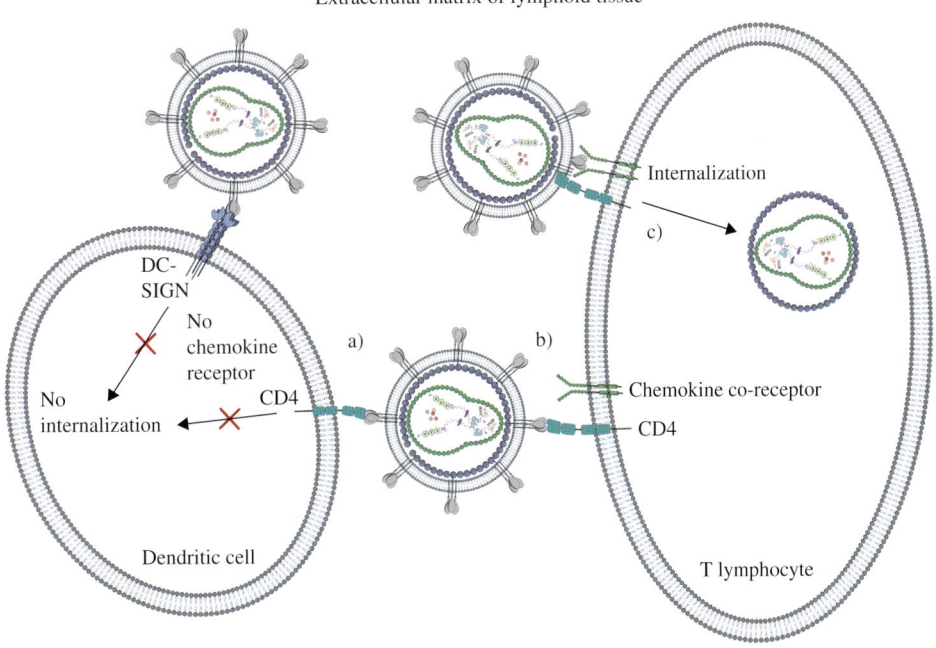

Dendritic cell delivers viral particle to T lymphocyte

Figure 8.8 Dendritic cell presents viral particle to T lymphocyte in lymphoid tissue. (a) HIV can attach to CD4 or DC-SIGN on the surface of a DC; however, internalization cannot occur for infection in the absence of a chemokine receptor. (b) DC can still present the viral particle attached to its surface to T lymphocytes in lymphoid tissue. (c) The interaction between gp41 of the virion and CD4 of the T lymphocyte triggers entry of HIV NC into the host cell cytosol. *Credit:* Figure partially created in BioRender.

mucosa-associated lymphoid tissue); bone marrow; lungs; kidneys; liver; adipose tissue; gastrointestinal tract; and urogenital system. Such persistent replication of competent HIV in these sites is the main barrier for treatment with potent antiretroviral therapy (ART) through HAART.

Pathogenesis and Clinical Manifestation

There are 3 clinical phases of HIV infection: (1) early phase, (2) middle phase, and (3) final phase. The early phase corresponds to the CDC classification group I. This is an acute primary infection where symptoms include fever, headache, fatigue, malaise, rash with small raised lesions, weight loss, pharyngitis, and gastrointestinal symptoms. These symptoms are nonspecific and can be caused by a number of different types of viral infections; therefore, HIV infection is sometime not suspected except in regions of the world that are known to have high incidences. The middle phase corresponds to the CDC classifications group II and group III. Group II is an asymptomatic infection. Group III is a persistent or chronic infection that usually leads to lymphadenopathy, which is exhibited by swollen or enlarged lymph nodes due to the infection itself, autoimmune disease, or malignancy. The final phase occurs when an HIV-infected individual has developed life-threatening conditions causing a condition called AIDS. The CDC classification of the final phase is further divided into 5 subgroups (A–E).

Subgroup A includes patients with constitutional disease. Most of these patients report fever, night sweats, myalgia (or muscle ache and pain), fatigue, anorexia (sometimes with nausea and vomiting), and weight loss. Subgroup B includes patients with neurologic disease, such as dementia, neuropathy, neuromuscular disorder, and neurosyphilis. Subgroup C includes patients with secondary infection of opportunistic pathogens such as cytomegalovirus, *Mycobacterium tuberculosis*, JCPyV (John Cunningham polyomavirus), bacterial and viral pneumonia (which are most common in the United States), oral thrush (or yeast infection developed inside the mouth), herpes simplex infection, and tuberculosis (which is most common in some parts of the world). Subgroup D includes patients with secondary neoplasms such as Kaposi sarcoma, non-Hodgkin lymphoma, and cervical carcinoma. Subgroup E includes patients with other conditions that are not listed above.

Diagnosis, Treatment, and Prevention

The following are recommendations and information about diagnosis, treatment, and prevention posted by the CDC, Mayo Clinic, and WHO. Various tests can be performed on blood or saliva samples to detect and diagnose an HIV infection. There are 3 reliable tests: antigen-antibody test, antibody test, and nucleic acid test. The antigen-antibody test detects the CA (or p24) in blood, usually taken from a vein. This test can only be performed at least 18 days after exposure and up to 90 days for more accurate results. The antibody test detects the presence of antibodies against HIV in blood or saliva. This is the most rapid HIV test. This test can be performed 23 to 90 days after exposure. The nucleic acid test detects the viral load in blood from a vein. This test can be performed 10 to 33 days after exposure and is usually only used when one has a high-risk exposure.

Additional tests for infections and/or complications are performed for individuals from known HIV-positive high-risk populations to reduce the possibility of the development of AIDS. The tests that are performed can be ones that detect tuberculosis, hepatitis B or C infection, cytomegalovirus, sexually transmitted infections (STIs), liver or kidney damage, urinary tract infection, cervical and anal cancer, and toxoplasmosis (caused by the parasite *Toxoplasma gondii*).

Even more tests may help identify or predict the phase and/or classification of HIV infection. The measurement of the patient's CD4 T-lymphocyte count is important since a cell count below 200 is diagnosed as HIV-induced AIDS. Results from testing for HIV RNA to determine the viral load can be used to predict one's chance of acquiring opportunistic infection and other HIV-related complications. A very low level of HIV RNA indicated as an "undetectable" viral load is, of course, preferred. Different HIV strains are resistant to different medications. Testing for medication resistance can possibly help determine the HIV strain that infected the patient.

Antiretroviral therapy is the method of treatment of choice for HIV/AIDS. This treatment requires daily intake of a combination of medicines. Unfortunately, there is no cure yet for HIV infection; however, this treatment can control the infection and help patients live better lives. ART aims at reducing the amount of HIV in the body as well as reducing the risk of transmitting the virus to others.

Classes of anti-HIV drugs used for ART include

- Nonnucleoside reverse transcriptase inhibitors inhibit protein necessary for HIV replication. This class of drugs includes efavirenz, rilpivirine (Edurant®), and doravirine (Pifeltro®).

- Nucleoside or nucleotide reverse transcriptase inhibitors are competitive inhibitors of substrates used for HIV replication. This class of drugs include abacavir (Ziagen®), tenofovir disoproxil fumarate (Viread®), emtricitabine (Emtriva®), and lamivudine (Epivir®). AZT (Retrovir) is used worldwide but not in the United States. These drugs are also available as a mix of the abovementioned drugs, such as emtricitabine–tenofovir disoproxil fumarate (Truvada®) and emtricitabine-tenofovir alafenamide fumarate (Descovy®).

- Protease inhibitors inhibit the activity of HIV PR, thus preventing virion maturation. This class of drugs includes atazanavir (Reyataz®), darunavir (Prezista®), and lopinavir-ritonavir (Kaletra®).

- Integrase inhibitors inhibit the activity of HIV IN, thus preventing the integration of the proviral DNA into host chromosome. This class of drugs includes bictegravir sodium–emtricitabine–tenofovir alafenamide fumarate (Biktarvy®), raltegravir (Isentress®), dolutegravir (Tivicay®), and cabotegravir (Vocabria®).

- Entry or fusion inhibitors block the entry of HIV into host cells. This class of drugs include enfuvirtide (Fuzeon®), maraviroc (Selzentry®), ibalizumab-uiyk (Trogarzo®), and fostemsavir (Rukobia®).

There are a number of treatment side effects from ART, such as nausea, vomiting or diarrhea, heart disease, kidney and liver damage, weakened bones or bone loss, abnormal cholesterol levels, higher blood sugar, and problems with thinking, emotions, and sleep.

Neither a cure nor an effective vaccine has been developed to date; however, HIV infection is very preventable. Preventive measures include both a behavioral adjustment and the use of medication. Behaviors that can help reduce the risk of HIV infection include using a male or female condom during sexual activity, frequent testing for HIV and STIs, abandoning the use of injected drugs or at least stop sharing needles, and even to the extreme measure of voluntary medical male circumcision. ART may be prescribed by physicians to prevent HIV infection.

Last, some medical devices can also be prescribed, such as dapivirine vaginal rings and injectable long-acting cabotegravir. There are medications to protect individuals who have been exposed or are at high risk for exposure such as pre-exposure prophylaxis and post-exposure prophylaxis. Finally, and very importantly, ART can be used to prevent transplacental transmission from mother to child.

Current Status

In 2022, it was estimated that there were between 33.1 and 45.7 million people who were infected with HIV in the world, either HIV positive or had AIDS; 1.2 million–2.1 million were children up to the age of 14 years, and 31.8 million–43.6 million were 15 years or older. Of those, 1.3 million were new reported cases in 2022, and between 480,000 and 880,000 individuals died from AIDS-related illnesses in this year.

From the start of the pandemic in the early 1980s until 2022, approximately 85.6 million cases of HIV infection had been reported, with 40.4 million dying from AIDS-related illnesses.

By the end of 2022, roughly 76% of all people living with HIV had access to ART. Of those people, 57% were children 0 to 14 years old and 77% were aged 15 years and older. It is a travesty to realize that roughly 9 million people are left without access to antiretroviral treatment.

Scientific Significance and Discoveries

Dr. Katherine D. McReynolds and her colleagues from the California State University, Sacramento, explored ways to reduce the infection rates of HIV-1 and SARS-CoV-2 (severe acute respiratory syndrome coronavirus 2) specifically and, more broadly, to control pandemics that cause damage to similar human tissues. Their target was the heparan sulfate proteoglycans (HSPGs) that are ubiquitously expressed on the surface of human cells. HSPGs are especially prevalent on epithelial cells that line the urogenital tract and respiratory system and serve as the susceptible host cell receptor for many viruses. Both glycoprotein complex gp41/gp120 of HIV-1 and spike protein S of SARS-CoV-2 attach to susceptible cells by interacting with HSPGs.

These researchers synthesized a series of hexavalent sulfoglycodendrimers (SGDs) that can mimic host cell HSPGs. They reported that these SGDs have the potential to act as broad-spectrum antiviral agents to be used as early interventions for patients with known exposure to a relevant virus, like HIV or SARS-CoV-2. SGDs may be effective in reducing symptoms and the duration of the particular illness.

The development of an HIV vaccine has been difficult and elusive the past 4 decades, going on 5. The very latest development on that front is a new phase 1 trial of a novel vaccine called VIR-1388, which began in September 2023. This trial is being conducted to assess the safety of the vaccine and its ability to elicit an HIV-specific immune response in individuals enrolled in this study. VIR-1388 is designed to induce the production of T lymphocytes that can prevent the establishment of a chronic viral infection.

Rous Sarcoma Virus

Historical Perspective

Dr. Francis Peyton Rous, a pathologist at the Rockefeller Institute, reported his discovery of an avian virus called Rous sarcoma virus (RSV) in the *Journal of Experimental Medicine* in 1911. RSV is the very first oncogenic retrovirus to have been described and studied. This remarkable finding opened the field of tumor virology, which showed that some viral infections can be the etiology for cancer. At the time, his report was ignored and even derided since it was a contradiction to what was thought to be known at the time, which is the fact that cancer was not infectious. Now, we know that it is true that cancer cells themselves are not infectious, but some viruses are associated with the de novo initiation of carcinogenesis in vertebrates, including humans. The study of this virus led to the identification of oncogenes, which in turn are the foundation for the molecular mechanisms of carcinogenesis. Dr. Rous received the Nobel Prize in 1966 for this discovery and his continued work on related research. Through using RSV as a study model, the noteworthy discovery of the RT earned Howard Temin and David Baltimore the 1975 Nobel Prize, and the identification of the first proto-oncogene c-Src (cellular Src) in the human genome earned Michael Bishop and Harold Varmus the 1989 Nobel Prize. In 1977, Raymond Erikson and Joan Brugge isolated the Src gene product. The discovery of the viral Src oncogene (v-Src) is especially profound when considering that the precursor form of this oncogene is encoded by the human genome. The human cellular homologue is referred to as the Src proto-oncogene (or c-Src). Both c-Src proteins and v-Src oncoprotein were also the first tyrosine kinases that were identified.

Classification and Structure of Rous Sarcoma Virus

Rous sarcoma virus is also a member of the Retroviridae family and the Orthoretrovirinae subfamily, like HIV, but belongs to the genus *Alpharetrovirus*. Because they are found in the same Baltimore classification, same family, and even same subfamily, RSV and HIV share a lot of common features. However, each also possesses unique characteristics and strategies to reproduce.

The structure of RSV is an enveloped virion with a mean diameter of 127 nm, but individual virion size varies significantly. Its CA has a faceted, almost polygonal, appearance in electron micrographs, with a tube-like, diamond-like, or coffin-like shape (Figure 8.9). Like in most other retroviruses, the genome includes 2 identical molecules of linear, plus-strand RNA molecules of 15,200 each. Each RNA molecule is capped at the 5' end (not an uncharged tRNA like the HIV genomic RNA) and polyadenylated at the 3' end, and possesses 3 standard ORFs found in all species of lentiviruses designated Gag, Pol, and Env.

There is an additional unique ORF toward the 3' end that encodes for the tyrosine kinase v-Src protein. The Gag gene encodes the structural proteins outer core membrane protein (MA or p17), CA (or p24), NC (or p7), and nucleic acid–stabilizing late assembly protein (LI or p6). The Pol gene encodes the RT (or p51) and RNase H (or p15), and PR (or p12). This gene does not encode IN (or p32), which is different from the HIV Pol gene. The RT itself possesses both RNase H activity and IN domain, which is absent in nonavian retroviral RTs. The Env gene encodes the envelope glycoprotein gp160, which ultimately gets cleaved into surface envelope glycoprotein gp120 (or SU) and

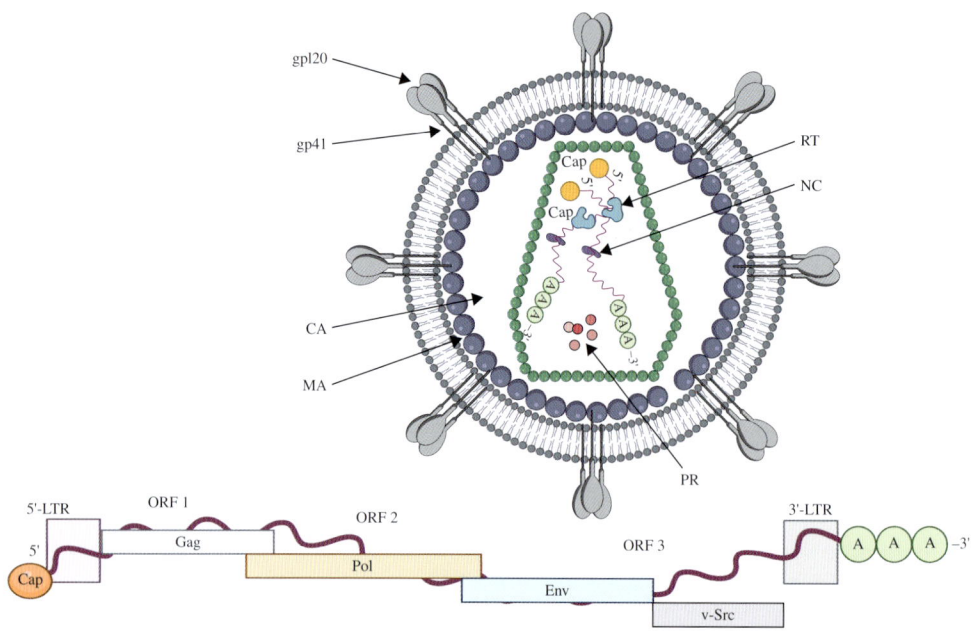

FIGURE 8.9 RSV virion structure and genome. RSV has an enveloped virion. Its 2 identical molecules of RNA genome are surrounded by a diamond-like CA. Each genome contains the 3 standard ORFs found in all species of lentivirus. It does not possess ORFs that encode for the 6 proteins unique to HIV. However, it does possess a gene that is not found in other lentiviruses known as the Src gene. *Credit:* Figure partially created in BioRender.

transmembrane protein gp41 (or TM) during the maturation process of the released viral particle. The v-Src gene that encodes a tyrosine kinase is not essential for RSV reproduction, but its presence greatly increases the virulence of RSV.

Infection Cycle of Rous Sarcoma Virus

Rous sarcoma virus can infect both dividing and nondividing cells, including avian macrophage, fibroblast, and muscle precursor cells. The mode of entry into the host cell and integration of the proviral DNA into the cellular chromosome are similar to activities carried out by HIV, with a few exceptions. RTs αβ and β from avian RSV harbor an IN domain that is absent in nonavian retroviral RTs. RSV IN contains a nuclear localization signal (NLS), which enables this enzyme to enter the nucleus to perform integration of the proviral DNA into the host chromosome. RSV does require the mitotic process of the host nondividing cell for the PIC to enter the nucleoplasm because its genome lacks the NLS and does not express Vpr. Uncoating, DNA replication, and PIC formation occur at the same rate in nondividing cells as in dividing cells; however, integration of the proviral DNA fails to occur in nondividing cells.

The integrated proviral DNA (or provirus) is transcribed by cellular RNA polymerase II, with resulting mRNAs capped and polyadenylated. RSV provirus, however, does not express the 6 gene products that are unique to HIV (Tat, Rev, Nef, Vif, Vpr, and Vpu/Vpx). The remainder of the RSV life cycle resembles that of the HIV life cycle.

Scientific Significance and Discoveries

The RSV genome encodes the oncogene Src, designated v-Src. It is recognized that v-Src originated from a eukaryotic gene that was mistakenly replicated and included in the RSV genome during viral RNA synthesis at some point during its evolution. The cellular version is designated c-Src.

The v-Src provides the tumorigenicity property of RSV. It is believed that RSV originally acquired its v-Src gene from a eukaryotic host cell during the process of viral genome replication. Before the viral plus-strand RNA genome can be replicated, the proviral DNA must first be integrated into the host cell chromosome. The thought is that as the viral RNA genome strand is being synthesized a considerable upstream portion of the c-Src gene is accidentally transcribed and included in the resulting viral genome (Figure 8.10). This portion of the c-Src gene is truncated, thereby missing the exon or exons that encode for the C-terminal domain, which can either block or expose the active/catalytic site. This truncated version of the gene is called v-Src.

In a eukaryotic cell, the c-Src protein kinase can be activated by a phosphorylation event in or near the catalytic site (Figure 8.11). This phosphorylated and thereby active

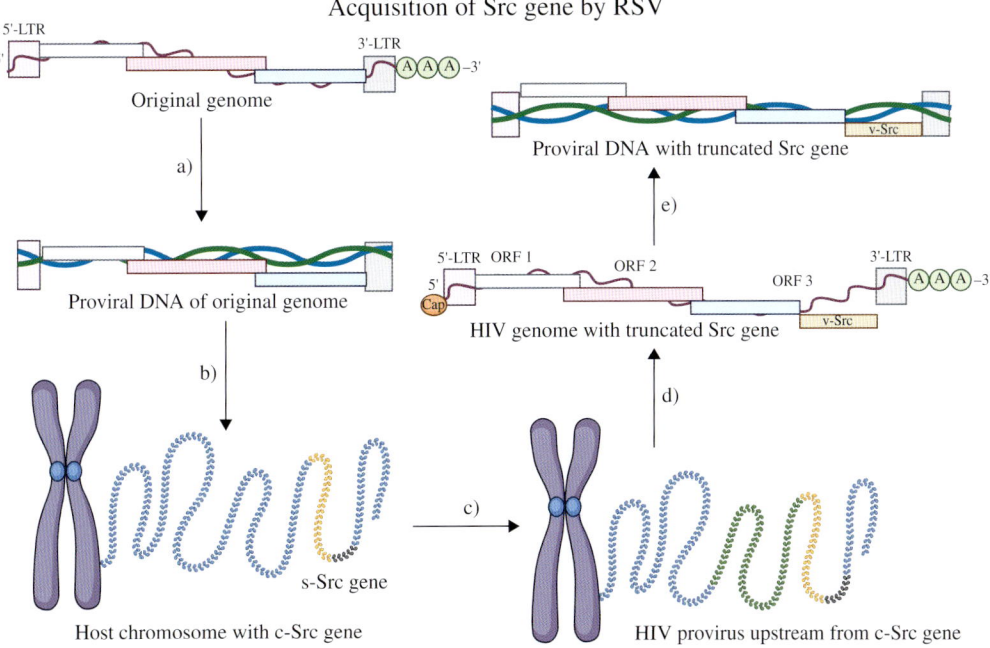

FIGURE 8.10   The acquisition of the Src gene by RSV. (a) The original HIV genome, which did not possess the Src gene, was used to produce a proviral DNA during an infection cycle. (b) The proviral DNA is targeted to a host chromosome that contains the Src gene (termed c-Src). (c) The provirus is located immediately upstream from the Src gene. (d) During viral genome replication, a large portion of the Src gene is included in the new viral genome. (e) During the next round of infection, the new viral genome containing the truncated form of the Src gene, termed v-Src, is used to produce a new proviral DNA, thus perpetuating the presence of v-Src in its genome. *Credit:* Figure partially created in BioRender.

Regulation of c-Src protein and v-Src protein activity

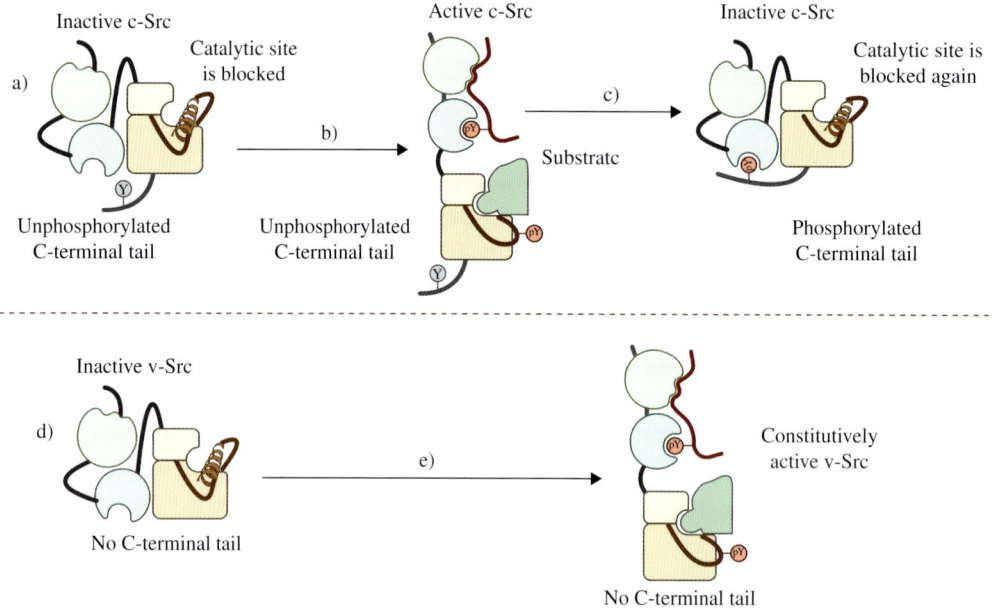

FIGURE 8.11 c-Src protein versus v-Src protein. a) Unphosphorylated c-Src protein is inactive where the catalytic site is blocked by a helix loop. (b) Autophosphorylation at a tyrosine residue (tyr416) in the loop activates c-Src, allowing substrate to bind to the catalytic site. (c) Phosphorylation at a tyrosine residue (tyr527) in the C-terminal tail inactivates c-Src again. (d) Unphosphorylated v-Src protein is inactive like c-Src. (e) Autophosphorylation at tyr416 causes v-Src to be constitutively active; inactivation is not possible since the C-terminal tail is missing. *Credit:* Figure partially created in BioRender.

form of c-Src protein kinase can be inactivated by a second phosphorylation event in the C-terminal domain, which possesses regulatory activity. Once this second site is phosphorylated, the C-terminal domain undergoes a conformation change that prevents substrates from entering the catalytic site. To reactivate c-Src protein kinase, a dephosphorylation event in the C-terminal domain must occur to allow the catalytic site to be exposed again.

Once activated, the function of both versions of the Src protein kinase is to induce the cell to enter the cell cycle, leading to upregulated cell proliferation. As the truncated version that is missing the C-terminal domain, v-Src kinase remains constitutively active once initially activated; therefore, the regulation of cell proliferation becomes uncontrolled in the RSV-infected cell from the initial activation. Uncontrolled cell proliferation has been shown to increase the chance of carcinogenesis. It is lucky for us that RSV is an avian virus and is not known to be zoonotic.

The Src family of kinases (SFK) consists of nonreceptor tyrosine kinases that regulate important cellular functions, such as cell proliferation, survival, differentiation, apoptosis, adhesion-dependent cell migration, and metabolism in a cell-autonomous manner. The SFK family is involved in supporting epithelial-mesenchymal transition prior to invasion and thus plays a critical role in tumor progression and metastasis.

The 11 members of vertebrate SFK are cellular sarcoma kinase (c-Src), lymphocyte-specific protein tyrosine kinase (Lck), B-lymphocyte kinase (Blk), hematopoietic cell kinase (Hck), Fyn (involved in the integrin signaling pathway), Gardner-Rasheed feline sarcoma virus oncoprotein (Fgr), Yamaguchi sarcoma virus oncoprotein (Yes), Lck/Yes novel tyrosine kinase (Lyn), Fyn-related Src family tyrosine kinase (Frk or Rak), Src-related kinase lacking C-terminal regulatory tyrosine and N-terminal myristoylation sites (Srms), and Yes-related kinase (Yrk).

The c-Src specifically can be activated by a variety of upstream cell surface or cytosolic receptors, such as receptor tyrosine kinase like the platelet-derived growth factor receptor, G-protein–coupled receptor like the β-adrenergic receptor, and an integrin bound to the extracellular MA. Once activated, this kinase can in turn modulate the activity of numerous effector functions; c-Src can affect downstream intracellular signaling pathways, such as the Ras/ERK/MAPK and Jak/STAT pathways. These pathways in turn modulate the activity and/or expression of a number of regulators, such as FAK, p130^{Cas}, and paxillin for cell adhesion and invasion; PI3K/Akt, Bcl-2, Bcl-xL, Mcl-1, and survivin for cell survival; VEGF and HIF1α for angiogenesis; and Myc, p21$^{WAF1/CIP1}$, and cyclin D, p27^{Kip1} and cdc2 for cell cycle progression.

Further Readings

Bagnato G, Leopizzi M, Urciuoli E, Peruzzi B. Nuclear functions of the tyrosine kinase Src. *Int J Mol Sci*. 2020;21(8):2675.

Baltimore D. RNA-dependent DNA polymerase in virions of RNA tumour viruses. *Nature*. 1970;226:1209–1211.

Brugge J, Erikson R. Identification of a transformation-specific antigen induced by an avian sarcoma virus. *Nature*. 1977;269:346–348.

Craigie R, Bushman F. HIV DNA integration. *Cold Spring Harb Perspect Med*. 2012;2(7):a006890.

Hatziioannou T, Goff S. Infection of nondividing cells by Rous sarcoma virus. *J Virol*. 2001;75(19):9526–9531.

Haura E. SRC and STAT Pathways. *J Thorac Oncol*. 2006;1:403–405.

Li G, Clercq. HIV genome-wide protein associations: A review of 30 years of research. *Microbiol Mol Biol Rev*. 2016;80(3):679–731.

Masenga S, Mweene B, Luwaya E, Muchaili L, Chona M, Kirabo A. HIV-host cell interactions. *Cells*. 2023;12(10):1351.

Ortiz M, Mikhailova T, Li X, Porter B, Bah A, Kotula L. Src family kinases, adaptor proteins and the actin cytoskeleton in epithelial-to-mesenchymal transition. *Cell Commun Signal*. 2021;19(67):2021.

Rous P. A sarcoma of the fowl transmissible by an agent separable from the tumor cells. *J Exp Med*. 1911;13:397–411.

Seitz R. Human immunodeficiency virus. *Transfus Med Hemother*. 2016;43(3):203–222.

Sharp P, Hahn BH. Origins of HIV and the AIDS pandemic. *Cold Spring Harb Perspect Med*. 2011;1(1):a006841.

Stehelin D, Varmus H, Bishop J, Vogt P. DNA related to the transforming gene(s) of avian sarcoma viruses is present in normal avian DNA. *Nature*. 1976;260:170–173.

Temin H, Mizutani S. RNA-dependent DNA polymerase in virions of Rous sarcoma virus. *Nature*. 1970;226:1211–1213.

Weiss R, Vogt K. 100 years of Rous sarcoma virus. *J Exp Med*. 2011;208(12):2351–2355.

Wells L, Vierra C, Hardman J, et al. Sulfoglycodendrimer therapeutics for HIV-1 and SARS-CoV-2. *Adv Ther (Weinh)*. 2021;4:2000210.

Wilen C, Tilton J, Doms R. HIV: Cell binding and entry. *Cold Spring Harb Perspect Med*. 2012;2(8):a006866.

Withers J, Beemon K. The structure and function of the Rous sarcoma virus RNA stability element. *J Cell Biochem*. 2011;112(11):3085–3092.

CHAPTER 9

DNA With Reverse Transcriptase

Hepatitis B Virus

Historical Perspective

In 1965, Dr. Baruch Blumberg discovered the hepatitis B virus (HBV). He shared the 1976 Nobel Prize for Medicine or Physiology with Dr. Carleton Gajdusek for their independent research on mechanisms that are involved in the origin and spread of infectious diseases. Dr. Gajdusek studied kuru, which is a slow-acting viral brain disease found in New Guinea. The discovery of kuru opened up a brand-new classification of infectious diseases, at the time termed *prion*. Anyway, back to HBV. The virus discovered by Dr. Blumberg was originally called the "Australia antigen" or "Au antigen" because the blood sample that was collected from an Australian aborigine reacted with antibodies in the serum of a patient with hemophilia. Of course, this antigen was later identified as the hepatitis B surface antigen (HBsAg). Persistent HBV infection may cause liver cirrhosis, leading to liver failure, where jaundice is a known symptom.

Hippocratic physicians described in the *Corpus Hippocraticum* a disease that manifested in a yellowish or greenish pigmentation of the skin and sclera, which we now know is due to high bilirubin levels. This condition is called jaundice. *Corpus Hippocraticum* is a collection of Ancient Greek textbooks, research notes, and philosophical reflections on medical topics written between the fifth and fourth century BCE. Hippocrates himself used the terms *ikteros* and *kirros* in his writing to describe this condition in 450 BCE. This condition was recorded in Babylonian clay tablets and even mentioned in the *Old Testament*. Incredibly, these Hippocratic physicians were able to make the connection between a liver problem and jaundice simply by observation without performing dissections. Diagnosis was based on symptoms, such as the color of the skin, urine, and feces. Recommended treatments at the time included the use of herbal medications, taking baths, dietary modification, and even bloodletting.

Using a large dataset of collected ancient viral genomes, researchers found HBV DNA in 137 ancient Eurasians and Native Americans dated as early as 10.5 thousand years ago. In fact, the most recent common ancestor of all HBV lineages was dated to between 20 and 12 thousand years ago, where evidence of the virus was found in remains of European and South American hunter-gatherers.

Outbreaks of catarrhal jaundice appeared multiple times in modern history. Outbreaks were recorded during multiple wars, including the American Civil War (1861–1865), the Franco-Prussian War (1870), World War I (1914–1918), and World War II (1939–1945). Information on an outbreak was published in 1885 involving 1289 shipyard workers in Bremen, Germany. In this case, HBV transmission occurred when these workers received a smallpox vaccine that was fortified with human lymph, which is the fluid that moves through the lymphatic system.

Between 1900 and the 1940s, outbreaks were caused by intravenous salvarsan inoculation and blood testing in diabetes clinics. Salvarsan inoculation was the first modern scientific treatment for syphilis. These outbreaks were associated with treatment of measles and mumps via convalescent plasma (therapy using blood from individuals who have recovered from an infection) in the 1930–1940s. In 1942, yellow fever vaccination led to an outbreak in the US Army.

In the 1930s, clinicians in Denmark who performed liver biopsies from individuals with catarrhal jaundice discovered a link to hepatitis that is characterized by inflammatory phenomena and thus associated with liver failure. Infectious hepatitis was shown to be caused by filterable agents in 1944, and, as described in the Historical Perspective section, the HBV was identified by Dr. Blumberg in 1965. By the late 1960s and early 1970s, chronic HBV infection had been linked to liver carcinogenesis. In 1970, HBV particles were visualized by electron microscopy. Drs. Blumberg and Irving Millman developed the first vaccine to protect against HBV in 1972. This HBV vaccine also has the distinction of being the first anticancer vaccine, which is used to reduce the chance of liver cancer in addition to its intended purpose to protect individuals from HBV infection. This is significant since chronic HBV and hepatitis C virus (HCV) are thought to induce roughly 80% of all liver cancers. In 1998, the Food and Drug Administration (FDA) approved lamivudine, which is the first direct-acting antiviral medication for chronic HBV infection.

Classification and Structure of Hepatitis B Virus

Hepatitis B virus belongs to the Hepadnaviridae family. This family of viruses includes 5 genera: *Parahepadnavirus* (eg, white sucker HBV), *Metahepadnavirus* (eg, bluegill HBV), *Herpetohepadnavirus* (eg, Tibetan frog HBV), *Avihepadnavirus* (eg, duck HBV), and *Orthohepadnavirus* (eg, HBV). The only other related family of virus is Caulimoviridae (eg, cauliflower mosaic virus).

The structure of the virus, which is one of the smallest viruses, is an enveloped virion with a diameter of 40–45 nm (Figure 9.1). The single relaxed, circular, partially double-stranded DNA genome (rcDNA) of 3300 base pairs is encased in an icosahedral capsid with a diameter of 32–36 nm. The HBV genome is very compact to fit inside such a small capsid. Structurally, the HBV genome includes a complete minus-strand DNA with a nick (or not closed), while the complementary plus-strand DNA is incomplete and shorter, thus leaving a gap. There are regions of short repeat sequences termed direct repeat (DR1 and DR2) at both ends of each DNA strand whose function is to hold the 2 strands together to preserve the circular structure of the genome. DR1 holds the 5′ end of the plus strand to the 3′ end of the minus strand, while DR2 holds the 5′ end of the plus strand to the 5′ end of the minus strand. Additionally, the 5′ end of the minus strand is covalently linked to the HBV Pol, while there's a capped RNA primer attached to the 5′ end of the plus-strand DNA.

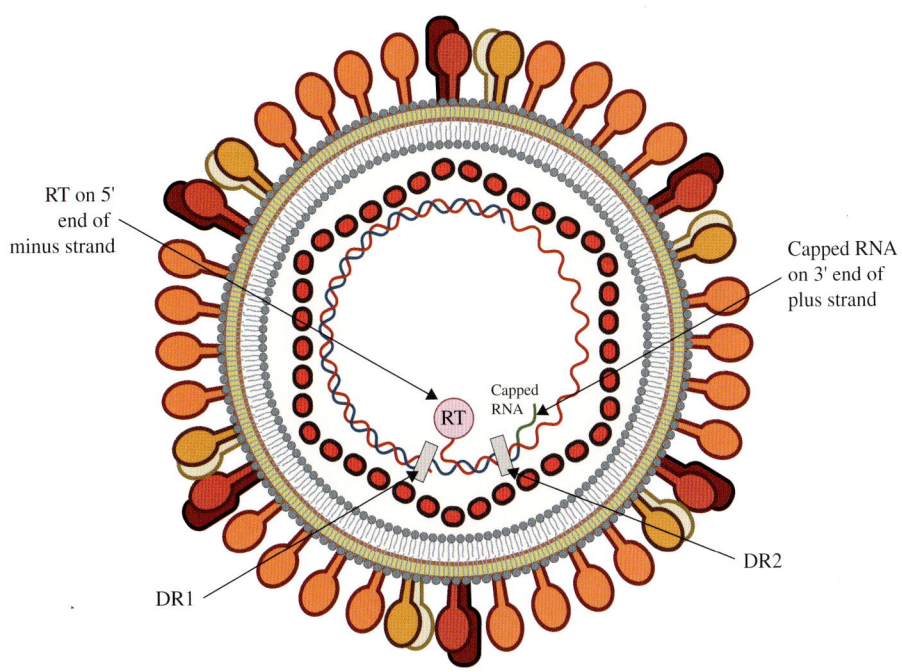

RT on 5' end of minus strand

Capped RNA on 3' end of plus strand

Capped RNA

RT

DR1

DR2

FIGURE 9.1   HBV virion structure and genome. HBV is an enveloped virion exhibiting 3 different surface proteins: L-HBsAg, M-HBsAg, and S-HBsAg. The single circular, gapped dsDNA genome is surrounded by an icosahedral capsid. The minus strand is complete but not closed. The synthesis of the plus strand is incomplete inside the capsid. A reverse transcriptase is attached to the 5' end of the minus strand, while a capped RNA is attached to the 5' end of the plus strand. *Credit:* Figure partially created in BioRender.

The HBV genome contains 4 overlapping open reading frames (ORFs; designated ORF S, C, P, and X). The P (Pol) ORF encodes a viral polymerase that consists of 4 domains: reverse transcriptase domain that functions as a RNA-directed DNA polymerase; a DNA-directed DNA polymerase, ribonuclease H (RNase H) domain that digests pregenomic RNA (pgRNA) that formed a hybrid with the minus-strand DNA after reverse transcription; a terminal protein (TP) domain that binds to the pgRNA to serve as a protein primer for the initiation of minus strand DNA synthesis; and a spacer domain with undefined function (Figure 9.2).

The C (core) ORF encodes the HBV core protein for the nucleocapsid (HBcAg [hepatitis B core antigen]), viral "e" antigen (HBeAg), and 22-kDa precore protein (p22cr). The S (surface) ORF encodes 3 viral surface glycoprotein antigens (HBsAg): large (L-HBsAg with preS1 domain), middle (M-HBsAg with preS2 domain), and small (S-HBsAg or Au antigen). The X ORF encodes the viral X protein (HBx), which has multiple functions that are discussed further in the "Scientific Significance and Discoveries" section of Hepatitis B virus in this chapter. HBx facilitates efficient viral reproduction by stimulating HBV gene expression. HBx also plays an important role in HBV-related hepatocellular carcinogenesis. This genome contains promoters and enhancers that are bound by tissue-specific transcription factors (TS-TFs) to regulate gene expression and encodes the polyadenylation site and signal for encapsidation during assembly.

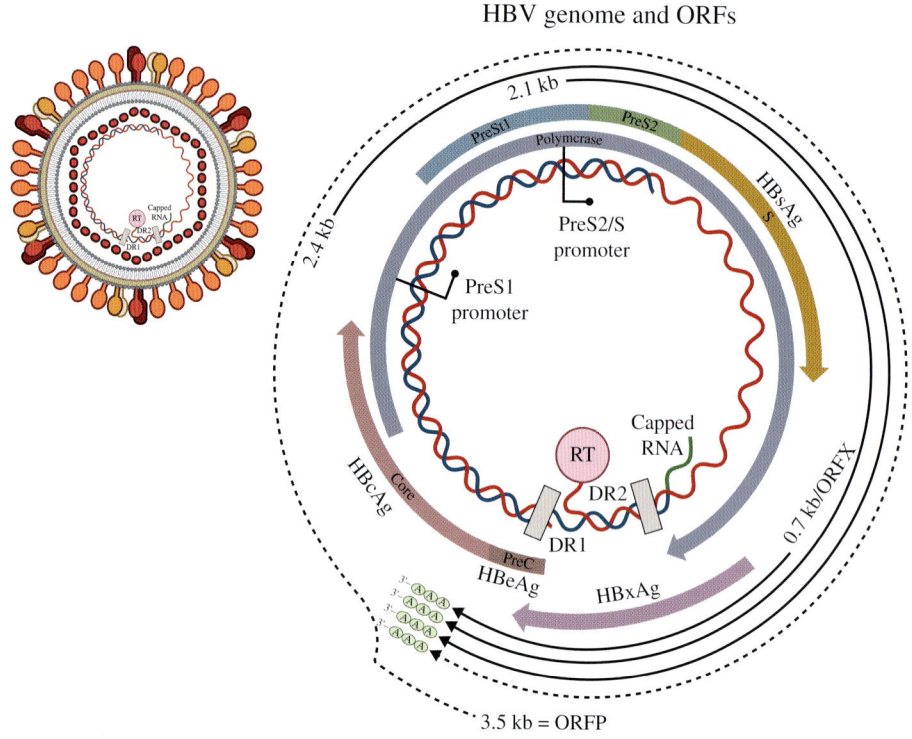

FIGURE 9.2 HBV genome, ORFs, and gene products. The HBV genome is a circular, gapped dsDNA. The 2 complementary strands are held together by 2 regions: DR1 and DR2. There are 4 ORFs, designated ORF P, S, C, and X. ORF P encodes the reverse transcriptase and RNAase H. ORF S encodes L-HBsAg, M-HBsAg, and S-HBsAg. ORF C encodes HBeAg and HBcAg. ORF X encodes HBxAg. *Credit:* Figure partially created in BioRender.

During transcription, 4 mRNA transcripts are synthesized. The 3.5-kb messenger RNA (mRNA) encodes HBcAg, HBeAg, and the viral polymerase. The 3.5-kb mRNA, also known as the pregenomic RNA (pgRNA), is also used as the template for genome replication. The 2.4-kb mRNA encodes L-HBsAg and M-HBsAg. The 2.1-kb mRNA encodes the S-HBsAg and M-HBsAg. The 0.7-kb mRNA encodes HBxAg.

Host Range, Transmission, Tropism, and Susceptible Host Cell

Hepatitis B virus species are known to infect humans and apes, such as chimpanzees, gorillas, orangutans, and gibbons, but are not known to be zoonotic. Even though HBV can attach to a variety of human cell types, tropism for HBV only includes host cells that exhibit sodium taurocholate cotransporting polypeptide (NTCP), which is a solute carrier protein encoded by the human sodium taurocholate cotransporting polypeptide (SLC10A) gene. Once bound, this receptor regulates the internalization of HBV through endocytosis, then membrane fusion. In humans, NTCP is expressed exclusively in the liver to uptake bile salts into hepatocytes. Therefore, HBV targets hepatocytes exclusively.

Transmission of HBV occurs through direct contact with blood or body fluids. Contaminated blood or body fluids can enter the body through sharing needles for intravenous drug, receiving infected blood or blood products through therapy, and sexual contact. HBV can also be transmitted transplacentally from an infected mother to her fetus. Open breaks in the skin and mucous membranes are easily accessible routes of entry for this virus, particularly in children.

Infecting the Host Susceptible Cell, Viral Particle Replication, and Tissue Damage

Initially, HBV attaches to heparan sulfate proteoglycan (HSPG) on the host cell surface with low affinity (Figure 9.3). This interaction leads to the subsequent interaction between the preS1 domain of L-HBsAg on the surface of the virion and SLC10A on the host cell surface with high affinity. NTCP induces the internalization of the virion. Internalization is further facilitated by the epidermal growth factor receptor (EGFR). HBV entry into the host cell occurs through clathrin-dependent, receptor-mediated endocytosis, followed by the fusion of the viral envelope and endosomal membrane. The nucleocapsid is released into the cytosol and transported toward the nucleus. Interaction between the nucleocapsid and nuclear protein complex around the nuclear pore induces uncoating, thus allowing the import to the viral genome into the nucleoplasm.

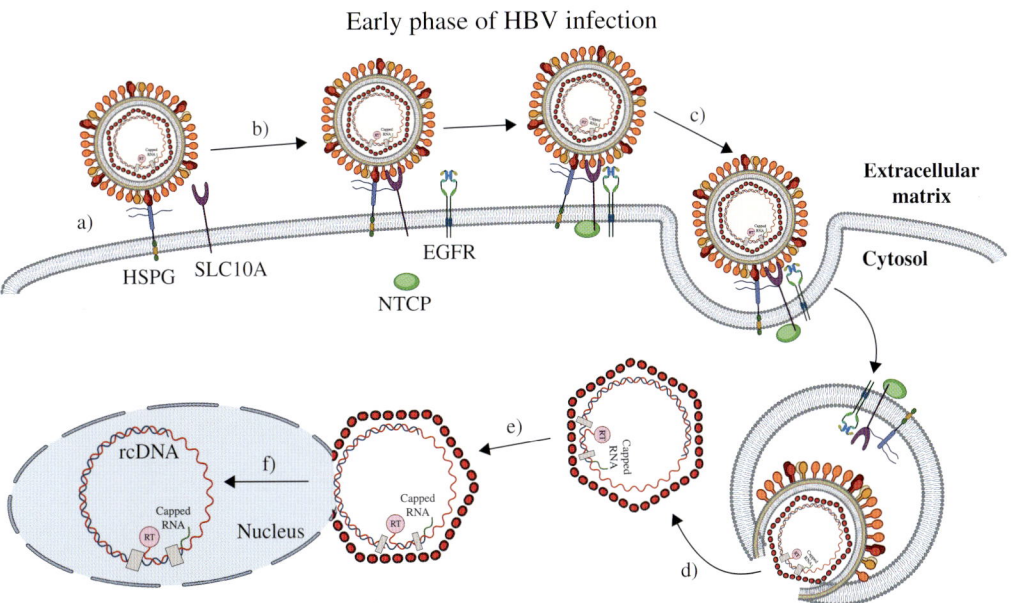

FIGURE 9.3 HBV life cycle leading to synthesis of rcDNA. The delivery of the HBV genome to the nucleoplasm is complex, involving many steps. (a) The virion attaches to HSPG with low affinity. HSPG is found on the surface of a number of cell types. (b) The range of tropism is greatly narrowed by the requirement for SLC10A, attached to NTCP on the host cell surface. (c) The presence of EGFR facilitates NTCP-induced endocytosis. (d) Once in the endosomal lumen, membrane fusion occurs, allowing the nucleocapsid to migrate into the cytosol. (e) The nucleocapsid is delivered to the nucleus, where uncoating occurs. (f) The rcDNA is then imported into the nucleoplasm. *Credit:* Figure partially created in BioRender.

The rcDNA is an incomplete (or gapped) circular dsDNA structure, therefore it must be remodeled into a covalently closed circular DNA (cccDNA) to be used as the template for transcription (Figure 9.4). This remodeling process occurs in the nucleoplasm and requires host DNA replication machinery and DNA repair mechanisms. DNA repair enzymes, tyrosyl-DNA phosphodiesterase 2 (TDP2) and flap structure-specific endonuclease 1 (FEN1), have been shown through in vitro assays to remove the reverse transcriptase and the terminal redundancy sequence DNA flap at the 5′ end of the minus strand of the rcDNA. The remaining nick is ligated by DNA ligase 1 (LIG1) or LIG3. These modifications on the minus strand result in a deproteinated rcDNA (ie, the rcDNA is not bound by histones).

The repair of the plus strand begins with DNA synthesis carried out by host DNA polymerase κ and α (Pol κ and Pol α) and topoisomerase I and II (TOPI and TOPII) to extend this strand to its full length using the minus strand as the template. FEN1 removes the capped RNA primer, then ligation of the nick is catalyzed by LIG1 and LIG3.

The resulting cccDNA is assembled with histones from the host cell to form nucleosomes and is also bound by viral factors such as HBcAg. Histone modifications within these nucleosomes enable the regulation of viral gene expression from the 4 overlapping ORFs. The cccDNA is a stable, nonreplicative, minichromosome (episome) and is considered the hallmark of an established HBV infection. This episome is not immediately integrated into the host chromosome.

Converting rcDNA into minichromosome

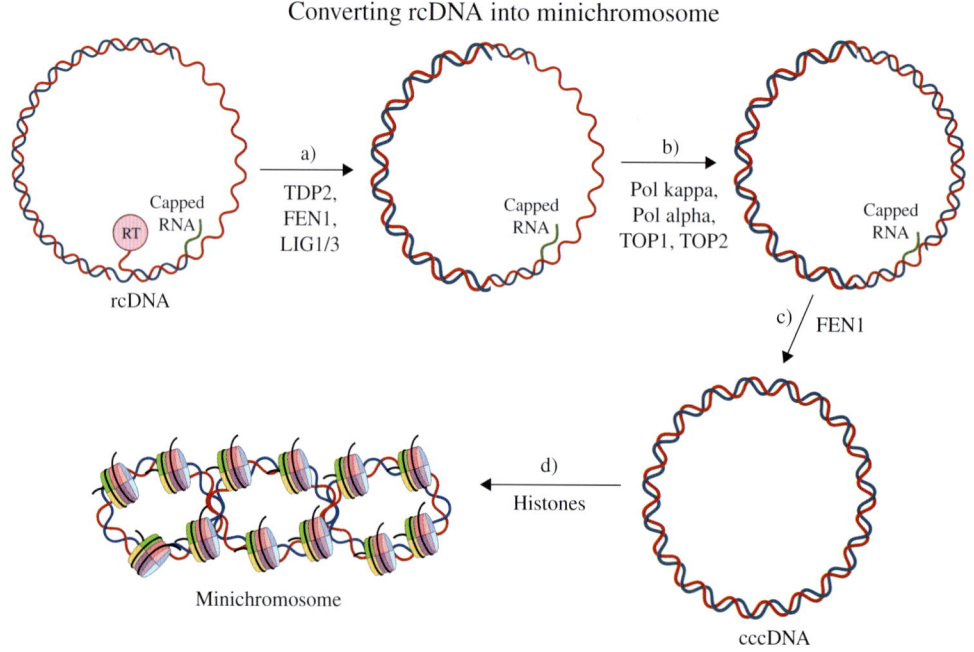

Figure 9.4   Modification of rcDNA into minichromosome. (a) TDP2 and FEN1 remove the RT at the 5′ end of the minus-strand DNA. LIG1 or LIG3 ligates the ends. (b) Pol κ, Pol α, TOP1, and TOP2 fill in the gap in the plus-strand DNA. (c) FEN1 removes the capped RNA at the 5′ end of the plus strand to form a cccDNA. (d) Histones bind to the cccDNA to form a minichromosome. *Credit:* Figure partially created in BioRender.

Host transcription factors, both general transcription factors (TFIIs, which regulate RNA polymerase II activity) and tissue-specific transcription factors (TS-TFs), are required to regulate transcription using the cccDNA as the template. Four viral mRNAs are produced from the 4 ORFs: P, C, S, and X. All mRNAs are capped and controlled to be polyadenylated from a common polyadenylation site. The mRNA transcribed from the P ORF not only encodes for the HBV Pol, but also acts as the template for viral genome replication. This mRNA corresponds to the full-length viral genome and is referred to as the pgRNA since it is also used as the template for minus-strand DNA synthesis for the progeny HBV rcDNA (Figure 9.5).

The other 3 mRNAs are referred to as subgenomic RNAs. TS-TFs that are involved in this process include peroxisome proliferator-activated receptor α, farnesoid X receptor, and retinoid X receptor α as nuclear receptors, and liver-enriched hepatocyte nuclear factor 3 and 4 (HNF3 and HNF4, respectively). Other transcription factors that augment the expression of pgRNA include CCAAT/enhancer-binding protein, nuclear factor 1, specificity protein 1, cyclic adenosine monophosphate (cAMP) response element binding protein (CREB), and liver receptor homologue 1 (LRH-1).

The transcription of pgRNA can also be suppressed by HNF6, prosperous-related homeobox protein 1, p53, and zinc finger and homeoboxes 2. HNF3 and HNF4 expression

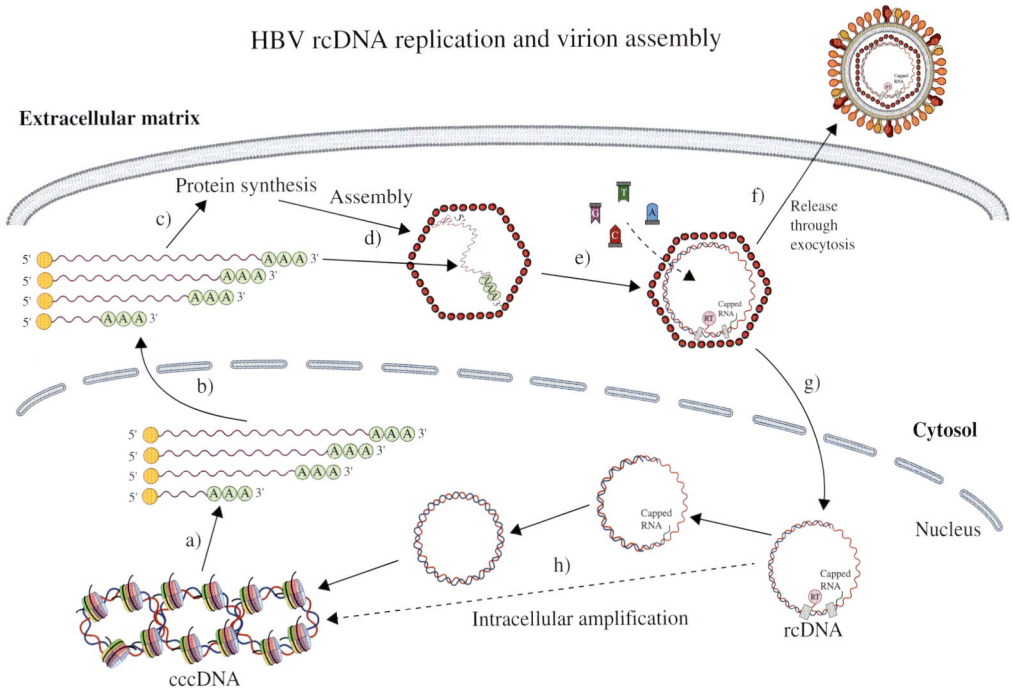

FIGURE 9.5 HBV genome replication and virion assembly. (a) The cccDNA is used as the template for transcription. (b) All 4 transcripts are exported to the cytosol. (c) Protein synthesis occurs in the cytosol. (d) Assembly of the capsid around the pgRNA. (e) Nucleotides are transported into the nucleocapsid for rcDNA replication. (f) The virion can be released from the host cell via budding. (g) The nucleocapsid can also migrate back to the nucleus, where the rcDNA is imported into the nucleoplasm. (h) More cccDNA is produced through intracellular amplification. *Credit:* Figure partially created in BioRender.

is especially high in hepatocytes, but they are also found in kidney and intestinal cells. The binding of HBV Pol and HBcAg to the pgRNA triggers its encapsidation within the hepatocyte's cytosol. During this process of assembly, deoxyribonucleotides are transported into the nucleocapsid to be used for DNA synthesis.

The virion maturation process occurs while the viral particle is still in the cytosol. HBV Pol (reverse transcriptase) uses its TP domain as a protein primer and the pgRNA as the template to synthesize the minus-strand DNA. The intermediate product is a hybrid containing the pgRNA and the newly synthesized minus-strand DNA. The RNase H domain of HBV Pol then digests most of the pgRNA, leaving only a capped RNA molecule to be used as the primer for the synthesis of the complementary plus strand catalyzed by the DNA-directed DNA polymerase domain of HBV Pol.

During genome replication, the capped RNA primer can initiate the synthesis of the plus-strand DNA at either the DR1 site or DR2 site. Initiation at DR1 produces a double-stranded, linear DNA (dslDNA), whereas initiation at DR2 produces rcDNA. HBV rcDNA is formed the majority of the time, while dslDNA synthesis is in the minority. In either case, the synthesis of the plus-strand DNA is incomplete once the source of deoxyribonucleotides inside the enclosed nucleocapsid is depleted.

Let's discuss the synthesis of nucleocapsids possessing an rcDNA first. After the plus-strand DNA is synthesized, the whole genome is circularized. Since the TP domain of HBV Pol is used to prime minus-strand synthesis and the capped RNA is used to prime plus-strand synthesis, both structures remain bound to the 5′ end of their respective strand. There are 2 possible paths for this rcDNA-containing mature nucleocapsid. It can either be enveloped and released from the hepatocyte as a progeny virion to complete the life cycle of HBV or be redirected back toward the nucleus, where the rcDNA can be imported into the nucleoplasm again to increase or replenish the intranuclear cccDNA pool.

Once in the nucleoplasm, of course, the rcDNA is converted into cccDNA, and the viral genome replication and virion assembly that is described above starts again. The biogenesis of cccDNA via this latter path is called intracellular amplification and is important in replenishing the cccDNA reservoir or concentration in the nucleus.

Why is intracellular amplification important? During the early stage of a primary HBV infection in a hepatocyte, the lifespan of cccDNA can be as long as 61 days. Its lifespan gradually shortens in later stages of infection and can be as brief as 3 days. With such truncated longevity, it is important for HBV to maintain an episomal pool of 5 to 10 molecules of cccDNA in the host cell nucleus to sustain viral persistence. Therefore, this pool of cccDNA is replenished through intracellular amplification.

Simultaneously, rcDNA-containing mature nucleocapsids can also be delivered to the plasma membrane to be released from the host cell via exocytosis. The mature nucleocapsid interacts with viral envelope polypeptides, specifically the preS1 domain of LBHsAg, that have been integrated into the endoplasmic reticulum (ER) membrane. This interaction induces the budding of the mature nucleocapsid into the ER lumen resulting in an enveloped virion (Figure 9.6). This enveloped virion is delivered through the ER and Golgi apparatus via vesicular transport. Envelope proteins undergo glycosylation during this journey. At the plasma membrane, the enveloped virion is released from the host cell through exocytosis.

Approximately 80%–95% of HBV virions possess the rcDNA, while only 5%–20% possess the dslDNA. HBV dslDNA is slightly longer than the genome length but contains all other features found in rcDNA, including the HBV Pol molecule at the 5′ end

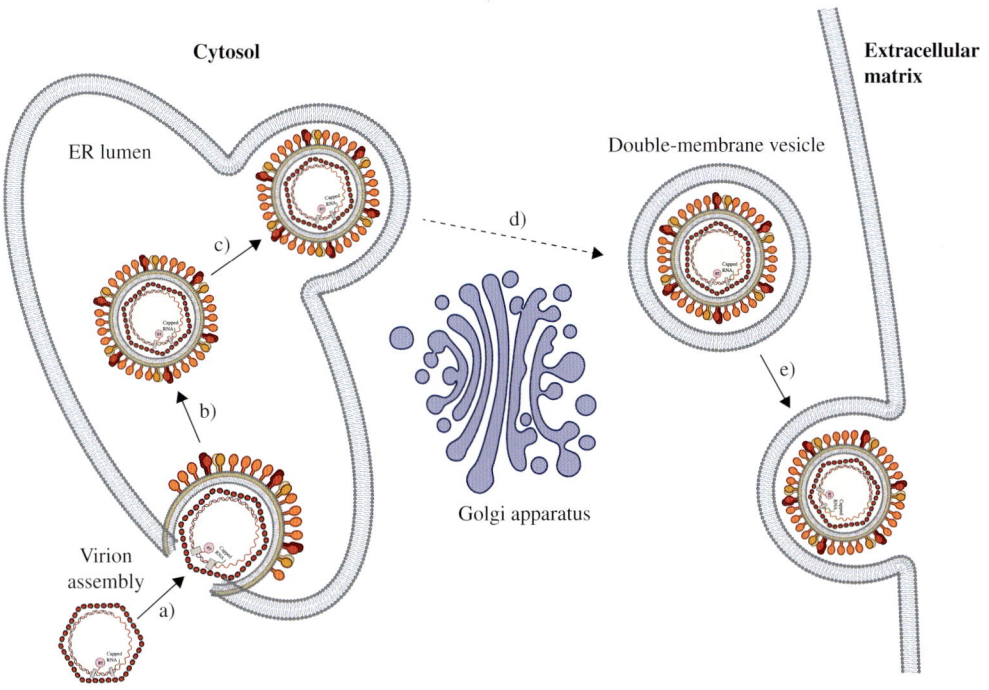

FIGURE 9.6   Release of HBV from host cell. (a) The nucleocapsid assembled in the cytosol buds into the ER lumen. (b) An enveloped viral particle is produced. (c) This enveloped HBV undergoes another budding event and is delivered toward the plasma membrane via vesicular transport. (d) A double-membrane viral particle is formed in the cytosol. (e) HBV released from the host cell via exocytosis is the mature single-enveloped viral particle. *Credit:* Figure partially created in BioRender.

of the minus strand, a capped RNA primer at the 5′ end of the plus strand, a DR1 site at the 5′ end, and a downstream DR2 site, and an incompletely synthesized plus-strand DNA. Similar to encapsidated rcDNA, dslDNA-containing nucleocapsid either can be enveloped and released from the host cell via exocytosis or can reenter the nucleus (Figure 9.7). In the nucleoplasm, dslDNA can either be modified into cccDNA or be integrated into the host genome. However, even though cccDNA produced from dslDNA can be used as the template for the synthesis of viral mRNAs, the pgRNA encoded by dslDNA in this case cannot be used for the synthesis of the full viral genome and thus is not supportive of additional rcDNA synthesis. This pgRNA, from dslDNA, is truncated, which renders it nonfunctional; therefore, dslDNA is deemed replication incompetent or replication defective.

HBV dslDNA can also be integrated into the host chromosome at random sites where double-stranded DNA breaks are located. Integration of the viral genome into host DNA is regulated by the host's nonhomologous end joining (NHEJ) or recombination repair mechanism, requiring the recruitment of Ku70:80 (X-ray repair cross-complementing protein dimer with 70-kDa and 80-kDa subunits) and DNA kinase complex. Ku70:80 dimers bind to ends of double-stranded DNAs to initiate nonhomologous end joining (NHEJ). Viral genome integration occurs early during host cell infection involving similar DNA repair mechanisms used to modify the rcDNA form into cccDNA.

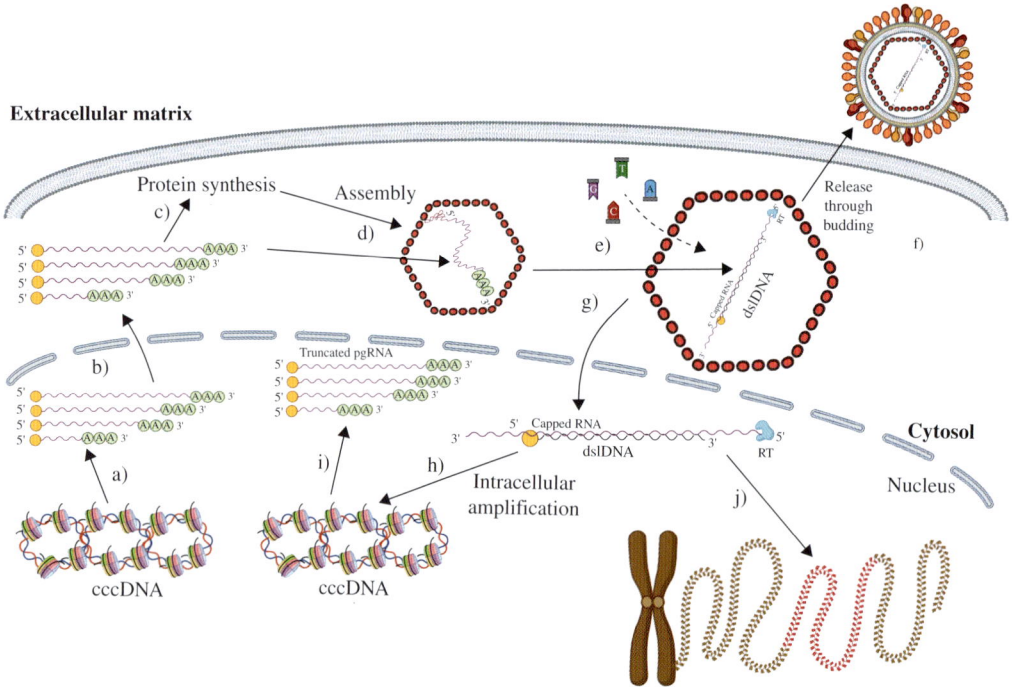

FIGURE 9.7 HBV dslDNA replication and integration into host chromosome. (a) The cccDNA is used as the template for transcription. (b) All 4 transcripts are exported to the cytosol. (c) Protein synthesis occurs in the cytosol. (d) Assembly of the capsid around the pgRNA. (e) Nucleotides are transported into the nucleocapsid for dslDNA replication. (f) The virion can be released from the host cell via budding. (g) The nucleocapsid can also migrate back to the nucleus, where the dslDNA is imported into the nucleoplasm. (h) More cccDNA is produced through modification of dslDNA. (i) However, the pgRNA synthesized from this cccDNA is truncated and thus nonfunctional. Virus replication is no longer possible. (j) The dslDNA can be integrated into the host chromosome, leading to chronic HBV infection. *Credit:* Figure partially created in BioRender.

Once integrated, the HBV dslDNA can still be expressed to produce viral proteins; however, truncated proteins are synthesized due to the absence of the shared polyadenylation site at the downstream region of the mRNAs. Additionally, since the pgRNA is nonfunctional, there would be no template strand for viral genome replication. However, because the HBV dslDNA is integrated, it would be replicated along with host chromosomes and transferred vertically to progeny cells, thus maintaining and even expanding the population of these cells. It is estimated that 1% of hepatocytes in the liver harbor HBV-integrated genome. The function of integration is still unknown; however, it has been proposed that this is a strategy for HBV to evade detection and eradication by the host's immune systems. Viral genome integration has been linked to chronic HBV infection.

Dissemination in the Host Body

Hepatitis B virus entry into the host can occur via direct transmission into the bloodstream or via open breaks in the skin and mucous membranes. Direct transmission into the bloodstream results in passive viremia, whereas virus that entered through breaks in the epithelial lining must undergo further hematogenous dissemination to enter the bloodstream, leading to active viremia. Once viremia is established, HBV particles eventually reach the liver since all the body's blood is filtered through the liver. HBV virions that are released from an infected cell remain in the liver since they have an exclusive tropism of hepatocyte. Hepatocytes pass the integrated viral genome to progeny cells through mitotic cell division.

Pathogenesis and Clinical Manifestation

Symptoms of an acute HBV infection range from mild to severe, including loss of appetite and abdominal pain, nausea and vomiting, a low-grade fever, weakness and fatigue, muscle and joint aches, dark urine, and/or jaundice (which is exhibited by yellowing of the skin and the whites of the eyes). Acute HBV infection is defined as one that lasts less than 6 months, where your immune system can likely clear the virus from your body, even with a more severe infection. Chronic HBV infection is one that can last a lifetime where no symptoms may be exhibited, but some chronically infected individuals may experience ongoing fatigue and mild symptoms of acute hepatitis.

Newborns (90% of 0 to 1 year old) and children under 5 years old (30%–50%) who are infected with HBV have a higher risk of chronic hepatitis. Of adults, 5%–10% may also develop chronic hepatitis. One's inflammatory responses to an HBV infection can cause acute and/or chronic hepatitis, which might lead to tissue damage. Continuous liver tissue damage caused by chronic hepatitis can lead to many serious complications, such as liver failure where vital functions of the liver shut down; periarteritis or polyarteritis nodosa, which is a serious inflammatory blood vessel disease; glomerulonephritis, which is damage to the kidney's glomeruli caused by extensive inflammation; and infantile papular acrodermatitis, which is a benign rash.

Late sequelae of chronic HBV infection include liver fibrosis, liver cirrhosis (extensive permanent scarring), and hepatocellular carcinoma (HCC). The major factor that can lead to hepatocarcinogenesis is likely the integration of dslDNA into the host cellular chromosome. Persistent expression of viral gene products, especially HBx, may stimulate oncogenic pathways. It is also possible that the integrated HBV DNA can modulate the expression of proximal host genes associated with carcinogenesis. There are viral promoters at the distal (3′) end of the integrated HBV DNA that can promote downstream cellular transcription.

Diagnosis, Treatment, and Prevention

It is estimated that 2 billion people globally have been infected with HBV, and more than 292 million of those individuals are living with chronic HBV infection. Therefore, individuals who are at risk, especially pregnant women and adolescents, are advised to be screened for the presence of HBV.

The initial assessment of HBV infection includes the recording of patient history and a physical examination, followed by serological assays to evaluate liver disease activity and the distinction between acute hepatitis versus chronic hepatitis. These assays detect antigens and antibodies that are present in the body either during or after an infection, which in turn can determine whether the patient had acquired immunity due to past infections or HBV vaccination or is susceptible to infection. Hepatitis antigen markers are HBsAg, HBcAg, and HBeAg, while hepatitis antibody markers are anti-HBc, anti-HBc immunoglobulin (Ig) M (IgM), and anti-HBe.

Samples may be collected from serum, plasma, and/or whole blood, even dried blood spot specimens and oral fluid specimens. Serological diagnostic assays include enzyme immunoassays, chemiluminescence immunoassays, and electrochemiluminescence immunoassays and result in high analytical sensitivity, specificity, and accuracy. A number of biochemical parameters can be measured to determine the severity of liver damage. These parameters include liver-damage-response enzymes aspartate aminotransferase (AST) and alanine aminotransferase (ALT), gamma-glutamyl transpeptidase, alkaline phosphatase, bilirubin, serum albumin gamma globulin, full blood count, and prothrombin time. However, if results from the above-mentioned tests are inconclusive, needle liver biopsy is used to detect the presence of liver cirrhosis. Unfortunately, this procedure is invasive, costly, and painful.

Molecular assays are additionally performed to detect and quantify the amount of viral DNA, for genotyping, and for identifying drug resistance and mutations analysis of the core region of the HBV genome. Quantifications of the viral load are performed using ultraviolet spectrophotometry, real-time polymerase chain reaction (RT-PCR), digital PCR, loop-mediated isothermal amplification, transcription-mediated amplification, nucleic acid sequence-based amplification, rolling circle amplification (RCA), electrochemical, quartz crystal microbalance, microcantilever, and/or surface plasmon resonance biosensors. Genotyping can be performed using reverse hybridization, restriction fragment polymorphism, multiplex nested PCR or real-time PCR, oligonucleotide microarray chips, reverse dot blot, restriction fragment mass polymorphism, and/or invader assay. Whole HBV genome sequencing, followed by phylogenetic analysis, may also be performed. The "gold standard" method for identifying mutations is the Sanger sequencing of nucleic acid fragments that were amplified via RT-PCR.

Currently, there is no known cure for HBV infection, although researchers believe that cccDNA is a potential therapeutic target. Therefore, a cure for hepatitis is thought to involve the targeting and destruction of cccDNA. In the meantime, a number of treatments have been shown to be effective in decreasing liver cancer mortality and reducing the need for liver transplantation. These medications may also slow or reverse liver disease progression and HBV infectivity. Current known effective anti-HBV drugs include lamivudine, telbivudine, entecavir, adefovir, and tenofovir.

Since prevention is the best method of treatment, HBV vaccination is the primary, most effective, and safest means to prevent hepatitis (Table 9.1). Depending on the particular FDA-approved vaccine used, an HBV vaccine can be administered as 2–4 injections separated by 1 or more months between injections. Avoid high-risk activities and behaviors whether or not you have been immunized. And, absolutely take precautions to prevent viral transmission if you are known to be infected.

Current Status

The Hepatitis B Foundation estimates that 1 out of 3 individuals around the world have been infected with HBV, which is approximately 2 billion people. Of these, 257–316 million individuals have chronic infection. There are an additional 1.5 million new cases each year, and the annual HBV infection-related deaths is around 820,000. HBV infection is a major global health concern. It is the primary trigger for HCC, which is the second leading cause of cancer deaths in the world. West Africa and China experience the highest HBV incidence rates, while in the United States approximately 2.4 million individuals have chronic HBV infection.

A.   U.S. infant (younger than 1 year old) Hepatitis B immunization schedules.

Vaccine	Dose 1 ("Birth Dose")	Dose 2	Dose 3	Dose 4
Engerix-B, Recombivax HB: 3-dose vaccine series	Within 24 hours of birth (HBV only vaccine)	1 month after dose 1 (HBV only vaccine)	6 months after dose 1 (HBV only vaccine)	
Vaxelis, Pediarix: 4-dose combination vaccine series	Within 24 hours of birth (HBV only vaccine)	6 weeks of age (combination vaccine against HBV + other diseases)	14 weeks of age (combination vaccine against HBV + other diseases)	6 months of age (combination vaccine against HBV + other diseases)

B.   U.S. children (1 year and older) and adult Hepatitis B immunization schedules.

Vaccine	Dose 1	Dose 2	Dose 3
Engerix-B, Recombivax HB, Twinrix (for adults 18 years and older): 3-dose vaccine series	Anytime (HBV only vaccine)	1 month after dose 1 (HBV only vaccine)	6 months after dose 1 (HBV only vaccine)
Heplisav-B (for adults 18 years and older only): 2-dose vaccine series	Anytime (HBV only vaccine)	1 month after dose 1 (HBV only vaccine)	

C.   Accelerated U.S. children (1 year and older) and adult Hepatitis B immunization schedules.

Vaccine	Dose 1	Dose 2	Dose 3	Dose 4
Engerix-B: 4-dose vaccine series	Anytime (HBV only vaccine)	1 month after does 1 (HBV only vaccine)	2 months after dose 1 (HBV only vaccine)	1 year after dose 1 (HBV only vaccine)
Twinrix (for adults 18 years and older only): 4-dose combination HAV and HBV vaccine	Anytime (combination vaccine against HBV + other diseases)	1 week after dose 1 (combination vaccine against HBV + other diseases)	1 month after dose 1 (combination vaccine against HBV + other diseases)	1 year after dose 1 (combination vaccine against HBV + other diseases)
Heplisav-B (for adults 18 years and older only): 2-dose vaccine series	Anytime (HBV only vaccine)	1 month after dose 1 (HBV only vaccine)		

D.   U.S. Hepatitis B immunization schedules for infants born to mothers who have Hepatitis B.

Vaccine	Dose 1 ("Birth Dose")	Dose 2	Dose 3	Dose 4
Engerix-B, Recombivax HB: 3-dose vaccine series	Within 24 hours of birth (HBV vaccine + HBV immunoglobulin)	1 month after dose 1 (HBV only vaccine)	6 months after dose 1 (HBV only vaccine)	

TABLE **9.1**   HBV Immunization Schedules

Vaccine	Dose 1 ("Birth Dose")	Dose 2	Dose 3	Dose 4
	(HBV vaccine + HBV immunoglobulin)	6 weeks of age (combination vaccine against HBV + other diseases)	14 weeks of age (combination vaccine against HBV + other diseases)	24 weeks of age (combination vaccine against HBV + other diseases)

E.   International Hepatitis B immunization schedules.

*Vaccine	Dose 1	Dose 2	Dose 3	Dose 4
3-dose vaccine series for infants younger than 1 year	Within 24 hours of birth (HBV only vaccine)	1 month after dose 1 (minimum 4 weeks of age) (HBV only vaccine)	6 months after dose 1 (minimum 24 weeks of age) (HBV only vaccine)	
3-dose vaccine series for children (1 year or older) and adults	Anytime (HBV only vaccine)	1 month after dose 1 (HBV only vaccine)	6 months after dose 1 (HBV only vaccine)	
4-dose combination vaccine for infants (younger than 1 year older)	Within 24 hours of birth (HBV only vaccine)	6 weeks of age (combination vaccine against HBV + other diseases)	10 weeks of age (combination vaccine against HBV + other diseases)	14 weeks of age (combination vaccine against HBV + other diseases)

* List of pre-qualified vaccine approved by the World Health Organization.

Vaccine	Vaccine Type	Date of Prequalification
ComBE Five (liquid)	Hepatitis B, Diphtheria-Tetanus, Pertussis (whole cell), Haemophilus influenzae type b	18May2012
Engerix	Hepatitis B	01Jan1987
Eupenta	Hepatitis B, Diphtheria, Tetanus, Pertussis (whole cell), Haemophilus influenzae type b	10Feb2016
Euvax B	Hepatitis B (pediatric and adult)	21Jan2020
Heberbiovac HB	Hepatitis B	11Dec2001
Hepatitis B vaccine (rDNA) (for adult)	Hepatitis B	12Nov2004
Hexasiil	Hepatitis B, Diphtheria, Tetanus, Pertussis (whole cell), Haemophilus influenzae type b, Polio (Inactivated)	21Mar2024
Hexaxim	Hepatitis B, Diphtheria, Tetanus, Pertussis (acellular), Haemophilus influenzae type b, Polio (Inactivated)	19Dec2014

TABLE 9.1   HBV Immunization Schedules (*Continued*)

Vaccine	Vaccine Type	Date of Prequalification
Pentabio	Hepatitis B, Diphtheria-Tetanus, Pertussis (whole cell), Haemophilus influenzae type b	19Dec2014

The Hepatitis B Foundation recommends different immunization schedules according to age groups. All vaccines require multiple doses.

TABLE 9.1   HBV Immunization Schedules (*Continued*)

Southeast Asia countries, China, Indonesia, Philippines, and the Pacific islands; the Middle East; Africa; and the Amazon basin are endemic regions showing high prevalence. The Mediterranean Basin and Eastern Europe are moderately endemic regions. Western Europe and North America are low endemic regions. The distribution of HBV genotypes differs among different global geographic regions and areas as well. There are 10 HBV genotypes classified as HBV A–J and 40 subgenotypes. Genotype A is predominant in Northwest Europe, North America, and Africa. Genotypes B and C are predominant in East Asia and Far East countries. Genotype D is predominant worldwide. Genotype E is, thus far, only detected in West Africa. Genotype F has been detected in Central and South America. Genotype G has been detected in Turkey, France, Canada, Vietnam, Germany, and America. Genotypes H and I have been detected in Central America, Mexico, Vietnam, and Laos. Genotype J has only been detected in Japan.

Most individuals become infected at birth (from contamination of the baby with blood during parturition rather than transplacentally) or during early childhood. HBV infection is an occupational hazard in certain professions as well. Individuals in these professions include dentists, surgeons, pathologists, mortuary attendants, and technicians; and scientists working in serology, hematology, and biochemistry; and people working in microbiology laboratories in hospitals or public health institutions, blood banks, or hemodialysis units. Body contact sports such as wrestling and rugby football may also pose a hazard.

In an effort to combat the HBV crisis, the World Health Organization (WHO) sets goals to be accomplished by the year 2030. They aim to reduce new infections by 90%, decrease mortality by 65% compared to 2015, detect 90% of all chronic infections, and reach the 90% vaccination status for children.

Scientific Significance and Discoveries

The ORF that encodes for HBx protein is found exclusively in the genome of the *Orthohepadnavirus* genus that infects mammals, but not in the genome of the *Avihepadnavirus* genus that infects birds. Even though HBx protein has not been found to be essential for the HBV life cycle, it appears to have transforming properties that can promote tumorigenesis and thus has clinical significance. It is believed that this viral protein plays a vital role in HBV persistence and liver pathogenesis.

What do we know about HBx? HBx has a dynamic subcellular localization associating with the nucleus and mitochondria and distributed throughout the cytoplasm. It is a transactivating inducer that wields its pleiotropic activities by regulating viral and cellular genes to control a variety of viral and cellular downstream events that are thought to be important for the successful establishment of an HBV infection and replication in the host cell. In fact, there is a direct correlation between the intracellular level of HBx and the development of liver cirrhosis that, in turn, increases the chance of cancer development. Cellular mechanisms that have been shown to be modulated by HBx include stimulating or inhibiting various signal transduction pathways, binding to cellular protein targets, modulating protein degradation, DNA repair, preventing the intrinsic apoptotic pathway, and inhibiting tumor necrosis factor α activity.

The following is a list of observed activities that are contributed by HBx pleiotropic function from in vitro studies:

- HBx plays an active role in controlling the cccDNA pool in infected hepatocytes.

- HBx promotes the recruitment of histone acetyltransferases to carry out epigenetic modifications of cccDNA minichromosomes to allow them to remain active and stable in infected hepatocytes. The stability of cccDNA in turn contributes to the increased expression of viral proteins, including HBx.

- HBx mediates abnormal DNA methylation, which correlates with enhanced HBV replication.

- HBx upregulates the expression of Yes-associated protein, which is a known oncogene that promotes hepatoma cell growth, which leads to the development of HCC.

- HBx contributes to the activation of the phosphoinositide 3-kinase and nuclear factor kappa B pathways to upregulate interleukin (IL) 34 expression. IL-34 in turn induces the activation of extracellular signal-regulated kinase (ERK) and Janus kinase–signal transducer and activator of transcription pathways, upregulation of B-cell lymphoma extra large antiapoptotic molecule, and expression of c-Myc (multifunctional transcription factor that promotes cell division). All these related activities contribute to the enhanced proliferation and migration of HCC cells.

- HBx binds to p53 to prevent its entry into the nucleus, thus inhibiting the induction of the apoptotic pathway.

- HBx regulates the levels of cell cycle regulatory proteins to activate cyclin-dependent kinase 2 (Cdk2), thus inducing quiescent hepatocytes to enter the cell cycle. Uncontrolled cell proliferation may lead to cancer progression.

- HBx interacts with cortactin, which is a substrate of the Src protein, to upregulate CREB1. Overexpression of CREB1 promotes the proliferation and migration of HCC cells.

- HBx contributes to the activation of the ERK/CREB signaling pathway to induce the expression of Forkhead box M1, which is associated with the progression of HBV-associated HCC.

- HBx activates p21 activated kinase to protect hepatoma cells from undergoing anoikis.

- HBx binds to the prebreast cancer 1 (AIB1) protein in human HCC cell lines to save it from being targeted for degradation. The synergistic activity of HBx and AIB1 promotes invasiveness of HCC cells.

- HBx upregulates beclin 1 expression to prevent hepatoma cells from undergoing autophagy in vitro.

- HBx promotes the activation of microRNA-3188 (miR-3188) that contributes to the activation of Notch signaling pathway, which may play a role in HCC progression.

- HBx promotes the expression of miR-5188 that is involved with the Wnt/ β-catenin-c-Jun signaling pathway. Wnt stands for wingless-type mouse mammary tumor virus integration site family. Increased activity of this pathway may result in promoting HCC stemness, metastasis, proliferation, and resistance to chemotherapy.

- HBx upregulates the expression of miR-21 to suppress the tumor suppressor programmed cell death 4, thus contributing to HCC development.

- HBx also suppresses the expression of a number of microRNAs, such as miR-148a, miR-520b, and miR-122, which in turn contributes to HCC development.

- HBx enhances the transcription of long noncoding RNA (lncRNA) DLEU2 (Deleted in Lymphocytic Leukemia 2), which in turn can possibly regulate the expression of cancer-related genes to promote hepatocytes transformation. DLEU2 is typically regulates cell proliferation, migration, and invasion as well as apoptosis. this lncRNA has been shown to be involved in cancer development and progression.

- HBx downregulates the tumor suppressor Dreh in an effort to promote HCC metastasis.

- HBx activates lncRNA HULC (Highly Upregulated in Liver Cancer), which in turn upregulates miR-539 to enhance HBV cccDNA stability and thus promotes HBV replication increased expression of HBx. HULC has been shown to promote tumor growth and metastasis by regulating certain metabolic pathways and autophagy.

- HBx upregulates lncRNA UCA1 (Urothelial Carcinoma Associated 1) in Hep3B cells to silence the Cdk inhibitor p27, leading to the promotion of cell growth and inhibition of apoptosis. UCA1 plays a role in cancer development and progression. Hep3B cells were initially isolated from the cancerous liver tissue of a young child, then developed into a human hepatoma cell line.

- HBx induces mitophagy by triggering the Pink1/Parkin-dependent pathway. Mitophagy is a process by which damaged mitochondria are selectively removed through autophagy. This leads to a diversion from apoptosis to promote host cell survival and virus persistence.

- HBx mediates reactive oxygen species (ROS) generation by interacting with either cyclooxygenase 3 or the voltage-dependent anion selective channel 3 mitochondrial protein, leading to mitochondrial dysfunction.

- HBx can interfere with the innate immune response via its interaction with and suppression of mitochondrial antiviral signaling protein and retinoic acid-inducible gene I.

- HBx-mediated elevation in intracellular ROS levels induces ER stress, which can lead to increased release of Ca^{2+} into the cytosol, which in turn contributes to increases viral replication.

Hepatitis Delta Virus

Historical Perspective

The first confirmed case of disease related to hepatitis delta virus (HDV) was reported in Canada in 2017. A 38-year-old male who was HBV positive entered the hospital exhibiting jaundice, fever, epigastric pain, and rash the past week. For 2 years, he had been receiving treatment to control the HBV infection. However, these symptoms were signs of progressive liver dysfunction. Assays were performed to determine the etiology of this sudden liver failure after years of seemingly successful treatment. All tests were negative except for those for HDV, which showed the presence of HDV antibody and RNA in the patient. This was a new development since he had tested negative for HDV as recent as 6 months prior. After having received treatment for the following 16 months, his liver function improved, and he was clinically stable.

It turned out that his wife was discovered to have chronic HBV infection that was diagnosed during a routine prenatal screening. However, this screening did not include testing for HDV at the time, which was only retrospectively uncovered during this episode for her husband. Furthermore, she was not receiving anti-HBV treatment; therefore, it was determined that this was a case of spousal transmission via cohabitation and sexual contact.

Mario Rizzetto, an Italian physician, discovered the hepatitis deltavirus (HDV) in 1977. He found a novel antigen (hepatitis delta antigen, HDAg) in liver biopsies and sera from patients with chronic HBV and severe liver disease. Initially, HDAg was thought to be one of the many HBV surface antigens that had yet to be identified at the time. The HDV genome was sequenced in 1986, and HDV was classified in its own genus, *Deltavirus*.

Only as recent as October 2022 did the very first international meeting on HDV infection take place in Milan, Italy. Experts in the field discussed discoveries and insights covering virology, epidemiology, pathogenesis, diagnosis, natural history, and antiviral treatment.

Classification and Structure of Hepatitis D Virus

Hepatitis delta virus belongs to the Kolmioviridae family of satellite virus. This family includes 8 genera: *Daazvirus* (eg, Chinese fire belly newt virus 1), *Dagazvirus* (eg, rhinotermitid virus 1), *Daletvirus* (eg, Swiss snake colony virus 1), *Dalvirus* (eg, dabbling duck virus 1), *Deevirus* (eg, ray-finned fish virus 1), *Deltavirus* (eg, HDV), *Dobrovirus* (eg, Chusan Island toad virus 1), and *Thurisazvirus* (eg, bat deltavirus).

The structure of the HDV is an enveloped virion with a diameter of 36–43 nm (Figure 9.8). The envelope is composed of HBV surface antigen lipoproteins (L-HBsAg, M-HBsAg, and S-HBsAg) that are required for the infection of hepatocytes. The covalently closed circular minus-strand RNA (cccRNA) of 1700 bases forms a ribonucleoprotein (RNP) complex/particle with the 2 forms of HDV antigen, large HDAg (L-DAg) and small HDag (S-DAg). This RNP particle has a diameter of 20 nm.

HDV contains the smallest known genome for a human virus. It is also the only known human satellite virus. Its genome consists of only 1 ORF, which is used to express 2 HDV antigen isoforms(L-DAg and S-DAg). There is a high degree of self-complementarity within the HDV RNA genome, which allows the cccRNA to fold into

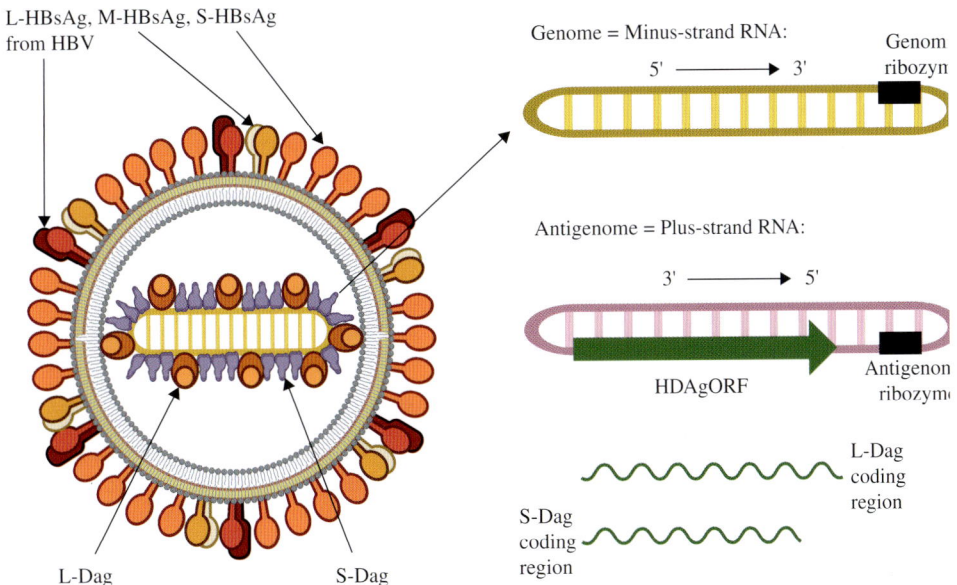

L-HBsAg, M-HBsAg, S-HBsAg from HBV

Genome = Minus-strand RNA:

5' ⟶ 3'

Genom ribozyn

Antigenome = Plus-strand RNA:

3' ⟶ 5'

HDAgORF

Antigenom ribozym

L-Dag coding region

S-Dag coding region

L-Dag   S-Dag

FIGURE 9.8 HDV virion structure and genome. HDV virion has an envelope that contains HBV membrane lipoproteins: L-HBsAg, M-HBsAg, and S-HBsAg. These lipoproteins are required for the attachment of HDV to its host cell. Its circular, minus-strand RNA is bound by multiple molecules of L-DAg and S-DAg, forming an RNP particle that has a rod-like structure. Its genome has only 1 ORF that encodes for both L-DAg and S-DAg. The genome also possesses a segment that functions as a ribozyme to carry out autocatalytic cleavage and autoligation. The plus-strand antigenome is the actual template used for translation of viral proteins. *Credit:* Figure partially created in BioRender.

unbranched, rod-like structures resembling viroids. Also similar to viroids, both the cccRNA and its complementary RNA strand possess ribozyme activity. L-DAg and S-DAg possess domains that mediate multiple functions, including oligomerization via a coiled-coil structure, host cell nuclear translocation via a bipartite signal sequence, and RNA binding via 2 arginyl-rich motifs. Additionally, L-DAg contains a prenylation site that is required for virion assembly.

Host Range, Transmission, Tropism, and Susceptible Host Cell

Humans are the only known natural host for HDV. Since HDV is a satellite virus and must co-infect a host with HBV, its mode of transmission mimics that of HBV. Transmission of HDV occurs through direct contact with blood or body fluids. Contaminated blood or body fluids can enter the body through sharing needles for intravenous drug use, receiving infected blood or blood products through therapy, and sexual contact. HDV can also be transmitted transplacentally from an infected mother to her fetus, but this is a rare occurrence. Open breaks in the skin and mucous membranes are easily accessible routes of entry for this virus. Inside the host organism, HDV infects hepatocytes predominantly due to the cellular tropism of HBV. In term of co-infection, if an individual is already experiencing chronic HBV infection (referred to as superinfection), then an eventual HDV infection typically leads to persistence. However, when both viruses co-infect a naïve host, then a transient infection of both HDV and HBV occurs.

Infecting the Host Susceptible Cell, Viral Particle Replication, and Tissue Damage

Hepatitis D virus must attach to the same host and cell types that are infected by HBV. Since these virions exhibit the same HBV surface antigens, the mode of attachment is also the same (Figure 9.9). HDV can only cause an infection in an individual who is already infected with HBV; otherwise, the HDV is avirulent and can be disposed of by the host immune system without being able to propagate. Attachment occurs via the interaction of the preS1 domain of L-HBsAg from both HDV and HBV to HSPG on the host cell first. The eventual interaction with NTCP on the host cell surface leads to the internalization of the virions, facilitated by the EGFR.

Once inside the host cell, the envelopes of HDV and HBV fuse with the endosomal membrane, which in turn induces uncoating of both viral species. The RNP complexes of both viruses exit the endosome and are transported into the cytosol, then trafficked and imported into the nucleoplasm.

The circular minus-strand RNA genome is used as the template by the host RNA polymerase II, which is a DNA-directed RNA polymerase, to produce a linear multimer of plus-strand complementary RNA via the rolling circle replication mechanism (Figure 9.10). This multimeric transcript undergoes autocatalytic cleavage catalyzed by

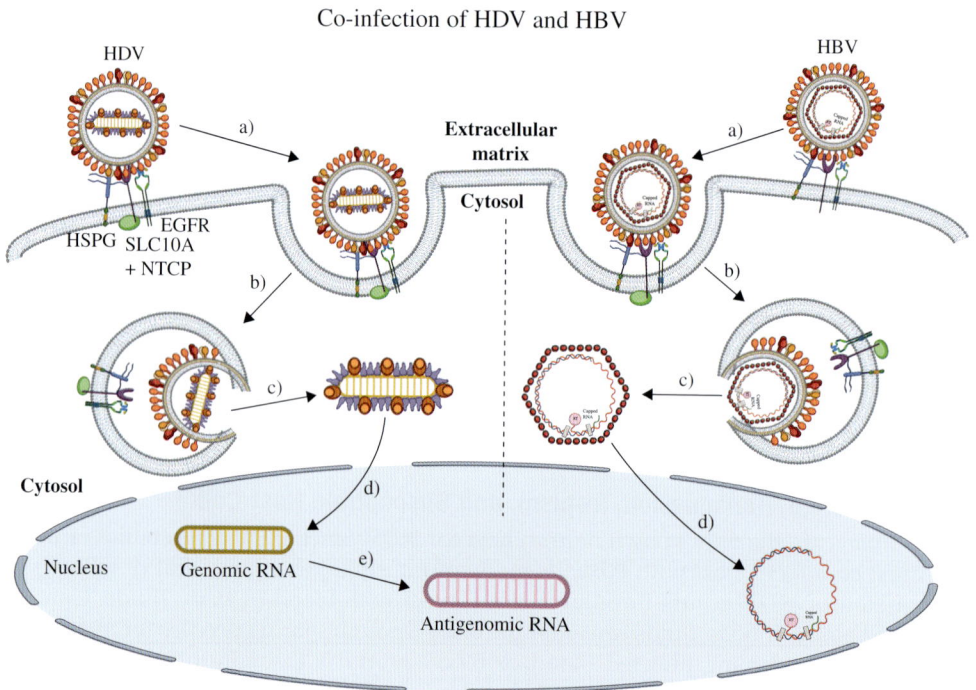

Figure 9.9 HDV entry into host cell and antigenome replication. (a) Both HDV on the upper left corner and HBV on the upper right corner bind to the similar set of surface receptors (HSPG, SLC10A + NTCP, and EGFR) before entering the host cell via endocytosis. (b) Once in the endosomal lumen, membrane fusion and uncoating occur. (c) Both nucleocapsids are then released into the cytosol. (d) Both RNP complexes are imported into the nucleoplasm. (e) The HDV genomic RNA is used as the template to produce HDV antigenomic RNAs to further the HDV life cycle, while the HBV genome is used for genome replication and protein synthesis to further the HBV life cycle. *Credit:* Figure partially created in BioRender.

Co-infection of HDV and HBV

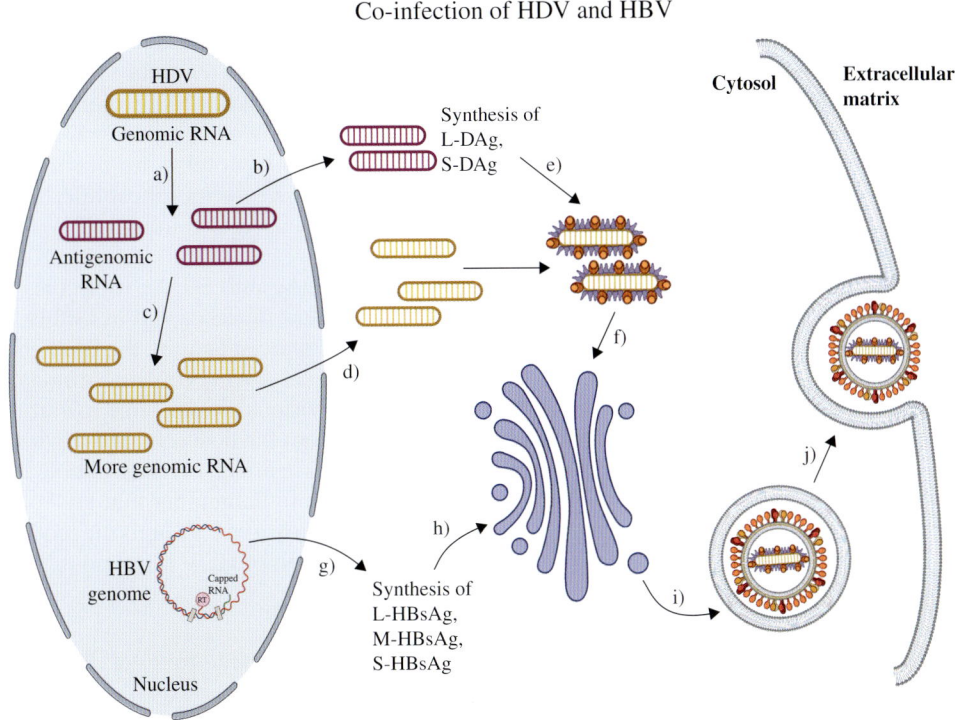

FIGURE 9.10 HDV genome replication, virion assembly, and release from the host cell. (a) HDV genomic RNA is used as the template to produce antigenomic RNA. (b) Sone antigenomic RNAs are exported to the cytosol for the synthesis of L-DAg and S-DAg. (c) Some antigenomic RNAs remain in the nucleoplasm to be used as the template for genomic RNA amplification. (d) Genomic RNAs are exported into the cytosol. (e) L-DAg and S-DAg molecules attach to each genome to produce an RNP complex. (f) RNP complexes are delivered to the Golgi apparatus and bud into its lumen to acquire the viral envelope. (g) In the meantime, the HBV genome is used as the template to synthesize mRNAs, which are exported to the cytosol. (h) L-HBsAg, M-HBsAg, and S-HBsAg are delivered and embedded into the ER membrane. They are then transported to the Golgi apparatus to adorn the envelope for HDV. (i) The enveloped viral particle is delivered toward the plasma membrane via vesicular transported. (j) HDV is finally released from the host cell via exocytosis. *Credit:* Figure partially created in BioRender.

its own element, called a ribozyme. Each unit of antigenomic RNA is then ligated to form a covalently closed circular plus-strand RNA. This structure resembles the cccRNA form of the viral genome.

These copies of circular plus-strand complementary RNA are pregenomic molecules that serve as the template to produce a linear multimer of minus-strand RNA genome, also via the rolling circle replication mechanism. This multimeric transcript also undergoes autocatalytic cleavage to form genomic RNAs, which in turn undergo self-ligation to produce HDV genomes (cccRNA) to be assembled. In summary, the HDV genome undergoes a double rolling circle replication mechanism to amplify its genome in the nucleus.

Plus-strand antigenome RNAs that were synthesized in the nucleus can also be exported into the cytosol to be used as templates for HDV protein synthesis. S-DAg functions as a transactivator of viral RNA replication. This population of these S-DAg RNPs is derived from molecules that were packaged with the genome in the original virion, as well as ones newly synthesized in the cytosol and imported back into the nucleus. L-DAg are also essential for assembly and, in some situations, can inhibit replication if necessary.

As stated previously, both L-DAg and S-DAg are encoded by the same ORF. L-DAg is the full-length form and S-Dag is identical to L-DAg except for being shorter by 19 amino acids. S-DAg is expressed constitutively, whereas the expression of L-DAg is stimulated by type I interferon (IFN-α and/or IFN-β). Both proteins are posttranslationally modified before virion assembly. S-DAg undergoes serine phosphorylation, while L-DAg is farnesylated. The farnesylated form of L-HDAg can inhibit HDV RNA replication and is essential for virion assembly.

Only circular minus-strand RNAs are bound by L-DAg and S-DAg molecules for assembly. HDV genome replication, protein synthesis, and nucleocapsid assembly are independent of HBV. However, once formed, this complex must acquire the lipid envelope by interacting with the Golgi apparatus membrane, which contains numerous HBsAgs, and budding into this organelle. The assembled virions are delivered toward the plasma membrane via vesicular transport and released from the host cell via exocytosis.

HDV possesses all the necessary surface components to attach to a susceptible host and induce internalization in the absence of HBV infection. Remember that HDV contains an envelope that exhibits many HBsAg molecules. However, without the required machinery from its cognate helper virus, HBV, reproduction of HDV is not possible.

Dissemination in the Host Body

HDV follows the same mode of entry into the human body, same route of dissemination, and same mode of exit as HBV.

Pathogenesis and Clinical Manifestation

Co-infection of HDV and HBV can cause mild-to-severe acute hepatitis. Three to 7 weeks after the initial infection, symptoms may include fever, fatigue, loss of appetite, abdominal pain, nausea, vomiting, dark urine, pale-colored stools, jaundice, and possibly, but rarely, fulminant hepatitis. Even though HDV infection is the most severe form of hepatitis, recovery is usually complete with supportive care to control and eliminate symptoms.

The good news is that symptoms caused by co-infection of HBV and HDV clear spontaneously after a short period of time in roughly 95% of individuals. Unfortunately, greater than 90% of individuals who have an HDV superinfection (HDV infection of an individual with chronic HBV) eventually develop chronic HDV infection, lasting more than 6 months. Chronic HBV-HDV co-infection is the most aggressive form of viral hepatitis, leading to a more rapid and higher rate of progression toward liver cirrhosis, liver cancer, and liver disease–associated mortality. There is a 10%–15% increase within 5 years from the initial infection and 80% increase after 30 years for the development of these diseases. HDV exacerbates symptoms of HBV infection by decreasing HBV replication while having minimal impact on the expression level of HBsAg to ensure the maintenance of a constant pool of this molecule for assembly into virulent agents but nothing else from HBV. Of course, decreasing HBV virion production means more HBsAg molecules are available for HDV reproduction. Additionally, HDV infection manages to evade most host immune responses, except for the production of IFN-α and -β, which leads to an increase in HDV viral load. Possible mechanisms of inhibition include the following:

- Epigenetic regulation of HBV cccDNA transcription may be regulated by HDAg. This activity causes an increase in the synthesis of HBV PreS/S mRNA while decreasing pregenomic mRNA synthesis.

- Both S-DAg and L-DAg have been shown to repress HBV enhancer sequences, which can cause a reduction in HBV replication.

- HDAg may bind to HBV mRNAs to affect their stability.

- In superinfected individuals, the genes in the integrated HBV DNA can be expressed to produce a lot of HBsAg without HBV replication.

Diagnosis, Treatment, and Prevention

For the diagnosis of HDV infection, several viral markers, such as HDV and HBV antigens, can be detected by anti-HDV antibodies using techniques such as enzyme-linked immunosorbent assay or radioimmunoassay. These assays are performed as the initial screening for HDV detection. Serological tests can also be used to diagnose and monitor HDV infection. Molecules that are to be detected through screening include liver enzymes, ALT, AST, HBsAg, HBV DNA level, anti-HDV antibody, and HDV RNA level. The detection of HDV RNA is performed to confirm the presence of an infection. Quantitative RT-PCR is always a useful tool to amplify target nucleic acids for detection. A chronic infection is the prognosis if there is positive detection of HDV RNA for more than 6 months. A quantitative microarray antibody capture assay to quantify the amount of anti-HDV IgG can be used for HDV diagnosis and has been shown to correlate with the detection of HDV RNA. The appearance of anti-HDV IgM correlates with an early acute infection as well as chronic infection. However, the detection and presence of anti-HDV IgM is unreliable. Immunohistochemistry and in situ hybridization can be used to detect HDAg in liver biopsy samples.

The primary recommended anti-HDV infection treatment is the use of pegylated IFN-α for therapy. Unfortunately, this treatment yields a low rate of response from the virus and is associated with significant side effects, such as fatigue, depression, reduced leukocytes, and reduced platelet counts. Individuals with decompensated cirrhosis (extreme liver damage and scarring), active psychiatric conditions, and autoimmune diseases are not recommended to go through this treatment. Fortunately, there are 2 promising drugs that are currently under phase III clinical trials: bulevirtide (BLV) and lonarfanib (LNF), as of a 2022 report. BLV inhibits the entry of HDV and HBV by interacting with and blocking the binding domain on NTCP. LNF prevents L-DAg from being farnesylated, thereby interfering with virion assembly. Last, a liver transplant is ultimately necessary for individuals with end-stage liver disease.

The best prevention for HDV infection is immunization against HBV since there is currently no anti-HDV vaccine. Unfortunately, options are limited, which means that an individual who is already HBV positive has no means for prevention. Other means of prevention include avoiding high-risk activities and behaviors whether or not you've been immunized. Avoiding high-risk activities and behaviors is crucial to prevent HDV and HBV transmission, especially if you've been infected. At risk individuals include those who inject drugs, sex workers, men who have sex with men, patients undergoing hemodialysis, people who have an HIV infection, and immigrants from areas of the world that have high levels of HDV infection.

Current Status

To date, 8 HDV genotypes have been identified. HDV1 is globally distributed but circulates predominantly in Europe, the Middle East, North America, and North Africa. HDV2 is common in Northern Asia, including Japan and Taiwan; and Middle Eastern countries like Iran and Egypt. HDV3 is mainly found in South America, especially the

Amazon Basin. HDV4 is predominant in China, Japan, and Taiwan. HDV5–8 circulate predominantly in Africa as well as various parts of Europe. Approximately 12–72 million people are infected with HDV worldwide. The most diverse variation of HDV genotypes is found in Central Africa, with the presence of HDV1, HDV5, HDV6, HDV7, and HDV8.

The WHO estimated that 5%–15% of the global population who have chronic HBV infection are affected by HDV infection. Roughly 20% of cases of liver disease and HCC are the result of HBV-HDV co-infection. Mongolia, the Republic of Moldova, and countries in western and central Africa are geographical hot spots for HDV transmission.

Further Readings

Block T, Alter H, London T, Bray M. A historical perspective on the discovery and elucidation of the hepatitis B virus. *Antiviral Res.* 2016;131:109–123.

Da B, Heller T, Koh C. Hepatitis D infection: From initial discovery to current investigational therapies. *Gastroenterology.* 2019;7(4):231–245.

Dandri M, Petersen J. Mechanism of hepatitis B virus persistence in hepatocytes and its carcinogenic potential. *Clin Infect Dis.* 2016;62(Suppl 4):S281–S288.

Datta S, Chatterjee S, Veer V, Chakravarty R. Molecular biology of the Hepatitis B virus for clinicians. *J Clin Exp Hepatol.* 2012;2(4):353–365.

Guvenir M, Arikan A. Hepatitis B virus: From diagnosis to treatment. *Pol J Microbiol.* 2020;69(4):391–399.

Kocher A, Barquera R, Key F, et al. Ten millennia of hepatitis B virus evolution. *Science.* 2021;374(6564):182–188.

Lampertico P, Degasperi E, Sandmann L, Wedemeyer H. Hepatitis D virus infection: Pathophysiology, epidemiology and treatment. *JHEP Rep.* 2023;5(9):100818.

Mentha N, Clement S, Negro F, Alfaiate D. A review on hepatitis D: From virology to new therapies. *J Adv Res.* 2019;17:3–15.

Mouzannar K, Schauer A, Liang T. The post-transcriptional regulatory element of Hepatitis B virus: From discovery to therapy. *Viruses.* 2024;16(4):528.

Rizzetto M. Hepatitis D virus: Introduction and epidemiology. *Cold Spring Harb Perspect Med.* 2015;5(7):a021576.

Rizzetto M, Canese MG, Aricò S, et al. Immunofluorescence detection of new antigen-antibody system (delta/anti-delta) associated to hepatitis B virus in liver and in serum of HBsAg carriers. *Gut.* 1977;18:997–1003.

Schollmeier A, Glitscher M, Hildt E. Relevance of HBx for Hepatitis B virus-associated pathogenesis. *Int J Mol Sci.* 2023;24(5):4964.

Stockdale A, Kreuels B, Henrion M, et al. The global prevalence of hepatitis D virus infection: Systematic review and meta-analysis. *J Hepatol.* 2020;73(3):523–532.

Tsukuda S, Watashi K. Hepatitis B virus biology and life cycle. *Antiviral Res.* 2020;182:104925.

Tu T, Budzindka M, Shackel N, Urban S. HBV DNA integration: molecular mechanisms and clinical implications. *Viruses.* 2017;9(4):75.

Tu T, Zhang H, Urban S. Hepatitis B virus DNA integration: In vitro models for investigating viral pathogenesis and persistence. *Viruses.* 2021;13(2):180.

Wei L, Ploss A. Mechanism of hepatitis B virus cccDNA formation. *Viruses.* 2021;13(8):1463.

Zhao F, Xie X, Tan X, et al. The functions of Hepatitis B virus encoding proteins: viral persistence and liver pathogenesis. *Front Immunol.* 2021;12:691766.

Zhao K, Liu A, Xia Y. Insights into hepatitis B virus DNA integration—55 years after virus discovery. *Innovation.* 2020;1(2):100034.

CHAPTER 10

Subviral Pathogens: Prion, Viroid, and Satellite

Prions

Historical Perspective

Scrapie is a disease displayed by sheep and goats; it is known as *rickets* in England, *scratchie* in Scotland, *der Trab* in Germany, *la maladie convulsive* in France, and *trzęsawka* in Poland. This condition was first reported in 1732 by Spanish shepherds, who observed merino sheep pathologically scraping against fences. In the spring of 1753, following an outbreak of scrapie in Lincolnshire, sheep farmers petitioned Parliament to introduce legislation to stop the movement of sheep. A report of this endemic in sheep was published in the Agricultural Improvement Society at Bath in 1788 and essentially stated that this was a serious issue. Benoit Fevrier from L'Institut Curie (Paris, France) suggested that this disease was caused by a "filterable agent" in 1899. Hans Creutzfeldt in 1920 and Alfons Jakob in 1923 described a fatal transmissible neurodegenerative condition, later to be named Creutzfeldt-Jacob disease (CJD). By the late 1930s, J. Cuille and P. L. Chelle demonstrated the transmissible nature of scrapie. Remember that the existence of virus, which is a transmissible, filterable agent, was reported by Dmitrii Ivanowski in 1892, and its presence was confirmed in 1898 by Martinus Beijerinck, which he referred to as *contagium virum fluidum*, meaning contagious living fluid.

The transmission of scrapie from sheep and goats to mice was demonstrated by J. A. Morris, D. C. Gajdusek, and R. L. Chandler in the early 1960s. The condition studied by Dr. Gajdusek is referred to as kuru, which won him the 1976 Nobel Prize, which he shared with Dr. Blumberg, who discovered the hepatitis B virus (HBV). Kuru is a slow-acting viral brain disease, with symptoms very similar to scrapie and CJD, found in Papua New Guinea that was later found to cause conditions similar to neurodegenerative diseases like Alzheimer disease, Parkinson, and Huntington. In 1967, Tikvah Alper, I. H. Pattison, and J. S. Griffith proposed that the infectious agent causing scrapie was an infective polypeptide. Dr. Pattison later showed evidence that the scrapie infectious agent was, in fact, of proteinaceous origin. Crucial experiments that supported their claim included the fact that this causative agent was resistant to exposure to

223

formalin, high temperature, and ionizing radiation, thus disproving that they were based on nucleic acid.

It was Stanley Prusiner who championed the hypothesis of this "proteinaceous infectious particle," which he coined "prion" in 1982. He demonstrated that scrapie infectivity is affected by mechanisms that traditionally destroy proteins but not affected by mechanisms that destroy nucleic acids. Interestingly, proin would have been the more appropriate name, but he thought that prion sounded better. For his work on prion, Prusiner received the Nobel Prize in 1997. In 1985, Bruce Chesebro and Richard Race isolated the messenger RNA (mRNA) transcript that encodes PrP 27–30, which is the protease-resistant prion particle or protein that was isolated by Dr. Prusiner. This mRNA was found in both infected and uninfected brain tissue. The first 2 cases of the bovine spongiform encephalopathy (BSE) outbreak were identified in the United Kingdom in 1985; however, this infection may have occurred as early as the 1970s. BSE is widely known as the "mad cow disease," which is a progressive and fatal disease of the nervous system in cattle. The first confirmed case of BSE in the United States was reported in late December of 2003 in an infected dairy cow imported to Washington State from Canada.

Prion Disease

Conditions caused by prions belong to a group of progressive, fatal, neurodegenerative disorders referred to as transmissible spongiform encephalopathies (TSEs) or simply prion diseases. In humans, this classification includes CJD, variant CJD (vCJD), variably protease-sensitive prionopathy (VPSPr), Gerstmann-Sträussler-Scheinker disease (GSS) syndrome, kuru, and fatal familial insomnia (FFI). Depending on the etiology of this disease, CJD is further classified as sporadic CJD, familial or genetic CJD, and acquired CJD. Sporadic CJD occurs when there is a spontaneous genetic mutation in the normal prion gene or the prion protein itself undergoes spontaneous misfolding after protein expression. Familial or genetic CJD occurs when an offspring inherits the mutated form of the gene from 1 or both parents. Acquired CJD is the result of consuming contaminated tissues.

Variant CJD is related to BSE when an individual consumes contaminated cattle product. VPSPr is extremely rare and occurs in individuals 70 years of age or older. GSS is also extremely rare and typically occurs in younger individuals at around 40 years old. Kuru was reported in Papua New Guinea and is caused by ritualistic cannibalism of infected human brain tissue. Once this ritual was discouraged and subsided, occurrences of kuru greatly declined, and kuru is now extinct. FFI is a rare genetic TSE, either sporadic or familial, that causes difficulty sleeping.

The TSEs in nonhuman animals include natural scrapie in sheep, goats, and mouflons; transmissible mink encephalopathy in ranch-reared mink; chronic wasting disease in mule deer and elk; BSE (or mad cow disease) in cattle; and feline spongiform encephalopathy in domestic cats.

Structure and Function of Nonpathogenic Prion Protein

The proteinaceous infectious particle named prion (PrP) by Dr. Prusiner is an alternative form of a prion protein encoded by a mammalian gene. The host, nonpathogenic prion protein is designated PrP^C, which has important cellular functions that are described in the next paragraph, whereas the mutated, infectious particle that causes diseases is designated PrP^{SC}. What is the function of PrP^C and how is PrP^{SC} produced? Let's take a look at this very interesting and unique mechanism for infection.

PrP^C is encoded by the PRNP gene in humans and mice that is conserved throughout vertebrates. PrP^C belongs to a mammalian prion gene family that includes SPRN (encoding a protein named shadoo or Sho) and PRND (encoding a protein named doppel or Dpl). PrP^C, Sho, and Dpl share structural domains that enable them to function as metal ion transporters. PrP^C itself is an exoprotein 30–35 kD in size and is modified by glycosylphosphatidylinositol (GPI) attachment. It is expressed in nerve cells and a number of peripheral cell types, particularly immune cells; however, its expression is most abundant in neurons of the brain and spinal cord. The highest concentration of GPI-modified PrP^C is localized in the lipid raft of the plasma membrane (Figure 10.1). Subcellular localization includes the endoplasmic reticulum ER, Golgi, and mitochondria.

In cultured cells, PrP^C expression can be induced by nerve growth factors, insulin, or insulin-like growth factor. PrP^C expression during ER stress, oxidative stress, and genotoxic stress has also been reported. PrP^C interacts with intracellular proteins to regulate signal transduction events. There are numerous proposed functions that are linked to PrP^C; these include:

Prion structures and subcellular localization

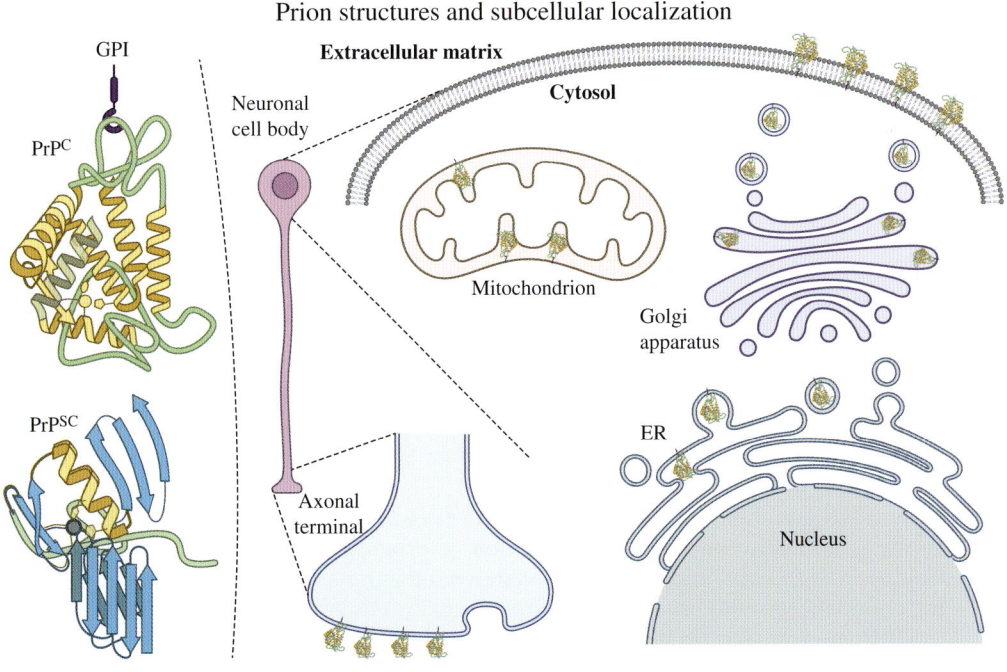

Figure 10.1 PrP^C versus PrP^{SC} structures and subcellular localization. PrP^C has a higher proportion of α-helices, whereas PrP^{SC} has a higher proportion of β-pleated sheets. PrP^C itself is modified by GPI attachment. There is a high concentration of PrP^C in the plasma membrane. PrP^C is also localized in the membrane of ER and Golgi apparatus as these molecules are being transported toward the plasma membrane via vesicular transport. PrP^C can also be delivered to mitochondria. *Credit:* Figure partially created in BioRender.

- neurite outgrowth
- antiapoptotic activities during oxidative stress–induced cell death
- proapoptotic activities during ER stress
- cellular uptake or binding of Cu^{2+} through its copper-binding domain
- formation and maintenance of synapses, adhesion to the extracellular matrix
- transmembrane signaling, regulation of myelin maintenance
- regulation of processes linked to cellular differentiation
- regulation of learning and memory as well as sleep patterns

PrP^C can also be released through exosomes (or extracellular vesicles), which are possibly involved in intercellular communication. The pathogenic agent PrP^{SC} is the misfolded form of PrP^C. As described, there are different origins for the production or the misfolded form.

Host Range, Transmission, Tropism, Susceptibility, Replication, and Dissemination

Prions are subviral particles that infect humans, sheep, goats, mouflons, mink, mule deer, elk, cattle, and domestic cats. The infectious agent that causes BSE is believed to have undergone interspecies transmission from cattle to humans, causing vCJD; this agent is the only known zoonotic prion. Species-specific infection depends on the transmission of different prion strains. Different prion strains are defined by 3 features: incubation period, pathology, and phenotype. Polymorphisms of the prion protein gene (PRNP) exist, which can produce different conformational variants or strains of PrP^{SC}, thus affecting variability in susceptibility. Different host organisms can be infected by the same prion strain. In turn, the same host organism can be infected by different prion strains.

Prions are transmitted via direct contact with bodily fluids such as saliva, blood, and urine or in feces. Transmission via indirect contact may also occur through contaminated soil, food, or water. Entry through the oral route is most common, leading to the amplification and accumulation of PrP^{SC} in gut-associated lymphoid tissues (GALTs). The prion that causes vCJD can persist in lymphoid tissues without undergoing neural dissemination, thus showing no signs of neurodegenerative symptoms, although the highest potential risk is from transcutaneous exposure such as through needlesticks, puncture wounds, and contamination of broken skin. Direct contact with any mucous membrane is also high risk.

In 1946, I. H. Plummer showed that transmission can occur from a scrapie-infected sheep to a healthy one by the inoculation of brain and cerebral spinal fluid. D. C. Gajdusek, C. J. Gibbs, and T. Alpers also showed the transmissibility of prions by infecting chimpanzees with brain material from humans who died of kuru or CJD.

PrP^C is expressed predominantly in neurons of the central nervous system (CNS), which means that amyloid formation can rise to a high concentration. In the CNS, PrPC is also found in astrocytes, oligodendrocytes, and microglia. In the peripheral nervous system (PNS), a high amount of PrPC can be found in the dorsal and ventral root ganglia of the spinal cord, sensory and motor axons, and Schwann cells., mast cells, macrophages, and dendritic cells as well as in organs, including the heart, intestine, kidneys, liver, pancreas, and spleen. Susceptibility is determined by several genes;

however, the expression of the prion protein gene (PRNP) is definitely the major genetic determinant.

PrP^C has a higher proportion of α-helices, while PrP^{SC} has a higher proportion of β-pleated sheets. This conversion from an α-helix-predominant form to a β-pleated sheet-predominant form may lower the Gibbs free energy of the prion protein. Proteins with lower Gibbs free energy are more stable; therefore, the reverse conformational change to the original PrPC form with higher Gibbs free energy is not likely. Thus, PrPSc is locked in this conformation and acts as a "seed" to produce more PrPSc to form amyloid fibrils. This process is also known as nucleation polymerization. When it encounters another PrP^C, PrP^{SC} forces this PrP^C molecule to misfold into the pathological conformation. This means that prions are capable of self-propagation or self-replication (Figure 10.2). The continuation of this process leads to an increase in the population of infectious prions. Since both PrP^C and PrP^{SC} can be efficiently secreted from the host cell within exosomes, both forms can be disseminated to noninfected neural cells or even remain in the extracellular matrix.

Nucleation polymerization, or the accumulation of PrP^{SC} molecules, yields the deposition of amyloid fibrils (or protofibrils) in the CNS, killing infected cells in the process, and leads to neurodegeneration causing TSEs. Amyloid fibrils are aggregates

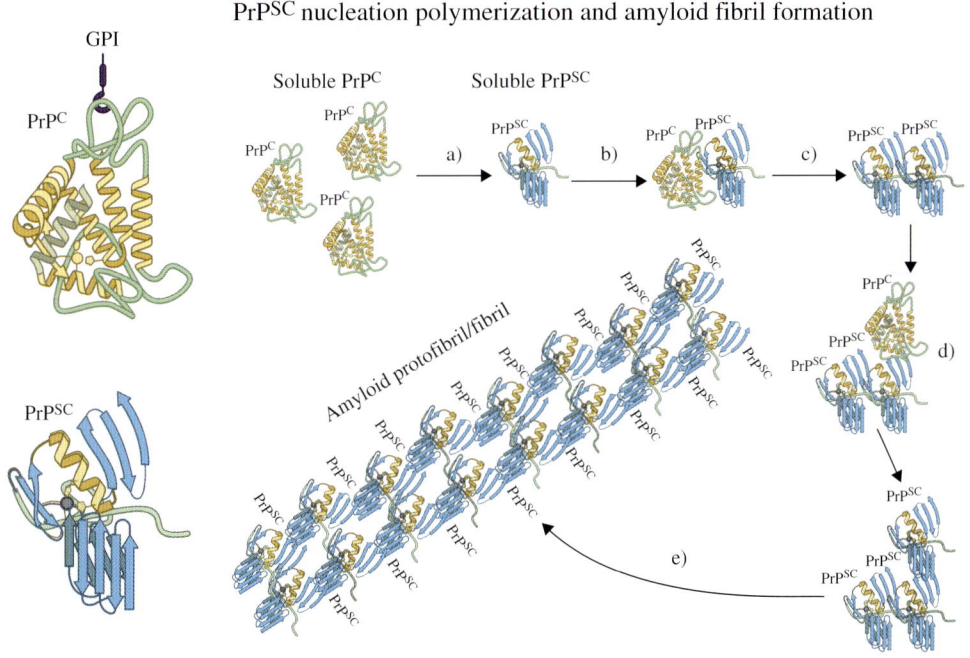

FIGURE 10.2   Growth of pathological PrP^{SC} seed and amyloid fibril formation. (a) A soluble PrP^C molecule is misfolded into PrP^{SC} molecule. This is the "seeding" or "nucleation" event. (b) This PrP^{Sc} can interact with another soluble PrP^{SC}. (c) Forcing the PrP^C to misfold as well. This is the start of self-replication. (d) As this process continues, it is known as nucleation polymerization. (e) The end result is the formation of an insoluble aggregate of proteins termed amyloid protofibril or amyloid fibril. *Credit:* Figure partially created in BioRender.

of soluble proteins that assemble to form insoluble fibers resistant to degradation. Amyloid aggregates appear as plaques in the brain of affected animals. These plaques are resistant to degradation by detergent, proteases, and an acidic environment.

As described in this section, GALTs are the principal target of a primary prion infection. Dissemination may occur via either the release of exosomes or through tunneling nanotubes (also referred to as membrane nanotubes). Nanotubes are cytoskeleton-dependent protrusions that extend from the plasma membrane to enable different animal cells to mediate the specific transfer of cellular materials, pathogens, and electrical signals between cells over long distances. Prions produced in the Peyer patches (in the ileum segment of the small intestine) can usually disseminate directly to the CNS, while prions from the spleen and lymph nodes often disseminate into the PNS before reaching the CNS.

Pathogenesis and Clinical Manifestation

Initial symptoms of nonhuman animals infected with prions include droopy ears, followed by shaking in the hindquarters. They can exhibit other abnormal behaviors, such as having altered gaits, licking excessively, and pathologically scraping against fences due to intense itching, hence giving rise to the name "scrapie."

In humans, typical symptoms include imbalance and problems with coordination. The individual may have difficulty in walking, changes in gait, and sudden and jerky movements. There is also a rapid development of dementia, which is exhibited by memory loss, impaired thinking, and personality changes. Other symptoms include trouble swallowing, blurry vision or even blindness, and insomnia and psychiatric symptoms like anxiety, depression, and hallucinations. These symptoms are associated with neuronal loss and a failure to stimulate inflammatory responses against a pathogenic infection or tissue damage. Fatality is often caused by falls, heart issues, lung failure, pneumonia, or other secondary infections.

The etiology of prion diseases can be sporadic, genetic, or acquired. Most prions in humans are from a rare aggregation of several PrP^C molecules that causes spontaneous conformational alteration into the PrP^{SC} form, which occurs at a very slow and limiting rate. This is the basis for sporadic prion disease etiology and accounts for 80% of all human cases. Approximately 10%–15%, however, are expressed from a genetically mutated PRNP gene, which can be heritable and is an autosomal dominant inheritance pattern. This is the basis for the genetic etiology and accounts for 10%–15% of human cases. Transmission of prions is the basis for the acquired etiology.

Diagnosis, Treatment, and Prevention

Diagnosis of TSEs typically consists of medical tests and physical examinations. Each prion disease requires its own specific set of diagnostic criteria. Blood tests are performed to detect the presence of PrP^{SC} in the patient's blood. There is also screening for possible neurological damage. Genetic testing is used to determine if the patient has gene variants. Magnetic resonance imaging is used to detect changes in brain structure. Lumbar punctures can be performed to screen the patient's cerebral spinal fluid for disease markers. Changes in brain waves can be measured with electroencephalography. Brain biopsies may also be obtained for histological screening.

For early disease diagnosis, a sensitive diagnostic assay called real time-quaking induced conversion can be used to amplify levels of PrP^{SC} that might otherwise be undetectable. Antisense oligonucleotides have been shown to be effective in inhibiting the expression of PrP^{C} and hence preventing the spontaneous conversion to PrP^{SC}, which is the etiology of 80% of all human TSEs. In terms of reducing the transmissibility of prions, hypochlorous acid (HOCL), a known antibacterial and antiviral agent, may be used to disinfect fomites, especially in hospitals. HOCL is a safe and very effective disinfectant that occurs naturally in the human body but, of course, should never be consumed.

Currently, there are no known treatments to cure TSE. The best approach is to treat symptoms that arise to impede the progress of the disease and prolong the patient's life. When developing therapeutic agents, there are 3 general approaches: (1) developing drugs that lower the expression of PrP^{C}; (2) developing drugs that inhibit the conversion from sPrP to PrP^{SC}; and (3) developing drugs that can block the downstream neurotoxic signaling cascade.

A number of medications have been shown to be effective. Flupirtine reduces cell death in neuronal cells in vitro. Treatment with flupirtine improves cognition but fails to increase survival. Pentosan polysulfate (PPS) reduces PrP^{SC} formation in vitro and in animal studies. PPS reduces abnormal prion deposition, reduces atrophy and neuronal loss, and reduces infectivity. However, this treatment does not improve neurological functions or affect brain postmortem pathology. Doxycycline binds to PrP^{SC} to hinder the process of nucleation polymerization to form amyloid fibrils. Doxycycline has been shown to prolong the incubation time of prions slightly.

Current Status

Creutzfeldt–Jakob disease is the most common form of human TSE. The reported incidence rate of CJD is 1–2 cases per million individuals globally, while about 350 cases are reported in the United States annually. Individuals who are 55 years of age and older are at higher risk of CJD, with a rate of 5 cases per million in the United States. However, there is no difference in the risk of sporadic CJD among the genders. This disease yields 100% fatality, with 70% of patients dying within a year of disease onset. The second most common type of TSE is genetic CJD, where patients usually inherit the autosomal-dominant PRNP gene mutations. The least common is acquired CJD, which occurs mainly in young adults in their late 20s. Variant CJD has also been detected in the United States, with a total of 4 cases reported since 2001.

Viroid

Historical Perspective

The potato spindle tuber disease was first reported by W. H. Martin in New Jersey in 1922, although the Department for Environment, Food, and Rural Affairs from the United Kingdom estimated that this disease probably originated from Central America much earlier. Eventually, this infection was spread throughout South America, Europe, Africa, and Asia and Oceania. Disease outbreaks have also been identified in tomato, pepper, and other citrus plants. In 1964, an outbreak of the potato spindle tuber disease occurred in the United Kingdom and caused yield losses of about 65% of potato plants. Theodor O. Diener reported in 1969 his discovery of a subviral particle that was identified as the causal agent for the potato spindle tuber disease; he later named this particle "viroid." The potato spindle tuber viroid (PSTVd) is the smallest and simplest known agent of infectious disease. Around the same time,

Joseph Semancik and Lewis Weathers reported a similar infectious agent, later named citrus exocortis viroid or ceVd, that causes diseases in citrus plants, including *Poncirus trifoliata*, Rangpur lime, and Swingle citrumelo and citrange. In 1977, Heinz Sänger reported that his group successfully isolated causative agents, viroids, in large quantities from diseased plants. Hans Gross sequenced the complete nucleotide sequence of PSTVd in 1978.

Two outbreaks on the tomato disease occurred in 2003 and 2011 in the United Kingdom. The 2011 outbreak caused almost 100% yield losses of tomato. The tomato planta macho disease was first observed in 1969 in the state of Morelos, Mexico; it was later shown to be caused by the tomato planta macho viroid (TPMVd). Four additional wild solanaceous plant species (*Solanum nigrescens, S. torrum, Physalis aff. foetens*, and *Jaltomato procumbens*) were found to be infected by viroids in Morelos.

Viroid Disease

Viroids are subviral particles that have been found to infect plants conclusively, although there are indications that they may also be able to infect phytopathogenic fungi. The International Committee on Taxonomy of Viruses initially classified viroids in the early 1990s. Viroids of the Pospiviroidae family target and cause diseases in the following plants:

- Apple scar skin viroid infects apple and pear;
- Apple dimple fruit viroid and apple fruit crinkle infect apple;
- Australian grapevine viroid, grapevine yellow speckle 1, and grapevine yellow speckle 2 viroid infect grapevine;
- Chrysanthemum stunt viroid infects chrysanthemum;
- Citrus bark cracking viroid infects citrus;
- Citrus bent leaf viroid and citrus dwarfing infect citrus;
- Citrus exocortis viroid infects citrus and tomato;
- Coconut cadang-cadang viroid infects coconut palm, African oil palm, and other monocots;
- Coconut tinangaja viroid infects coconut palm;
- *Coleus blumei* 1 viroid infects *Coleus* and *Mentha* spp.;
- *Coleus blumei* 3 viroid infects *Ocimum basilicum* and *Melissa officinalis*;
- *Columnea* latent viroid infects *Columnea, Brunfelsia, Nemathanthus*, and tomato;
- Hop stunt viroid (HSVd) infects citrus, grapevine, *Prunus* spp., hop, and cucumber;
- Hop latent viroid infects hop;
- *Iresine* viroid infects *Iresine*;
- Mexican papita viroid infects *Solanum cardiophyllum*;
- Pear blister canker viroid infects pear and quince;
- PSTVd infects tomato and avocado;
- Tomato apical stunt viroid; and
- TCDVd and TPMVd infect tomato.

Viroids of the Avsunviroidae family target and cause diseases in the following plants:

- Avocado sun blotch viroid infects avocado;
- Chrysanthemum chlorotic mottle viroid infects chrysanthemum;
- Eggplant latent viroid infects eggplant; and
- Peach latent mosaic viroid infects peach and nectarine.

Structure and Function of Viroid

Viroids are classified into 2 families: Pospiviroidae and Avsunviroidae. Pospiviroidae includes 5 genera with 39 members: *Pospiviroid* (eg, potato spindle tuber viroid), *Hostuviroid* (eg, hop stunt viroid), *Cocadviroid* (eg, citrus bark cracking viroid), *Apscaviroid* (eg, apple dimple fruit viroid), and *Coleviroid* (eg, coleus blumei viroid 1). Avsunviroidae includes 3 genera with 5 members: *Avsunviroid* (eg, avocado sunblotch viroid), *Pelamoviroid* (eg, chrysanthemum chlorotic mottle viroid), and *Elaviroid* (eg, eggplant latent viroid). The 2 families are identified on the basis of their biological, biochemical, and structural features. Further classification of species within these 2 families is based on multiple criteria, including subcellular localization of the viroid, whether they are transported into the nucleus or into the plastid (mainly in the chloroplast); whether rolling circle replication (RCR) is the mode of viroid replication, either symmetric or asymmetric; the presence of one or all conserved sequences and motifs like terminal conserved sequence, the presence of terminal conserved hairpin, and ribozyme motif. Species differentiation is based on the threshold of 90% conservation of nucleotide sequence and host specificity.

The structure of viroids is a naked, circular, single-stranded RNA (ssRNA) that is not encapsidated. The size of the genome is 246–434 nucleotides and does not contain open reading frame (ORFs) that encode for proteins (Figure 10.3). The viroid itself possesses some enzymatic activities. Viroid ssRNAs contain high amounts of self-complementary sequences that form multiple secondary structures, which contribute to enzymatic activities and produce compact, robust, rod-like structures. Viroids are not only thermodynamically stable but also include metastable elements such as secondary hairpins (hairpins I and hairpins II), loops (loop E), and pseudoknots. These elements are in kinetically favored conformations to provide essential regulation of viroid replication, intracellular movement, as well as intercellular movement. They are also involved in the induction of diseases and facilitate entry and exit of the viroid molecule from the plant vascular system.

The structure of viroid in the Pospiviroidae family is a circular ssRNA genome that can form rod-shape, quasi rod-shape, or branched (with secondary structures) structures consisting of 5 domains: central conserved region (CCR), pathogenicity-determinant domain (P), domain with high sequence variability (V), and 2 terminal domains (TL and TR). Additionally, the ssRNA molecule is composed of an unbranched series of short helices and small loops. It is localized in the nucleus and utilizes host nuclear DNA-dependent RNA polymerase II (DdRPII) to carry out replication via the asymmetrical RCR mechanism.

The structure of viroids in the Avsunviroidae family is a circular ssRNA genome that can form a branched-rod shape consisting of an intrinsic hammerhead ribozyme motif. It is localized in the chloroplast. This ssRNA molecule lacks the CCR found in

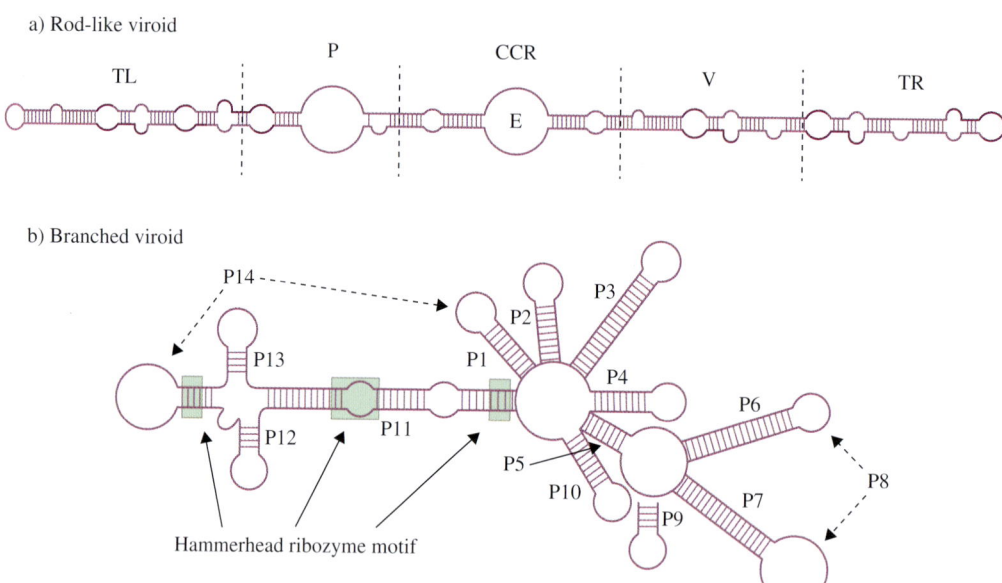

a) Rod-like viroid

b) Branched viroid

Hammerhead ribozyme motif

FIGURE 10.3 Viroid structures. (a) Rod-like viroid structure. The 5 domains of the viroid molecule are the CCR, pathogenicity-determinant domain (P), domain with sequence variability (V), and 2 terminal domains (TL and TR) at each end of the rod-like structure. (b) Branched viroid structure contains 14 loops with 3 regions of hammerhead ribozyme motif. *Credit:* Figure partially created in BioRender.

viroids of the Pospiviroidae family. It utilizes host plastid RNA polymerase to carry out replication via a symmetrical RCR mechanism.

Host Range, Transmission, Tropism, Susceptibility, Replication, and Dissemination

The host range for viroids includes higher plants such as monocots and dicots, vegetable crops, woody perennials, and agronomic and ornamental plants. As of today, diseases caused by these infectious agents are globally distributed. There are 4 modes of viroid transmission: vegetative propagation, mechanical transmission, infected seed and pollen, and aphid or vector transmission.

1. Vegetative propagation (vegetative reproduction) is a form of asexual reproduction in plants. New plants arise from a fragment of the parent plant called vegetative propagules. The most common mechanisms of vegetative propagation are grafting, cutting, layering, suckering, and tissue culture and tuber, bulb, or stolon formation. New growths are clones of the parental vegetation; therefore, DNA assortment, DNA recombination, or horizontal transfer does not occur.

2. Mechanical transmission occurs as the result of direct contact with viroid-infected sap and contaminated farming tools that are used during normal cultivation activities under favorable conditions. The mode of transmission

occurs prevalently in vegetations that are grown along rows like potatoes and tomatoes.

3. Seed and pollen transmission occurs among germplasm collections, which include primitive landraces and wild species related to particular crops, and developed varieties and breeders' lines. Germplasm is used to develop new plant varieties for food, feed, fiber, turf, forages, and ornamentals. Germplasm is also used for forestry, industrial, and medicinal purposes. This mode of transmission is a major contributor to epidemic and pandemic outbreaks due to international seed and pollen trade and exchange.

4. Vector transmission occurs when viroids are transmitted from 1 plant to another via a vector. The vector can be insects like larvae of a codling moth and aphids (small sap-sucking insects). These insects may have ingested viroid-contaminated sap from an infected plant using mouthparts with a needle-like stylet used to access and feed on plant cells. Fungal transmission may also occur through infection by phytopathogenic ascomycetes such as apple scab, rice blast, the ergot fungi, black knot, and the powdery mildews. These ascomycetes themselves may have been targeted by microbes like *Nostoc* sp. PCC 7120 (filamentous blue-green cyanobacterium) and even other fungi like viroid-infected *Saccharomyces cerevisiae* (unicellular yeast fungus) before transmission to vegetations. The vector can also be bacteria or viruses that introduce viroid-modified genetic material to the plant.

Since viroids do not encode for proteins, they require host mechanisms for intracellular as well as intercellular movement. Once inside the host cell, these RNAs are delivered to either the nucleus (mechanism used by the Pospiviroidae family) or plastid (mechanism used by the Avsunviroidae family). It's been proposed that Virp (viroid RNA-binding protein1) is a plant protein that is responsible for intracellular trafficking by forming a ribonucleoprotein complex with viroids. Virp is a bromodomain-containing protein that possesses 2 important domains: an RNA-binding domain and a nuclear-localization signal. Import into the nucleus may also require additional binding of TFIIIA and ribosomal protein L5 (RPL5). TFIIIA is a general transcription factor that regulates the synthesis of 5S ribosomal RNA.

Viroid replication is divided into 3 steps: (1) transcription, (2) cleavage, and (3) ligation. Viroids are nonprotein-coding RNAs and thus do not encode for proteins (Figure 10.4).

1. Replication of viroids in the Pospiviroidae family is initiated at the TL loop of the original circular positive-polarity RNA, which serves as the template to be used by DdRPII to produce an oligomeric (multimeric) intermediate of the opposite polarity. This positive-polarity RNA is repeatedly transcribed to amplify viroid synthesis greatly.

2. The linear, multimeric intermediate strand in turn serves as the template to be used by DdRPII to produce a linear, multimeric concatemer of the positive polarity. The participation of transcription factor IIS, TFIIIA, and RPL5 in viroid replication of both positive and opposite polarity have been reported.

3. Viroids of the Pospiviroidae family do not possess intrinsic ribozymes, so they must rely on a host ribonuclease for cleavage. Multimeric RNAs are in turn cleaved to proper unit length, forming monomeric circular RNAs of the original

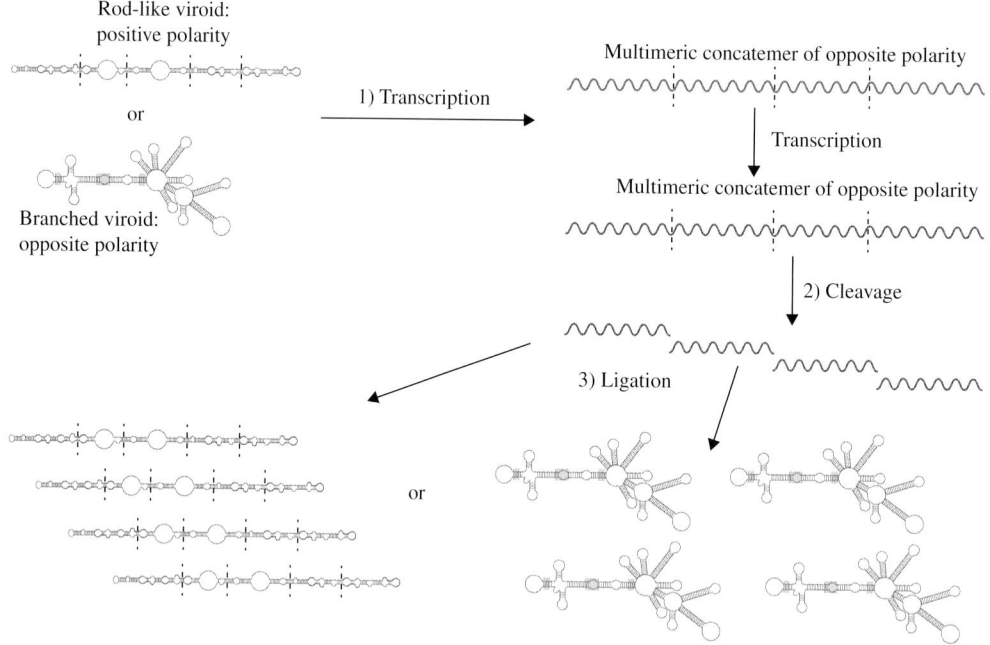

FIGURE 10.4   Viroid replication and amplification. (1) The viroid of positive polarity is used as the template for the synthesis of a linear, multimeric intermediate of opposite polarity. Transcription is catalyzed by cellular DNA-dependent RNA polymerase II. This multimer of opposite polarity is in turn used as the template for the synthesis of a linear multimer of positive polarity. (2) The multimer of positive polarity is then cleaved. The rod-like viroid relies on cellular ribonuclease for cleavage, while the branched viroid employs its intrinsic ribozyme activity. (3) Ligation is the final step producing many more viroid molecules. *Credit:* Figure partially created in BioRender.

polarity. The CCR contains a cleavage-ligation site that is targeted by the host RNase III type enzyme for cleavage and host DNA ligase 1, which is repurposed as an RNA ligase, for ligation into individual units of circular RNA.

Replication of viroids in the Avsunviroidae family carries out symmetric RCR for amplification following these 3 steps: (1) The host plastid RNA polymerase uses the original circular positive-polarity RNA as the template to produce an oligomeric intermediate of the opposite polarity. (2) The intermediate is cleaved and ligated into the circular RNA form of the opposite polarity. (3) Each circular RNA of the opposite polarity undergoes the same RCR process again to produce linear, multimeric concatemers of the original polarity to be cleaved and ligated, producing viroids. Members of the Avsunviroidae family, unlike the Pospiviroidae family, do possess an autocatalytic RNA ribozyme in each polarity to carry out self-cleavage, which produces monomeric circular RNAs of the original polarity. The efficiency of the ribozyme activity is enhanced by host factor PARBP33 (Plant Adaptation RNA Binding Protein). PARBP33 binds to chloroplast RNAs to regulate their maturation, stabilization, and editing. Ligation is catalyzed by chloroplast transfer RNA ligase.

After amplification, viroids are transported back into the cytosol. These infectious agents are rapidly disseminated from the mesophyll to the bundle sheath via plasmodesmata that are formed between the 2 structures. From the bundle sheath, viroids enter the phloem, then spread into the upper leaves (Figure 10.5). Systemic viroid transport involves the aid of host proteins such as phloem protein 2 (PP2) and α-helical protein Nicotiana tabacum (Nt)-4/1. PP2 is a dimeric chitin-binding lectin that has an RNA-binding domain. It has a defense-related function and is the most abundant protein in phloem exudate. The specific function of Nt-4/1 is still unknown.

Pathogenesis and Clinical Manifestation

Some viroid-infected plants exhibit symptoms; however, other viroids can replicate inside the host cell without eliciting disease symptoms. Symptoms vary depending on the species of plant. These symptoms include stunting; epinasty (flower petals are bent outward or downward); vein discoloration and clearing; leaf distortion and mottling; chlorotic mosaic (discoloration or albinism); necrotic spots; cankers; scaling and cracking of bark; malformation (of tubers, flowers, and fruits); and even death of the plant. A major cytopathic symptom includes the presence of accumulated paramural bodies, termed plasmalemmasomes, in the cytoplasm. These structures are invaginations of the plasmalemma. The presence of plasmalemmasomes is associated with leaf epinasty and blistering. Other cytopathic effects include abnormal plastid development, irregular thickening of the cell wall, and electron-dense deposits in the cytoplasm as well as chloroplasts.

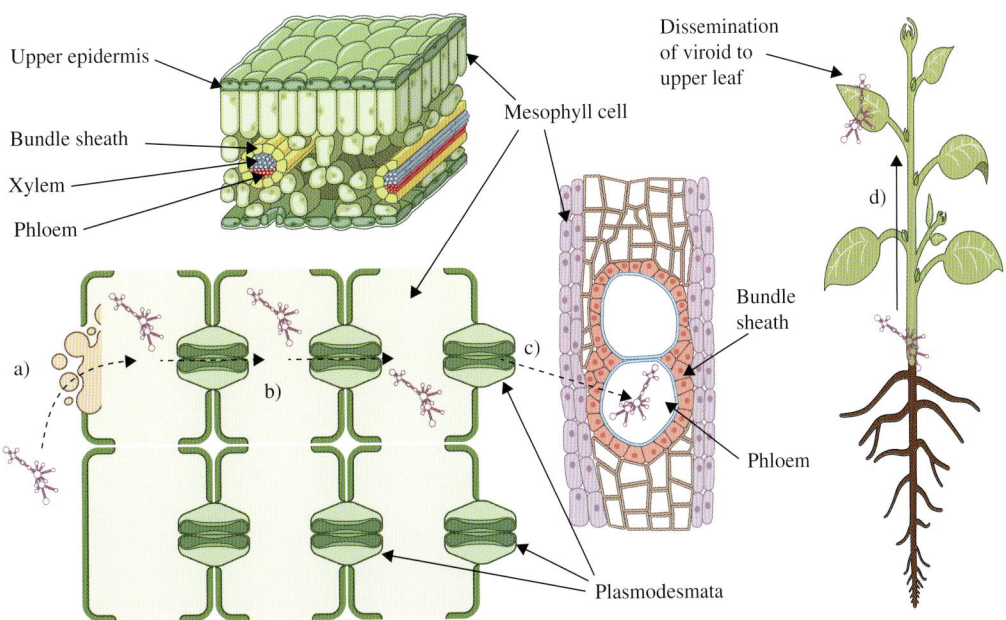

Figure 10.5 Dissemination of viroids from original damaged plant cell to cell in upper leaves. (a) Viroids enter plant cell through damaged cell wall. (b) Viroids spread to neighboring plant cells by transport through channels between cells called plasmodesmata. (c) Viroids eventually migrate to the phloem and sieve tissue that carries water and nutrients up toward the tip of the plant. (d) Viroids are carried to upper leaves along with water and nutrients. *Credit:* Figure partially created in BioRender.

For potatoes, PSTVd causes smaller but elongated tubers that are misshapen and cracked. The eyes of the tubers are larger and become knob-like protuberances. These outgrowths may themselves develop into small tubers. Leaves on the potato plant are smaller as well. Growth of infected potatoes as well as tomato plants is reduced at first, then develops into stunting. There's chlorosis of leaves on the top of these tomato plants, initially turning yellow, then red and/or purple. Leaves become brittle, and flower and fruit initiation ceases. For infected pepper plants, there are only very mild symptoms, which include growth reduction and abnormal margin of leaves located near the top of the plant. Infected vine plants are smaller and produce smaller leaves. Viroid infection of many solanaceous ornamentals is often asymptomatic.

In the infected host cell, viroids alter expression of genes that are involved in defense and stress response, cell wall structure, chloroplast function, protein metabolism, and hormone signaling. One of the mechanisms for viroid pathogenesis is via RNA silencing of host mRNAs. Since viroids are non–protein-coding RNAs, they require host machinery for this function; therefore, they hijack cellular processes to damage host RNAs. There are 2 pathways for RNA silencing: (1) Dicer-like 1 (DCL1) regulates the synthesis of microRNAs (miRNAs) and (2) DCL2, DCL3, and DCL4 regulate the synthesis of small interfering RNAs (siRNAs). Both miRNAs and siRNAs can be further amplified through the activity of RNA-directed RNA polymerases (RdRPs). Both pathways can be hijacked by the invading viroid to produce large amounts of viroid-derived small RNAs (vd-sRNAs), which are 21–24 nucleotides long. The vd-sRNAs can guide the RNA-silencing–mediated degradation of complementary and physiologically relevant host mRNAs, such as chloroplastic heat shock protein 90 mRNA, transcription factor StTCP23 (spindle tuber teosinte branched1/Cycloidea/Proliferating cell factor 23), and thylakoid translocase subunit required for chloroplast development. Large amounts of vd-sRNAs have been detected from specific hot spot regions within both positive- and opposite-polarity RNA strands.

Viroid pathogenesis may also involve viroid-associated manipulation of alternative splicing and epigenetic mechanisms such as DNA methylation and histone modification. In fact, the activity of RNA-directed DNA-methylation discovered in PSTVd-infected cells is a major regulator of epigenetic alteration leading to overexpression of certain host genes. Last, the impairment of host translational machinery has also been associated with viroid infections.

Intracellular Host Defense Against Viroids

Interestingly, host RNA silencing is the primary mechanism by which the host cell defends itself against a viroid infection. In other words, posttranscriptional RNA silencing is used by both the host and infectious agent to damage each other's RNAs. What's even more fascinating is the fact that vd-sRNAs drive sequence-specific degradation of both viroids and host mRNAs. However, only viroid intermediates are susceptible to RNA-silencing machinery, while mature, circular viroid RNAs are ignored. A number of key regulators of RNA silencing are upregulated in viroid-infected plants, including DCLs, Argonaute protein family, and RdRP6. The high amount of base-paired, stem-loop structures of viroid RNA are unsurprisingly targeted by the various DCLs for cleavage. The activity of DCL2 and DCL3 appear to be antiviroid to defend the host cell, while DCL4 (found in both the nucleoplasm and cytosol) plays a key role in processing vd-sRNAs to carry out host RNA silencing.

Diagnosis, Treatment, and Prevention

The primary tool to diagnose a viroid infection of plants is visual inspection. As described in the Pathogenesis and Clinical Manifestation section, the following visible symptoms can be easily detected: dwarfing, stunting, leaf chlorosis, epinasty and necrosis, vein discoloration, bark cracking, distortion of tubers, flowers and fruits, delays in flowering, foliation and ripening, and, in some cases, death of the plant. Detection of the presence of viroids can include biological indexing (or bioassays) and direct detection of the genomic viroid RNA, for which the most common method is the molecular (specifically dot-blot) hybridization technique. The reliable polymerase chain reaction (PCR) technique is universally used since it's a highly rapid, versatile, specific, and sensitive molecular method to detect nucleic acid. Reverse transcription PCR and real-time PCR assays, biosensors, rolling circle amplification, metagenomics, and deep sequencing are also used for viroid diagnosis. Most recently, 2-dimensional, nondenaturing/denaturing polyacrylamide gel electrophoresis were developed to purify low-molecular-weight nucleic acids.

The application of either heat therapy (35–37°) or cold therapy (5–20°) had been used with mixed results as treatment to eliminate viroid in host plants. Cryotherapy involving the incubation of shoot tips in liquid nitrogen (−196°C) yielded much better results. Shoot-tip grafting technology has also been applied with high success. However, this is a difficult technique that requires a high level of expertise. Other techniques such as chemotherapy, tissue culture, thermotherapy, and electrotherapy have been used to protect plants against viroid infection.

Viroid disease management is divided into 2 parts: (1) prevention of an infection and (2) viroid eradication. (1) Deterrence of the introduction of infected plant material is the most effective method to control viroid disease. Prevention of an infection requires careful and complete preclusion of the introduction of viroid into a specific crop. Planting material, including tubers, seeds, and plants, must be tested to ensure that they are viroid free. Workers must adhere to hygiene best practice to ensure that their hands, gloves, clothes, and equipment are viroid free since this infectious agent is mechanically transmissible. Chemical substances (eg, 1%–5% sodium hypochlorite, 6% hydrogen peroxide, and 2% sodium hydroxide with 2% formaldehyde) can be used as disinfectants to clean equipment used in the field. The number of different plant species must be limited when grown in a small and/or enclosed area like a greenhouse. Aphid population must be controlled or eradicated since they are a major vector for viroids. (2) Viroid eradication can be achieved through the destruction of viroid-infected plants. All equipment and material used during the handling of infected plants must be thoroughly cleaned, disposed by incineration, or taken to a refuse dump. Crop rotation in the field can reduce or eliminate infected volunteer plants.

Current Status

In the 21st century, a number of outbreaks occurred. Two tomato disease outbreaks occurred in 2003 and 2011 in the United Kingdom. The 2011 outbreak caused almost 100% yield losses of tomato. There was an outbreak of hop stunt disease (HSD) caused by HSVd in Washington State in 2004, and another outbreak was confirmed as recent as 2017 in Ohio State. In fact, HSD epidemics have been reported in the past 2 decades worldwide, such as in Xinjiang, China, and Slovenia in Europe. New PSTVd variants have also been discovered from potatoes, tomatoes, and other ornamental plants during the past 2 decades in Canada, China, Europe, India, Japan, New Zealand, Russia, Turkey, and United States.

Satellite

Historical Perspective

The origin of satellite genomes is still unknown; however, possible natural original sources of these nucleic acids could have been from either the genome of the cognate helper virus or even other co-infecting satellites, the host organism genome, or transmission vectors. The first satellite was identified in 1962, which was a satellite virus associated with the tobacco necrosis virus. In 1969, a satellite RNA associated with the tobacco ringspot virus was also identified. By 1978, the discovery of another satellite RNA associated with the cucumber mosaic virus (CMV) named CARNA 5, the causal agent of lethal tomato necrosis disease, was reported at a conference in Munich. CARNA 5 is a double-stranded satellite RNA whose size is similar to a viroid and can induce pathogenic effects. However, it differs from a viroid in that CARNA 5 is dependent on the helper CMV to replicate. Infection of CARNA 5 along with CMV is shown to have necrogenic properties, whereas infection with CMV alone yields marginal symptoms. This is to say that CARNA 5 appears to enhance the infectivity of CMV.

Satellite Nucleic Acid Disease

Most satellites are associated with plants but can be found in bacteria and animals such as mammals and arthropods. They don't actually cause specific diseases, but they can modulate symptoms triggered by their cognate helper viruses. This modulation can be increasing, decreasing, or having no effect. Example of pathogens that can be co-infected with satellites include: malaria, dengue virus, Zika virus, Vibrio cholerae, and blood flukes causing Schistosomiasis (also known as snail fever). Additionally, satellite nucleic acids have been found to be associated with various cancers such as breast, ovarian, and prostate cancer.

Structure and Function of Satellite Nucleic Acid

Satellites are divided into many classifications: single-stranded satellite (ss-satellite) DNA, double-stranded satellite RNA, single-stranded satellite RNA (large ss-satellite RNA, small linear ss-satellite RNA, circular ss-satellite RNA), satellite-like RNA, and satellite virus.

There are 2 groups of ss-satellite DNA: alphasatellite with 40 members (eg, *Ageratum* yellow vein, milk vetch dwarf C1 alphasatellite) and betasatellite with 62 members (eg, *Ageratum* leaf curl Cameroon betasatellite, zinnia leaf curl betasatellite). There are 24 members of double-stranded satellite (ds-satellite) RNA (eg, M satellites of *Saccharomyces cerevisiae* L-A virus, *Bombyx mori* cypovirus 1 satellite RNA). Members of ss-satellite RNA are divided into 3 groups: large ss-satellite RNA with 11 members (eg, *Arabis* mosaic virus large satellite RNA, bamboo mosaic virus satellite RNA); small linear ss-satellite RNA and satellite-like RNA with 15 members (eg, artichoke mottled crinkle virus satellite RNA, pea enation mosaic virus satellite RNA); circular ss-satellite RNA (also known as virusoid) with 10 members (eg, *Arabis* mosaic virus small satellite RNA, velvet tobacco mottle virus satellite RNA).

Satellite-like RNAs include 1 member of hepadnavirus-associated satellite-like RNA (eg, *Deltavirus*-like satellite-like RNA) and 3 members of polerovirus-associated RNA (eg, beet western yellows virus ST9-associated RNA, tobacco vein distorting virus-associated RNA). They are regarded as components that can rescue a deficiency in the cognate helper virus. Additionally, they are often identified as a part of the helper virus genome. However, satellite-like RNAs can be dispensable since they are not always associated with the helper viral genome.

Satellites are viral or subviral particles that require the co-infection of a host cell with a helper virus. They are, in fact, parasites of parasites. The cognate helper viruses provide special viral proteins for the replication, encapsidation, movement, and transmission of these satellites. Satellites are genetically distinct from their helper viral genome. They are not a part of the helper viral genome or required for the helper virus infection cycle.

There are 2 distinguishing classes of satellite: satellite nucleic acid and satellite virus. Satellite nucleic acids encode either nonstructural proteins or no proteins at all and therefore require coat proteins of the helper virus to encapsidate their genetic material. Satellite viruses encode their own structural proteins, which are distinct from those of their helper viruses, to encapsidate their genetic material. A related category is satellite-like RNA, which, as the name indicates, resembles satellite nucleic acids. It requires co-infection with a helper virus for propagation; however, unlike satellite nucleic acids, they encode a protein that contributes to the biological success of their helper viruses. In fact, they are sometimes found within the viral genome, but not always, and thus are considered to be an integral element of the helper virus.

Single-stranded satellite DNAs are usually circular molecules that are 680–1300 bases long and typically encode a nonstructural protein that may play a role in the pathogenicity of the helper virus. Double-stranded satellite RNAs are linear molecules of 500–1500 base pairs. These molecules contain G + C contents of roughly 50%.

Single-stranded satellite RNAs are the most common type of satellites. Nucleic acids of satellites range from 194 bases to 1500 bases. All satellites have 4 features in common. (1) They regulate, often exacerbate, the symptoms caused by their cognate helper viruses. (2) They require the RdRP of their cognate helper viruses for genome replication. (3) They alter the production of their cognate helper viral particles, usually to decrease it, in the host cell in order to reserve materials and effort for their own replication. (4) They can accumulate to high concentrations in the host cell and be used as vectors for foreign genes.

Large satellite RNAs are linear molecules of 900–1300 bases that encode a protein required for either the replication of the satellite RNA or efficient systemic movement of the satellite RNA. Some encode other proteins with varying function, such as altering symptom responses and suppressing the accumulation of their cognate helper viruses. Small satellite RNAs are either linear or circular molecules of 220–390 bases with a high degree of internal base pairing, similar to viroids; however, they do not possess enzymatic activities. These molecules can affect the production of cognate helper viruses as well as degree of symptoms exhibited in the host.

Structure and Function of Satellite Virus

Satellite viruses include: chronic bee-paralysis virus-associated satellite virus (CBPSV) with 1 member (eg, chronic bee-paralysis satellite virus); satellites that resemble tobacco necrosis satellite virus with 4 members (eg, maize white line mosaic satellite virus, tobacco mosaic satellite virus); nodavirus-associated satellite virus with 1 member (eg, *Macrobrachium rosenbergii* nodavirus XSV); adenovirus-associated satellite virus with 1 member (eg, *Dependovirus*); and Mimivirus-associated satellite virus with 1 member (eg, *Acanthamoeba castellanii* Mamavirus–associated satellite virus; also known as Sputnik, which is a virophage). A virophage is a small, ds-satellite virus that co-infects with a giant virus.

There are only 2 satellite viruses that are associated with humans: hepatitis delta virus (HDV) of the Deltaviridae family and adeno-associated virus (AAV). HDV is a satellite ssRNA virus whose cognate virus is hepatitis B virus. A detailed discussion of HDV can be found in Chapter 9 of this textbook under the hepatitis delta virus section. AAV is a satellite ssDNA virus whose cognate virus is adenovirus.

CBPSV is a small viral particle with 17 nm in diameter. Its genome includes 3 single plus-strand RNA molecules totaling 1100 bases. This is the only known satellite virus with a segmented genome. Increased replication of CBPSV appears to decrease the accumulation of its cognate helper virus, chronic bee-paralysis virus.

Satellites that resemble tobacco necrosis satellite virus are small viral particles 17–22 nm in diameter. Its genome is a single plus-strand RNA of 824–1240 bases. The size of a nodavirus-associated satellite virus is 15–25 nm in diameter, which includes a single plus-strand RNA genome of 800–900 bases.

ASV is a small viral particle 25 nm in diameter. Its genome is a linear, ss-DNA molecule of 4500 bases. Its known cognate helper virus is not only the adenovirus, but also herpesvirus and papillomavirus.

Mimivirus-associated satellite virus (also known as Sputnik) is a fairly large viral particle 74 nm in diameter. Its genome is a single, circular double-stranded DNA molecule of 17,000–33,000 base pairs. It possesses a 73% A-T content and 16–34 ORFs. One of these proteins is the DNA polymerase ε; therefore, Sputnik does not require the host DNA replication machinery, although it does depend of the transcriptional machinery of the cognate helper virus. Sputnik is classified as a virophage, which is a bona fide parasite of the giant viruses Mimivirus and Mamavirus (for a more deltailed discussion, see Chapter 11 under the Virophage section).

Pathogenesis and Clinical Manifestation

Satellite nucleic acids are associated with not only mostly plants but also some bacteria. Satellite virus, on the other hand, can be found in bacteria and animals such as mammals and arthropods. The synergistic and/or antagonistic interactions among satellites, their cognate helper viruses, and the host cell yield 3 possible types of symptom modulation: attenuation, exacerbation, and no significant effect. Symptom attenuation is the most common result. Attenuation occurs when the satellites reduce the replication and accumulation of their cognate helper virus, therefore decreasing the pathogenicity and symptoms caused by the helper virus. This action is highly beneficial to the host. Some satellites may exacerbate symptoms caused by their cognate helper virus and thus are detrimental to the host. Satellites may be benign and have no effect on either helper virus or host cell activity; therefore, all symptoms observed come from helper virus activity.

The host cell's RNA-silencing mechanisms are involved in not only the defense of itself against infections, but also the use by both the satellite and its cognate helper virus to induce or eliminate pathogenicity and symptoms. This trilateral competition for the targeting of RNA-silencing mechanisms is a means for each of these 3 entities to enhance its reproducibility while suppressing the competition. Of course, RNA silencing is 1 of the most effective host antiviral defense mechanisms, especially in plants. The host cell uses this mechanism to silence satellite RNAs and helper viral genome, thus blocking their replication. Satellites and their cognate helper viruses use this mechanism not only to block each other's reproduction, but also to silence physiologically relevant genes of the host cell. The exhibition of symptoms occurs when the helper virus targets host genes. The exacerbation of symptoms can occur when host genes are additionally targeted by the satellite. The following are examples of this trilateral competition:

- The capsid proteins of turnip crinkle virus (TCV) have been shown to inhibit the DCL2/DCL4 silencing pathway of *Nicotiana benthamiana*, a very important plant platform for biopharmaceutical protein and vaccine production. However, TCV capsid proteins are most effective in the free, cytosolic form and are weak suppressors when assembled into a capsid. TCV satellite RNA can decrease the assembly of TCV, thus reducing the accumulation of virions. This reduction results in the increased level of free TCV capsid proteins, which in turn can strongly suppress host RNA-silencing ability. The downregulation of the RNA silencing actually benefits the helper virus, allowing it to survive inside the host cell. However, viral symptom-inducing capability is also limited due to the significant reduction in virion production.

- Cymbidium ringspot virus (CymRSV) competes with CymRSV satellite RNA to reproduce in *Nicotiana benthamiana*. However, to eliminate its competition, CymRSV expresses an RNA-silencing suppressor P19 to control and reduce the accumulation of its cognate satellite.

- A satellite can sometime cause symptoms in the host. For example, CMV-associated satellite RNAs can induce systemic necrosis in tomato plants and a yellowing symptom in some Nicotiana species. In both cases, the satellite RNA's siRNAs guide the cellular RNA-silencing machinery to degrade host mRNAs.

- CMV-associated satellite RNA acts as an RNA-silencing suppressor by sequestering siRNAs. This sequestration prevents siRNAs derived from CMV from entering into the RNA-silencing machinery, thus preventing the degradation of CMV mRNAs.

- The 2b protein of CMV is a silencing suppressor protein that can induce symptoms in *Nicotiana benthamiana* and arabidopsis. However, CMV-associated satellite RNAs have been shown to use the RNA-silencing machinery to target the 2b protein, thus attenuating the pathogenicity of CMV.

Further Readings

Baiardi S, Mammana A, Capellari S, Parchi P. Human prion disease: molecular pathogenesis, and possible therapeutic targets and strategies. *Expert Opin Ther Targets*. 2023;27(12):1271–1284.

Castle A, Gill A. Physiological functions of the cellular prion protein. *Front Mol Biosci.* 2017);4:19.

Diener T. Origin and evolution of viroids and viroid-like satellite RNAs. *Virus Genes*. 1996;11(2–3):119–131.

Di Serio F, Owens R, Navarro B, et al. Role of RNA silencing in plant-viroid interactions and in viroid pathogenesis. *Virus Res*. 2023;323:198964.

Gnanasekaran P, Chakraborty S. Biology of viral satellites and their role in pathogenesis. *Curr Opin Virol*. 2018;33:96–105.

Hadidi A, Sun L, Randles J. Modes of viroid transmission. *Cells*. 2022;11(4):719.

Hu C-C, Hsu Y-H, Lin N-S. Satellite RNAs and satellite viruses of plants. *Viruses*. 2009;1(3):1325–1350.

Katzourakis A, Aswad A. The origins of giant viruses, virophages and their relatives in host genomes. *BMC Biol*. 2014;12:51.

Koonin E. Viroids and viroid-like circular RNAs: Do they descend from primordial replicators? *Life*. 2022;12(1):103.

Kovalskaya N, Hammond R. Molecular biology of viroid–host interactions and disease control strategies. *Plant Sci*. 2014;228:48–60.

Lee B, Koonin EV. Viroids and viroid-like circular RNAs: Do they descend from primordial replicators? *Life*. 2022;12(1):103.

Liberski P. Historical overview of prion diseases: A view from afar. *Folia Neurophathol*. 2012;50(1):1–12.

Mahabadi H, Taghibiglou C. Cellular prion protein (PrPc): Putative interacting partners and consequences of the interaction. *Int J Mol Sci*. 2020;21(19):7058.

Mougari S, Sahmi-Bounsiar D, Levasseur A, Colson P, Scola B. Virophages of giant viruses: An update at eleven. *Viruses*. 2019;11(8):733.

Sano T. Progress in 50 years of viroid research—Molecular structure, pathogenicity, and host adaptation. *Proc Jpn Acad Ser B Phys Biol Sci*. 2021;97(7):371–401.

Steger G, Riesner D, Prusiner S. Viroids, satellite RNAs and prions: Folding of nucleic acids and misfolding of proteins. *Viruses*. 2024;16(3):360.

Venkataraman S, Badar U, Shoeb E, Hashim G, AbouHaidar M, Hefferon K. An inside look into biological miniatures: Molecular mechanisms of viroids. *Int J Mol Sci*. 2021;22(6):2795.

Wang Y. Current view and perspectives in viroid replication. *Curr Opin Virol*. 2021; 47:32–37.

Zabel M, Reid C. A brief history of prions. *Pathog Dis*. 2015;73(9):ftv087.

CHAPTER 11

Giant Virus and Virophage: Discoveries in the 21st Century

Acanthamoeba polyphaga Mimivirus

Historical Perspective

La Scola et al. reported the identification of the first "giant virus" (also referred to as "megaphage" or "girus") that infects the amoeba *Acanthamoeba polyphaga* in 2003. The giant virus is named *Acanthamoeba polyphaga* Mimivirus (APMV). The name Mimivirus for "mimicking microbe" refers to the fact that these pathogenic agents resemble a small gram-positive coccus when Gram stained. For about a decade, this infectious agent was thought to be an intracellular bacterium of amoebas until the publication in 2003. This Mimivirus was actually isolated from a power-plant cooling tower during a search for the causative agent of a pneumonia outbreak in Bradford, England, in 1992. No amplification products were obtained when low-stringency polymerase chain reaction (PCR) was performed with universal 16*S* rDNA bacterial primers. In other words, the absence of ribosomal DNA confirmed that this entity was not bacteria.

In 2014, Legendre et al. reported the discovery of *Pithovirus sibericum* in a more than 30,000-year-old radiocarbon-dated ice core harvested from permafrost in Siberia, Russia. Giant viruses have only been discovered in recent years because of their size. Traditionally, viruses are filterable infectious agents that can pass through filters with pores of 200 nm. Researchers searching for viruses failed to identify these viruses in the filtration since they remain in the residue. Mimiviruses are so large that they can be viewed with light microscopy. In 2008, an even larger giant virus strain was isolated in another water-cooling tower in Paris, France. This virus is a relative of APMV and was named *Acanthamoeba castellanii* Mamavirus (ACMV), which infects the amoeba *Acanthamoeba castellanii*. This particular giant virus not only contains a slightly larger genome than Mimivirus but also hosts a parasite virus named Sputnik. Sputnik is classified as a virophage, defined as a virus that parasitizes a virus (for a more detailed discussion, see Chapter 10 under the Structure and Function of Satellite Virus section), that is associated with both Mimivirus and Mamavirus. APMV and ACMV belong to the same genus of giant virus.

Giant Virus Overview

Giant viruses are classified under the phylum Nucleocytoviricota. Members of this phylum are also known as nucleocytoplasmic large DNA viruses (NCLDVs). Giant viruses are so-named because they are the largest known viruses, in terms of both virion and genome size (Figure 11.1). Giant viruses are considered to be ones with genomes larger than 300 kilobase pairs and with capsid diameters of about 200 nm or more. This is a lineage of eukaryotic viruses that includes many animal and protist pathogens; however, amoebas are the main hosts of giant viruses.

The size of giant viruses ranges from 200 nm to as big as 1.5 μm with genomes that range from 300 to 2500 kilobase pairs (or 2.7 megabase pairs). Pithovirus has the distinction of being the largest known virus and measures 1.5 μm in length and 0.5 μm in diameter. The genomes of giant virus contain hundreds, even thousands, of open reading frames (ORFs) that encode for proteins involved in functions such as DNA maintenance (DNA replication, DNA repair, and recombination); transcription and RNA maturation; and translation and proteostasis. Recent astounding data report the existence of giant viral genes that regulate cellular metabolic pathways involving in glycolysis, fermentation, gluconeogenesis, tricarboxylic acid cycle, photosynthesis, and lipid β-oxidation. These genes are found to be orthologues of those

Size comparison of giant virus with HIV and *E. coli*

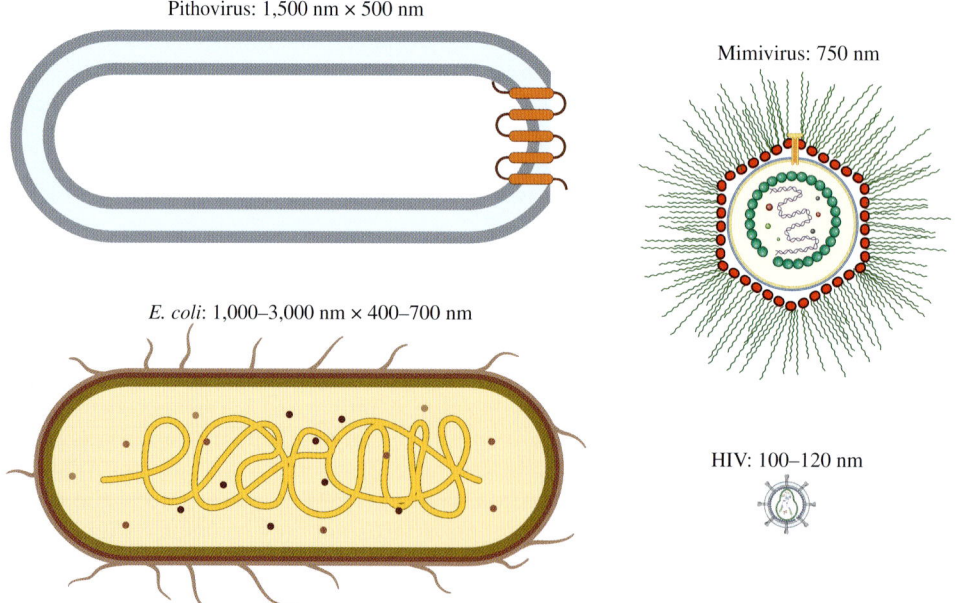

FIGURE 11.1   Size comparison of giant virus with another virus and bacteria. The giant virus *Pithovirus* is roughly the same size as the bacterium *Escherichia coli*, twice as large Mimivirus, and 10–15 times larger than the human immunodeficiency virus. *Credit:* Figure partially created in BioRender.

found in eukaryotes, such as aminoacyl transfer RNA (tRNA) synthetases. Marine giant viral genes can also include those involved in the recycling of carbon metabolism, nitrogen, and nutrients as well as soil organic matter degradation. Ultimately, 5 genes have been found to be common to all NCLDV viruses and thus to all giant viruses.

Giant viruses are found in fresh and marine waters, including thermal waters and deep-sea vents. They have also been detected in soil and plants, as well as animals like humans and ruminants. They mimic microbial prey that are internalized by the host cell (mostly amoebas) via phagocytosis or endocytosis. They are released from the host cell via either exocytosis or cell lysis. Transmission is via the natural external fluid environment to the next host. A discussion of APMV, in the Classification and Structure of Mimivirus section of this chapter, is used as a sample of the life cycle of a giant virus.

Presently, 32 families of NCLDV that reside in diverse ecosystems, most abundantly in the ocean, have been identified, including several well-known families that infect vertebrates, for example, Asfaviridae (eg, African swine fever virus) and Poxviridae (eg, smallpox virus); and invertebrates, for example, Ascoviridae (eg, *Heliothis virescens ascovirus*), Iridoviridae (eg, lymphocystis disease virus), and Phycodnaviridae (eg, *Paramecium bursaria* chlorella virus 1). Giant viruses include families that infect algae and other protists are Marseilleviridae (eg, tunisvirus and insectomime virus), and Mimiviridae (eg, APMV). Other giant virus families include Allomimiviridae (eg, *Pyramimonas orientalis* virus), Molliviridae (eg, *Mollivirus sibericum*), Pithoviridae (eg, *Pithovirus massiliensis*), Pandoraviridae (eg, *Pandoravirus braziliensis*), and Phycodnaviridae (eg, *Chrysochromulina brevifilum* virus PW1).

Classification and Structure of Mimivirus

Acanthamoeba polyphaga Mimivirus belongs to the Mimiviridae family and genus *Mimivirus*. Members of this genus also include Mamavirus (second giant virus ever discovered), Mimivirus Bombay, Mimivirus gilmour, Mimivirus golden, and Samba virus.

The structure of APMV is a spherical, multilayered virion with an overall diameter of 750 nm. The linear double-stranded DNA (dsDNA) genome of 1200 kb is associated with proteins and messenger RNAs (mRNAs) to form the core compartment (or viral core or nucleocapsid) (Figure 11.2). These proteins include 12 enzymes involving with the regulation of transcription, 5 involved in DNA repair mechanism, 2 involved regulating RNA modification, and 5 involved in translational modification. The core compartment is, in turn, enclosed by 2 monolayers, inner and outer, of lipid membrane. The diameter of this whole structure thus far is 340 nm. This spherical membrane-bound structure is encapsidated inside an icosahedral capsid of 450–500 nm in diameter. The capsid itself is covered with a layer of closely packed collagen-like fibrils that radiate from the center, giving the virion a total diameter of 750 nm. The fibrils themselves are about 120–140 nm long and 1.4 nm thick and are extensively glycosylated. Mimivirus is the only giant virus and thus member of NCLDV that is known to possess this peripheral fiber layer. Another interesting feature is the absence of fibrils at 1 of the 12 vertices of the capsid. This region is instead covered by a starfish-shaped proteinaceous structure named "stargate."

The APMV genome contains approximately 979 ORFs, and gene expression is regulated through early and late gene promoters. The ORFs are delineated into 4 main groups:

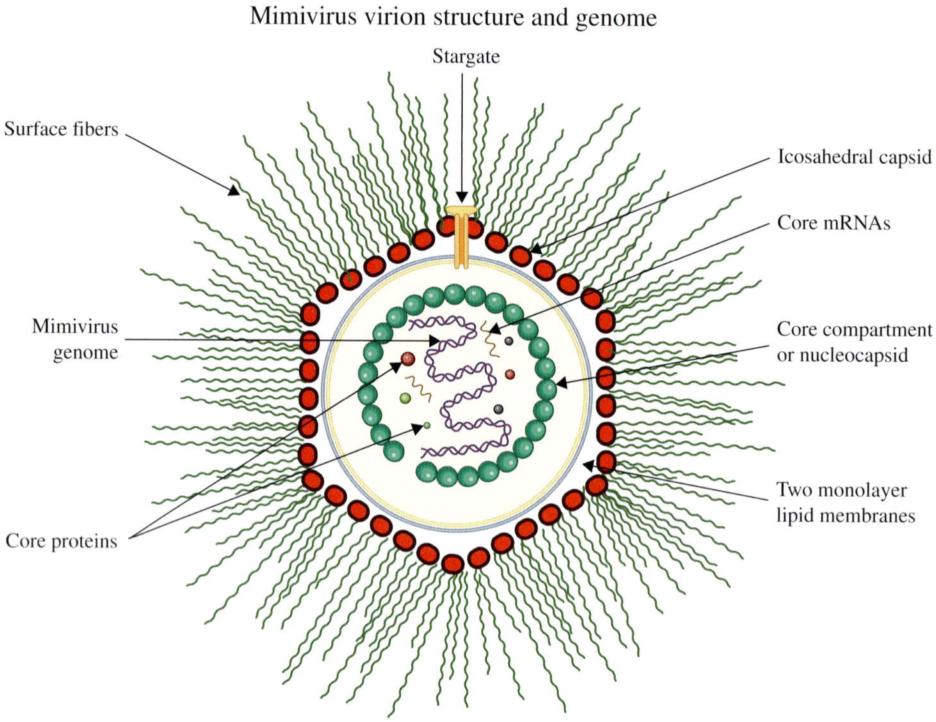

Mimivirus virion structure and genome

FIGURE 11.2 AMPV virion structure and genome. Schematic of Mimivirus illustrating the multilayered virion that is surrounded by a layer of fibers. *Credit:* Figure partially created in BioRender.

1. Megavirales core genes (9 of them): D5-lie helicase-primase/ATPase (adenosine triphosphatase) (superfamily III helicase); DNA polymerase elongation subunit family B; A32-like virion packaging ATPase; A18 helicase (superfamily II); transcription factor VLTF2 (viral late transcription factor 2); major capsid protein (MCP) D13L; thiol oxidoreductase; D6R/D11L-like helicase (superfamily II); and serine/threonine protein kinase. *Note:* The first 5 listed are found in the genome of all giant viruses.

2. Genes involved in lateral gene transfer.

3. Duplicated genes.

4. ORFans: Orphan genes or orphan ORFs are new genes that enable an organism to adapt to its specific living environment.

APMV proteins include those that are involved in DNA repair mechanisms, cell motility, membrane biogenesis, translation (eg, amino-acyl tRNA synthase and translation factors), protein folding, secondary metabolite biosynthesis, transport and catabolism, and metabolism of amino acid, lipid, and polysaccharide. The viral genome also contains a mobile DNA element or mobilome, which corresponds to pro-virophages and transpovirons (dsDNA episomes within the giant virus or even in the virophage capsid). Some genes encode for introns and inteins. There is even a Mimivirus virophage resistance element (MIMIVIRE) that helps defend against virophage infection of the giant virus. MIMIVIRE is analogous to the CRISPR (clustered regularly interspaced short palindromic repeats)-Cas (CRISPR-associated) pathway. The CRISPR-Cas pathway is a bacterial immune system against invading viruses and plasmids.

Infection Cycle of APMV in Amoeba

The Acanthamoebae appear to be the only confirmed host of APMV; however, sponges and corals have been indicated to be potential hosts. Additionally, there are signs that APMV may infect vertebrates, like mice and even humans, as well. There have been reports of the presence of APMV in mice after intracardiac infection and in human phagocytic cells, peripheral blood mononuclear cells, and the gut. The AMPV genome has also been detected in monkeys and bovines.

The glycosylated fiber layer enables APMV to attach to the host cell with yet to be identified specific receptor (Figure 11.3). The virus is internalized into the host cell via phagocytosis. Once in the cell, uncoating occurs first, leading to the fusion between the 2 viral monolayer lipid membranes with the phagosomal membrane. The viral core is then released into the cytosol through the channels of the stargate structure. These channels are open as the result of phagocytosis. The viral core contains the dsDNA genome

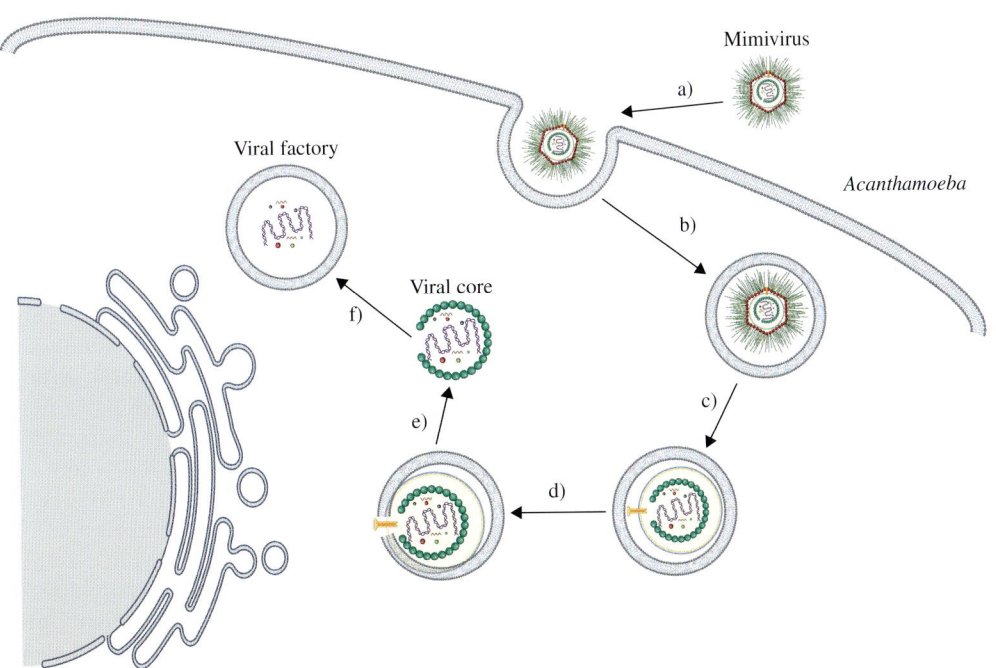

FIGURE 11.3   AMPV entry into host cell. (a) AMPV enters the host amoeba via phagocytosis, a process used to internalize food for protists. (b) The virion is placed in a phagosome. (c) Uncoating occurs. (d) The 2 monolayer lipid membranes fuse with the phagosomal membrane. (e) The viral core is released into the cytosol through the "stargate" complex. (f) The viral genome is enveloped in a viral factory where virion production occurs. *Credit:* Figure partially created in BioRender.

with associated proteins and mRNAs and is surrounded by a core wall composed of viral proteins. This integrity of this structure is maintained until structures referred to as "viral factories" are constructed. Viral factories are specialized endoplasmic reticulum (ER)–derived compartments formed in the cytosol where virion production occurs. Once in the cytosol, expression of early genes to regulate viral genome replication and the construction of viral factories begins immediately.

Since APMV encodes for most of its own DNA replication and transcription mechanisms, there is less requirement for host machineries from the nucleus. It's worth noting that not all giant viruses encode for their own DNA replication and transcription machinery, like the pandoraviruses and molliviruses. In this case, the genome of these viruses does need to be imported into the host cell nucleus to gain access to the necessary functions. The Marseilleviruses constitute yet a third type of giant viruses; they encode for some functions but still require additional help from nuclear activities.

For APMV, virion production occurs within viral factories and is initiated with the formation of the inner and outer monolayer lipid membranes around viral proteins and mRNAs, but without the genome itself. These monolayer membranes originated from a lipid bilayer that is derived from host ER membrane. It is thought that this lipid bilayer ruptures after the enclosure, leading to the formation of 2 monolayer lipid membranes. Eventually, multiple viral factories fuse into a large compartment where capsid assembly occurs for virion production. Capsid protein, L425, is used as the scaffolding protein for capsid assembly to enclose the 2 membranes. The viral genome is then deposited into this structure through a transient pore in the capsid. Release of APMV is due to the lysis of the host amoeboid cell during the rapid and high-level propagation of virions.

Infection Cycle of APMV in Humans

APMV appears to target human myeloid cells, specifically phagocytic cells like circulating monocytes, monocyte-derived macrophages, and THP (Tohoku Hospital Pediatrics)-1 myelomonocytic cells. THP-1 is a monocyte isolated from peripheral blood from an acute monocytic leukemia patient used in immune system disorder research. Only phagocytes can carry out phagocytosis, which is a mechanism used to target and destroy pathogens like viruses and bacteria. APMV is found mostly in the gut, although they've also been detected in human feces and from respiratory samples in patients suspected to have tuberculosis. Water is likely a source of human exposure to these viruses. In addition to the fecal-oral route, transmission may also occur through direct contact, consumption, and/or inhalation of contaminated water present in sources such as fresh and marine water, drinking water, shellfish, and air. Aerosols, aerosolized water, and mechanical ventilation can be contaminated with amoebas infected with APMV. Fomites such as walls and equipment in hospitals are potential sources of transmission.

Once in the host body, attachment occurs when the highly glycosylated fibrous exterior of the virus binds host cell surface. This interaction does not require specific host receptors, which is possibly a reason for the broad host range and tropism. Internalization of APMV occurs through either phagocytosis, mediated by cytoskeleton reorganization and the phosphatidylinositol 3-kinase pathway, which is known to be involved in phagocytosis and macropinocytosis.

Once inside the host cell, the APMV life cycle in human cells is similar to its life cycle in amoebas. Virion maturation occurs in the cytoplasm, and virions are released by exocytosis or lysis of the host cell.

It is well known that *Acanthamoeba*, a genus of free-living amoebas, can be a part of the normal microbiota in humans as well as other vertebrates. These protists are ubiquitous and can be found in a number of environments that are connected with humans, including soil; air; aquatic environments; sewage treatment systems; contact lenses; hospital environments (ie, hospital floors and emergency rooms, nurseries, and kitchens); and ventilation and air conditioning systems. Protists that are transmitted into human hosts may themselves be infected by giant viruses like APMV. These viruses can in turn disseminate and infect target human cells.

APMV is resistant to fairly high temperatures up to 65°C; however, they can be thermally inactivated at 75°C and are sensitive to exposure with active chlorine at 3% for 5 minutes. They can possibly withstand an acidic environment and therefore are able to survive in the human gut. Due to these resilient features, their ability to infect humans is only natural and possibly prevalent. APMV presence has been linked to severe and chronic diseases by themselves, such as granulomatous amoebic encephalitis, cutaneous acanthamoebiasis, amoebic keratitis, primary amoebic meningoencephalitis, and pneumonia. It is unclear, however, if the presence of APMV in the human gut yields any clinical or biological significance. It is possible that they are only transient passengers and contribute to disease symptoms caused by the host local inflammatory responses.

Detection involves using antibodies against giant viral components with techniques such as immunohistochemistry on tissue specimens as well as respiratory samples and immunofluorescence on blood and also tissue specimens and as well as respiratory samples. Universal PCR systems, often used to amplify nucleic acid in extremely small quantities, have been proven ineffective in detecting giant virus, mostly due to low viral loads and the quite varied genetic diversity among giant virus species. Data on genome sequences are currently insufficient.

Current Status

Eight giant viruses were discovered in melting Siberian permafrost and were revived. These giant viruses are dubbed "zombie virus" because they are still infective after being revived (or resurrected to stay with the theme). (1) *Pithovirus sibericum* was discovered in a 30,000-year-old sample in 2014. It is 1 of the largest known viruses at about 1.5 μm in length. This virus was revived when permafrost samples were exposed to amoebas. *Pithovirus sibericum* poses no harm to humans or other animals. (2) *Mollivirus sibericum* was found in the same 30,000-year-old Siberian permafrost sample as *Pithovirus sibericum*. This virus also poses no harm to humans or other animals. (3) *Pithovirus mammoth* was recovered from a clump of petrified mammoth wool unearthed on the banks of the Yana riverbank. It is estimated that this virus is 27,000 years old and measures 1.8 μm in length. (4) *Pandoravirus mammoth* was also discovered in the 27,000-year-old petrified mammoth wool from the Yana riverbank. It was found again in petrified mammoth stomach content in the Lyakhovsky Islands. This virus is estimated to be 28,600 years old. (5) *Pandoravirus yedoma* is the oldest known virus resurrected from permafrost and was discovered in 48,500-year-old icy deposits under a lake in Yukechi Alas. It measures 1 μm in length. (6) *Megavirus mammoth* was isolated from the same 27,000-year-old clump of ice discovered on the Yana riverbank where *Pandoravirus mammoth* and *Pandoravirus yedoma* were found. (7) *Pacmanvirus lupus* was discovered in the 27,000-year-old intestinal remains of a Siberian wolf (*Canis lupus*). (8) *Cedratvirus lena* was isolated from permafrost on the muddy banks of the Lena River. It measures 1.5 μm in length, similar to *Pithovirus sibericum*.

Virophage

Historical Perspective

In 2008, *Acanthamoeba castellanii* Mamavirus (ACMV) was the second giant virus discovered and was isolated in a water-cooling tower in Paris. ACMV is a close relative of APMV. The discovery of this giant virus came with a surprise. It hosts its own parasite virus, named Mimivirus-associated satellite virus (also known as Sputnik). Sputnik is not classified as a satellite virus but as a virophage, which is defined as a virus that parasitizes virus host. Virophages act as bona fide parasites that can inhibit giant virus replication. The study of parasite-on-parasite or "hyperparasitism" is an emerging field, along with discovering giant viruses.

Classification of Virophage

Virophage is used to describe any virus containing a dsDNA genome that can co-infect a eukaryotic host cell with a giant virus. Sputnik was the first virophage discovered and was named because of its satellite-like characteristic. Once both viruses had been internalized, the virophage would appropriate a portion of the transcription and DNA replication machinery of the giant virus to carry out its own reproduction. Of course, this sounds very much like a satellite virus except, unlike a satellite virus, a virophage actually interferes with the giant viral activities to inhibit their reproduction and possibly induce the synthesis of defective viral particles.

Currently, only a handful of virophages have been identified. They belong to the Lavidaviridae family and are classified into 2 identified genera, *Sputnikvirus* and *Mavirus*, along with a number of unclassified species. The genus *Sputnikvirus* is further divided into 2 groups of strains: species Mimivirus-dependent virus Sputnik and species Mimivirus-dependent virus Zamilon.

1. Mimivirus-dependent virus Sputnik includes Sputnik, Sputnik2, and Sputnik3.
   - Sputnik was isolated from a water-cooling tower in Paris, France, along with its giant virus host ACMV. Even though originally found in a sample with Mamavirus, Sputnik can infect both ACMV and APMV.
   - Sputnik2 was isolated from lens liquid with its giant virus host *Acanthamoeba polyphaga* Lentille virus.
   - Sputnik3 was isolated from a soil sample collected in Marseille, France, with a Mimivirus reporter but without its natural giant viral host.
   - Rio Negro virophage was isolated in the waters of the Negro River in Brazil with Samba giant virus.
2. Mimivirus-dependent virus Zamilon includes Zamilon1 and Zamilon2.
   - Zamilon1 was isolated from a soil sample in Tunisia.
   - Zamilon2 was isolated in poplar sawdust from a bioreactor in New York State.

The genus *Mavirus* was the second genus of virophage to be discovered and includes *Cafeteriavirus*-dependent mavirus that parasitizes the giant virus *Cafeteria roenbergensis* virus (CroV). *Cafeteriavirus*-dependent mavirus and its giant virus host can co-infect the marine phagotrophic flagellate *Cafeteria roenbergensis*. Studies of this virophage led researchers to determine that the virophage is linked to eukaryotic mobile genetic elements of the Polinton (also known as Maverick) class. Mobile genetic elements in eukaryotes are either retrotransposons and DNA transposons that are divided into 5 major classes: long terminal repeat (LTR) retrotransposons, non-LTR retrotransposons, cut-and-paste DNA transposons, rolling circle DNA transposons (or Helitrons), and self-synthesizing DNA transposons (or Polintons/Maverick). It has been suggested that these transposons may be able to form virions under appropriate conditions, hence the connection with the virophage.

The most recent virophage that was isolated in 2022 is *Chlorella* virus virophage SW01 (CVv-SW01), which parasitizes a large green alga virus called *Chlorella* virus XW01 (CV-XW01) from the waters of Lake Dishui near Shanghai, China. CV-XW01 belongs to the Mimiviridae family and is closely related to the giant virus CroV; however, CVv-SW01 is phylogenetically distant from the virophage mavirus.

Structure of Virophage

Virophages are nonenveloped with small icosahedral capsids with a diameter of 35–74 nm. They possess a closed, circular dsDNA genome of 17,000–19,000 base pairs in length and containing 16–34 ORFs. The genome of all identified virophages contain 4 conserved genes encoding for (1) major capsid protein (MCP), (2) minor capsid protein (mCP), (3) ATPase involved in DNA packaging, and (4) cysteine protease (PRO), which is a cysteine protease possibly involved in capsid maturation. MCP and mCP are involved in morphogenesis.

Sputnik (or Mimivirus-associated satellite virus) is a large viral particle 74 nm in diameter (Figure 11.4). Its dsDNA genome of 18,343 base pairs contains 21 partially overlapping

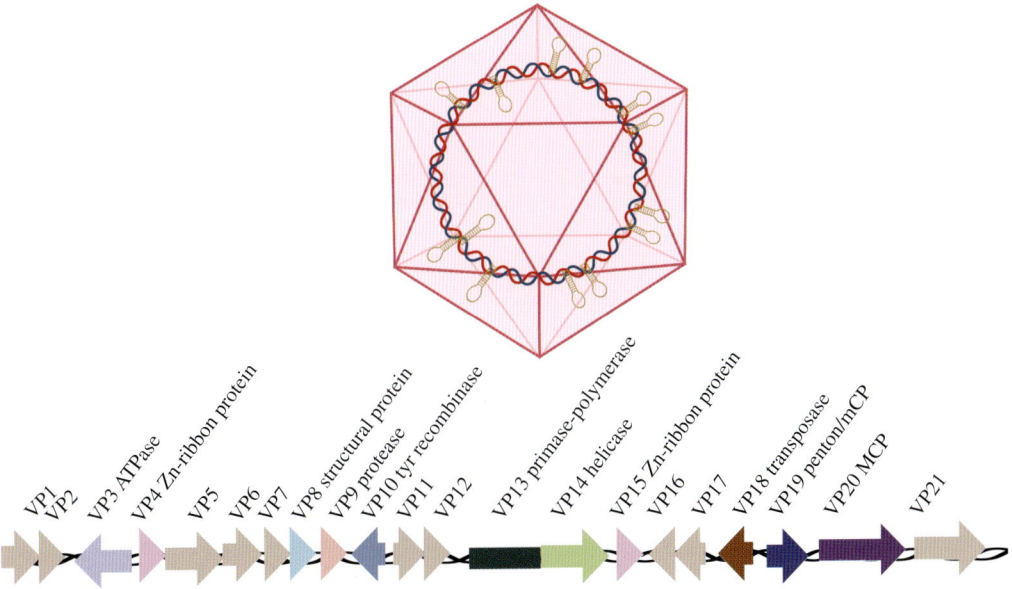

FIGURE 11.4 Sputnik structure and genome. Sputnik is a simple non-enveloped virus with a circular dsDNA genome surrounded by an icosahedral capsid. Its genome contains 21 ORFs. Proteins encoded by 11 of these ORFs have been identified, while 10 are still unknown. *Credit:* Figure partially created in BioRender.

ORFs encoding for proteins such as DNA-packaging ATPase, zinc-ribbon protein, structural protein, PRO, recombinase, helicase-primase, transposase, mCP, and MCP.

Zamilon is a spherical virophage with a diameter of 60 nm. Its dsDNA genome of 17,276 base pairs contains 20 ORFs. This genome has a 76% homology in sequence and 15 predicted gene products that are similar to those of Sputnik. Zamilon gene products include transposase, mCP, MCP, helicase, recombinase, PRO, structural protein, zinc-ribbon protein, and DNA-packaging ATPase. Mavirus has a diameter of 75 nm. Its dsDNA genome of 19,063 base pairs contains 20 ORFs that encode for interesting proteins, such as a helicase, retroviral integrase, protein-primed DNA polymerase β, zinc-ribbon protein, DNA-packaging ATPase, PRO, mCP, and MCP.

Infection Cycle of Virophage

Virophages that parasitize Mimivirus can make use of the fiber layer of this giant virus. For instance, Sputnik strains are thought to attach to, or be trapped in, the capsid fibrils of its giant virus host, then enter the host cell when the Mimivirus is internalized by the host cell (Figure 11.5). Virophages that parasitize giant virus that do not have a fiber

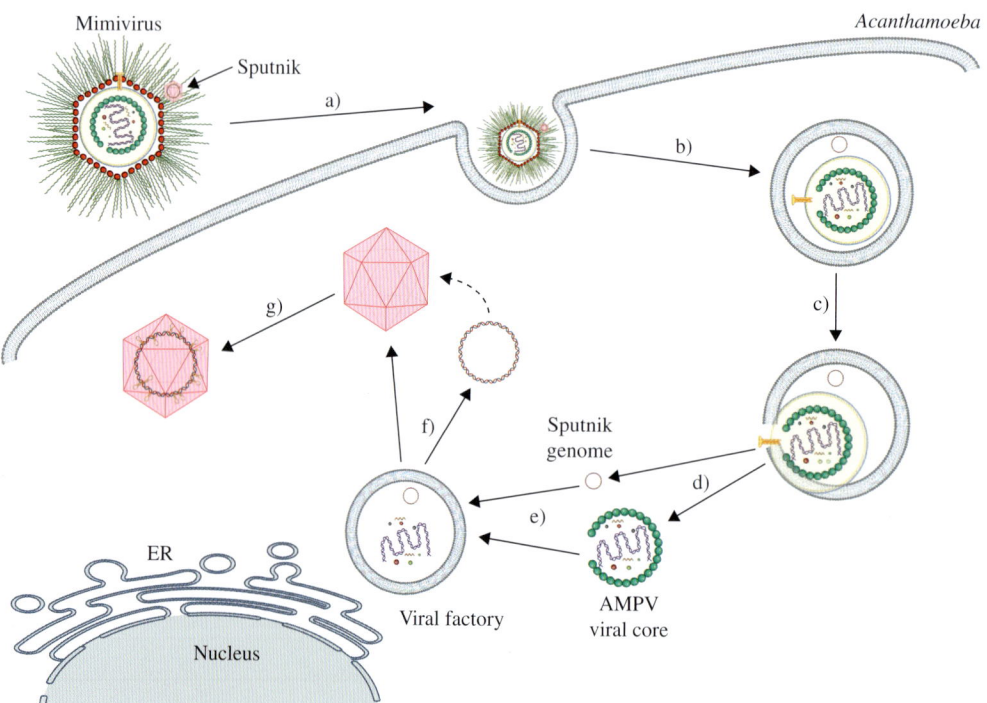

FIGURE 11.5   Sputnik life cycle. In the upper left corner, the schematic shows the actual size difference between AMPV and Sputnik. Sputnik is trapped in the capsid fibrils of the AMPV. (a) Both the giant virus and its cognate virophage are phagocytosed by the host protist. (b) Uncoating of both viruses occurs in the phagosome. (c) A pore in the phagosomal membrane is produced as the result of membrane fusion. (d) Both the AMPV viral core and Sputnik genome are released into the cytosol. (e) Sputnik uses the same viral factory created by AMPV for its genome and protein synthesis. (f) An empty Sputnik icosahedral capsid and its genome are formed independently. (g) The viral DNA-packaging ATPase inserts the circular dsDNA genome into the empty capsid, producing a complete, mature virophage. *Credit:* Figure partially created in BioRender.

layer can infect the host cell independently from its giant virus host. For example, mavirus and its giant virus host CroV are independently internalized via clathrin-mediated endocytosis by the green alga *Chlorella*.

Once in the host cell, uncoating of the virophage occurs to release its genome into the cytosol. Virophages sequester their giant virus transcription machinery to express their genomes as late genes instead of using the host cell's transcription machinery. The virophage replication occurs alongside its giant virus host in viral factories. MCPs and mCPs are used to assemble the icosahedral virophage capsid but do not appear to require either giant viral or cellular initiation factors. Once the empty capsid is formed, the DNA-packaging ATPase inserts the dsDNA genome into the virion. Then, MCP molecules are processed by the cysteine protease to provide the stability of the capsid in an acidic environment.

Virophage replication siphons off resources that are also required by the giant virus, therefore reducing giant virus reproduction while increasing host cell survival. A 70% reduction in Mimivirus production is observed during a co-infection with Sputnik. In addition, a high amount of defective giant viral particles is detected.

Zamilon is different; however, this virophage does not appear to affect the propagation or increase the number of defective particles of its giant virus host. This therefore suggests that Zamilon does not affect the Mimivirus's ability to cause lysis of the host cell.

Mavirus, on the other hand, is unique in that it can integrate its own DNA at several loci in the nuclear genome of its cellular host *Cafeteria roenbergensis* regulated by the retroviral integrase. Thus, mavirus can remain latent in this provirophage form until this same host cell is superinfected by its giant viral host CroV again. Prophage expression exploits the transcription machinery of CroV for mavirus propagation in the viral factories created by its giant virus host. Mavirus production, however, does not reduce the replication of CroV in this scenario and thus does not appear to have a protective role for the host cell. CroV can still accumulate inside the host cell and cause cell lysis, releasing both the giant virus and its cognate virophage. Once released, mavirus can co-infect neighboring cells with CroV, where now the virophage can inhibit giant virus replication to protect the host cell from cell lysis.

Current Status

In the past 2 decades, researchers performed metagenomic sequence-based analyses for virophage signatures in eukaryotic genomes of protists, fungi, and basal metazoans in search for new virophages. DNA-packaging ATPase, a cysteine protease, an MCP, and an mCP are the 4 core proteins that were used as signature genes for virophage. There were 328 high-quality genomes (either complete or near complete) that had been identified as of 2023. Some of the unclassified viruses in the Lavidaviridae family that were discovered include the following:

- Ace Lake mavirus was isolated from the waters of Antarctic Lake Ace. Its giant virus host is still unknown but is likely a member of Mimiviridae family.
- Dishui Lake virophages 1–8 were isolated from the waters of Dishui Lake in Shanghai, China, as well as in other freshwater bodies. Its giant virus host is likely a member of the Phycodnaviridae family.
- Guarani virophage was isolated from freshwater samples in the Pampulha lagoon in Belo Horizonte, Brazil. Its giant virus host is likely a member of the Mimiviridae family.

- Organic Lake virophage was isolated from samples from Organic Lake, Antarctica. Its giant virus host is likely a member of the Mimiviridae family.
- *Phaeocystis globosa* virus virophage (PgVV) was isolated in the Dutch coastal waters of the North Sea. Its giant virus host is *Phaeocystis globosa* virus, which infects algae of the genus *Phaeocystis*. PgVV is closely related to mavirus.
- Plantovirus Saccamoebae "comedo" was isolated in sycamore trees with its giant virus KSLT-5. This giant virus likely belongs to the Mimiviridae family and genus *Mimivirus*.
- Qinghai Lake virophage was isolated in the planktonic microbial community of Tibetan mountains. Its giant virus host is likely a member of the Phycodnaviridae family.
- Rumen virophage (RVP) was isolated from a wastewater bioreactor and the rumen of sheep. Its giant virus host most likely belongs to the Mimiviridae family.
- Yellowstone Lake virophages 1–4 were isolated from Yellowstone Lake with members of viruses of the Phycodnaviridae family.

Further Readings

Abrahao J, Dornas F, Silva L, et al. *Acanthamoeba polyphaga mimivirus* and other giant viruses: An open field to outstanding discoveries. *Virol J.* 2014;11:120.

Alempic J-M, Lartigue A, Goncharov A, et al. An update on eukaryotic viruses revived from ancient permafrost. *Viruses.* 2023;15(2):564.

Belhaouari D, De Souza G, Lamb D, et al. Metabolic arsenal of giant viruses: Host hijack or self-use? *eLife.* 2022;11:378674.

Brandes N, Linial M. Giant viruses—Big surprises. *Viruses.* 2019;11(5):404.

Colson P, Aherfi S, La Scola B. Evidence of giant viruses of amoebae in the human gut. *Hum Microb J.* 2017;5–6:14–19.

Colson P, La Scola B, Levasseur A, Caetano-Anolles G, Raoult D. Mimivirus: Leading the way in the discovery of giant viruses of amoebae. *Nat Rev Microbiol.* 2017;15:243–254.

Diesend J, Kruse J, Hagedorn M, Hamm C. Amoebae, giant viruses, and virophages make up a complex, multilayered threesome. *Front Cell Infect Microbiol.* 2017;7:527.

Ghigo E, Kartenbeck J, Lien P, Pelkmans L, Capo C, M J-L, Raoult D. Amoebal pathogen Mimivirus infects macrophages through phagocytosis. *PLoS Pathog.* 2008;4(6):31000087.

La Scola B, Audic S, Robert C, et al. A giant virus in amoebae. *Science.* 2003;299(5615):2033.

Legendre M, Bartoli J, Shmakova L, et al. Thirty-thousand-year-old distant relative of giant icosahedral DNA viruses with a pandoravirus morphology. *Proc Natl Acad Sci U S A.* 2014;111(11):4274–4279.

Mougari S, Sahmi-Bounsiar D, Levasseur A, Colson P, La Scola B. Virophages of giant viruses: An update at eleven. *Viruses.* 2019;11(8):733.

Roux S, Fischer M, Hackl T, Katz L, Schulz F, Yutin N. Updated virophage taxonomy and distinction from Polinton-like viruses. *Biomolecules.* 2023;13(2):204.

Tokarz-Deptula B, Chrzanowska S, Gurgacz N, Stosik M, Deptula. Virophages—Known and unknown facts. *Viruses.* 2023;15(6):1321.

CHAPTER 12

Bacteria, Archaea, and Plant Viruses

Viruses That Infect the Bacterial Host

Ernest Hanbury was the first to observe bactericidal activities against cholera bacteria from the Ganges River in 1896. Two decades later in 1915, William Twort discovered bacteria-killing viruses, and D'Herelle coined the name "bacteriophage," meaning "bacteria-eater," in 1917 to describe viruses that infect bacteria host cells. Bacterio-phages replicate by following similar steps that have been described regarding human viruses in previous chapters: adsorption, penetration of the viral genome (instead of whole viral particle entry), replication, maturation, and release.

Strategies for entry into and release of virus from bacterial host cells are quite differ-ent from those infecting human cells. The first step of the bacteriophage life cycle is still adsorption (or attachment); however, attachment of the first viral protein to its specific host receptor is reversible because there is no permanent change in virion morphol-ogy during this initial encounter. A second, stronger interaction with a secondary host receptor is required to cause a more stable conformational change in virion morphology, thus leading to an irreversible attachment. Virion morphological change is necessary to induce uncoating to allow the viral genome to penetrate into the host cell. Endocytosis is not a viable strategy for bacteriophage entry since the phage genome must bypass several layers of bacterial surface structures, such as the glycocalyx, S-layer, and pep-tidoglycan cell wall. Some bacteriophages rely on their ability to inject their genome into the bacterial cytoplasm through the use of tail fibers and core proteins. Others take advantage of the formation of the F-pilus that is created during mating between 2 bacterial cells. Through either mechanism, a combination of mechanical and enzymatic action is required for genome injection into the host cell cytosol.

Virus replication involves the lytic cycle or virulent infection, resulting in host cell lysis for the release of newly produced virions. However, some viruses can enter the lysogenic cycle or temperate infection, where the viral genome may remain dormant. For example, lambda (λ) phage has a choice of either going through the lytic cycle directly or remaining in a dormant lysogenic cycle. This type of virus is referred to as a temperate phage. Temperate phage DNA that had been integrated into the host genome is called a prophage. A host cell that carries a prophage is called a lysogen or lysogenic cell. The lysogen continues to perform normal functions and is not affected by the pres-ence of the prophage. Once integrated into the cellular chromosome, the prophage is transferred horizontally to progenies when the lysogen undergoes binary fission, thus increasing the population of bacterial clones that carry the prophage.

255

The prophage may or may not impose any virulent activity on the host cell until the lytic cycle is stimulated again. Certain external conditions, such as exposure to radiation or mutagenic chemicals, production of large amounts of DNA damage, or extreme changes in the environment, can induce a shift to the lytic cycle. Genes from some prophages may be expressed to produce toxins that are harmful to the host. In such a case, the prophage becomes active through gene expression to initiate lytic infection. Interestingly, a lysogen is immune to additional or secondary infection by the same virus.

In August 2022, the International Committee on Taxonomy of Viruses (ICTV) updated the classification system for bacteriophage and archaeal virus. Of all bacteriophage and archaea viruses, 96% possess a double-stranded DNA (dsDNA) genome and are classified under the class Caudoviricetes and order Caudovirales (tailed phages). Viruses in this order that had been discovered increased from 1359 in 2015 to 4483 in 2022. The remaining 4% of viruses possess either a single-stranded DNA (ssDNA) or plus-strand RNA genome. So far, no minus-strand ssRNA bacteriophages or archaeal viruses have been isolated. In total, 8437 bacteriophage and archaeal genomes have been sequenced. The following families are composed of bacteriophage and archaeal virus members:

- Ackermannviridae: non-enveloped, contractile tail, linear dsDNA
- Atkinsviridae: non-enveloped, isometric, linear ssRNA
- Autographiviridae (formerly known as Siphoviridae): non-enveloped, contractile tail (long), linear dsDNA (eg, Lambda, T5, HK97, N15)
- Autolykiviridae: non-enveloped, isometric, circular dsDNA
- Blumeviridae: non-enveloped, isometric, linear ssRNA
- Chaseviridae: linear dsDNA
- Clavaviridae: non-enveloped, rod shaped, circular dsDNA
- Corticoviridae: non-enveloped, isometric, circular dsDNA (eg, PM2)
- Cystoviridae: enveloped, spherical, linear dsRNA (eg, Φ6)
- Demerecviridae: linear dsDNA
- Drexlerviridae (formerly known as Podoviridae): non-enveloped, noncontractile tail (short), linear dsDNA (eg, T7, T3, Φ29, P22)
- Duinviridae: non-enveloped, isometric, linear ssRNA
- Fiersviridae (formerly named Leviviridae): non-enveloped, isometric, linear ssRNA (eg, MS2, Qβ)
- Finnlakeviridae: non-enveloped, isometric, circular ssDNA (eg, FLiP)
- Globuloviridae: enveloped, isometric, linear dsDNA
- Guenliviridae: linear dsDNA
- Herelleviridae: non-enveloped, contractile tail, linear dsDNA
- Inoviridae: non-enveloped, filamentous, circular ssDNA (eg, M13)
- Matshushitaviridae: enveloped, isometric, linear dsDNA
- Microviridae: non-enveloped, isometric, circular ssDNA, (eg, ΦX174)
- Paulinoviridae: non-enveloped, filamentous, circular ssDNA

- Picobinrnaviridae: non-enveloped, isometric, linear ssDNA
- Plasmairidae: enveloped, pleomorphic, circular dsDNA
- Pleolipoviridae: enveloped, pleomorphic, circular ssDNA, circular dsDNA, or linear dsDNA
- Plectroviridae: non-enveloped, filamentous, circular ssDNA
- Portogloboviridae: enveloped, isometric, circular dsDNA
- Rountreeviridae: linear dsDNA
- Salasmaviridae: linear dsDNA
- Schitoviridae: linear dsDNA
- Simuloviridae: enveloped, isometric, linear dsDNA
- Solvpiviridae: non-enveloped, isometric, linear ssRNA
- Sphaerolipoviridae: enveloped, isometric, linear dsDNA
- Steitzviridae: non-enveloped, isometric, linear ssRNA
- Stravoviridae (formerly known as Myoviridae): non-enveloped, contractile tail, linear dsDNA (eg, T4, Mu, P1, P2)
- Tectiviridae: non-enveloped, isometric, linear dsDNA
- Turriviridae: enveloped, isometric, linear dsDNA
- Zobellviridae: linear dsDNA

Lambda (λ) Phage

Infection Cycle of λ Phage

Lambda (λ) phage (also known as enterobacteria phage λ, coliphage λ, *Escherichia* virus lambda) is a temperate bacteriophage belonging to the Autographiviridae (formerly known as Siphoviridae) family and genus Siphoviridae. Other members of this genus include T5, HK97, N15. Its host is *Escherichia coli* bacteria. The λ phage linear dsDNA genome of 48,502 base pairs, which contains 73 open reading frames (ORFs), is encapsidated in a non-enveloped icosahedral protein shell or capsid with a diameter of 50–60 nm (Figure 12.1). This capsid "head" is attached by a long, flexible, noncontractile tail that is about 150 nm and provides a route for the delivery of the viral genome into the *E. coli* host cell cytosol. Six tail fibers 35 nm long are attached to the tail, whose function is to allow the viral particle to adsorb to the host cell surface.

Adsorption occurs when the glycoprotein J (gpJ) on the tip of each tail fiber interacts with LamB, a porin molecule and subunit of the mannose phosphotransferase system (PTS) permease complex (a sugar-transporting system), on the host cell surface. The gpJ molecules not only adhere to the host cell surface and seal the tail tips but also extend the central tail fiber toward the mannose PTS permease complex. Lambda phage genomic dsDNA is then injected through the hollow tube in the tail and through the mannose PTS permease complex into the cell's cytoplasm. Once inside the host cell, the linear dsDNA is immediately circularized. There is a 12-base 5′ overhang at each end of the dsDNA called the "cos end." These *cos* (cohesive) ends can base pair to produce a circular dsDNA genome once it enters the host cell. The resulting site of the base pair is called a cos site. Ligation is catalyzed by host DNA ligase, producing

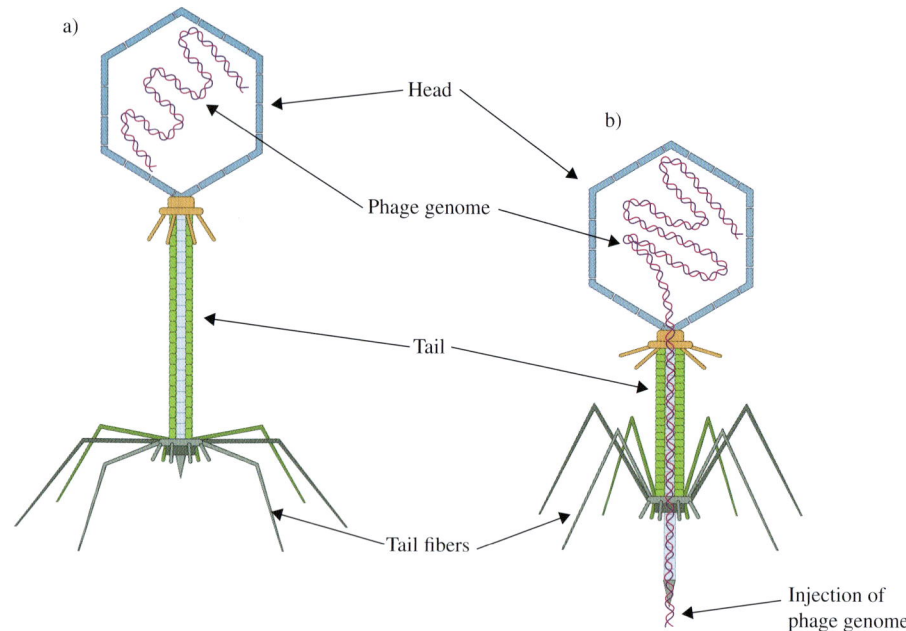

FIGURE 12.1 Lambda (λ) phage structure and genome. (a) Schematic of λ phage showing a non-enveloped icosahedral capsid head with a long, flexible, noncontractile tail. The phage head encapsidates are linear dsDNA genome. Six tail fibers are attached at the end of the tail. (b) After attachment to the host cell surface, the central tail fiber extends and punctures the cell surface. The genome is then injected into the cytoplasm through the hollow tube in the tail. *Credit:* Figure partially created in BioRender.

the *cos* site. This *cos* site can later be recognized and bound by the enzyme λ terminase, which has a function analogous to a restriction endonuclease, to form *cos* and ends again at the end of the phage life cycle. The λ terminase holoenzyme is a heteroligomer composed of small (gpNu1) and large (gpA) subunits. The cos ends can also be involved with the formation of concatemers as well as integration into the cellular chromosome.

The λ phage is a temperature phage, which means that the circular dsDNA genome can enter either the lytic infection/cycle or the lysogenic infection/cycle, depending on the extracellular conditions and intercellular controls (Figure 12.2). The lytic cycle is a period of rapid virion propagation leading to host cell lysis to release virions into the extracellular matrix. The lysogenic cycle occurs when the phage DNA is integrated into the bacterial chromosome for long-term dormancy.

Lytic Infection

Let's discuss lytic infection first. After the circularization of the genome, expression of λ phage genes occurs temporally where the immediately early genes *Rho* and *N* are transcribed first by the host RNA polymerase. The temporal nature of λ phage transcription is controlled by the activation of its 3 promoters: promoter leftward (pL) and promoter rightward (pR) control early gene transcription to regulate the lytic cycle,

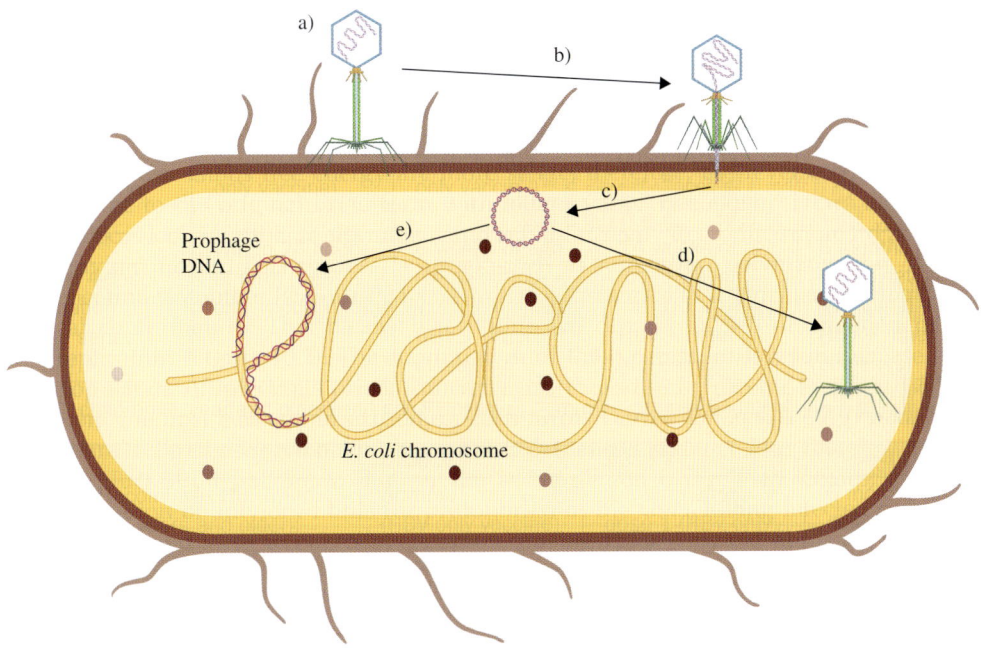

FIGURE 12.2 Lambda phage life cycle. (a) Lambda (λ) phage attaches to surface of host bacterial cell. (b) The flexible tail extends through the cell wall and plasma membrane. (c) Phage genome is injected into the cytoplasm and circularized. (d) Lytic cycle leads to rapid virion propagation and eventually release from the cell when the host cell is lysed. (e) Lysogenic cycle occurs when the prophage DNA is integrated into the host chromosome. *Credit:* Figure partially created in BioRender.

while promoter for repressor maintenance (pRM) is the promoter of the lysogenic cycle. As the lytic cycle progresses, transcription is regulated by Rho and N proteins. Rho protein is known to cause termination of transcription of the *cro* gene, while N protein causes the termination of immediate early transcription to fail, thus allowing the expression of downstream early genes. This function earned N protein the name "transcription antiterminator." So, the expression of N protein inhibits Rho protein function, allowing the *cro* gene to be expressed. Cro protein is a weak repressor that regulates gene expression during early phage development, which in turn is required for normal late-stage lytic growth. Cro is known to repress the transcription of cI from the promoter pRM to prevent the progression into the lysogenic cycle. The function of cI is to silence lytic functions. Lambda phage cI protein directly turns off all λ phage lytic genes and is required for the stimulation of the lysogenic cycle.

Early genes are expressed before DNA replication. Regulatory proteins expressed by the lambda *O* and *P* early genes from the pRM promoter are required to initiate viral DNA synthesis by the *E. coli* DNA polymerase. During the early stages of infection, DNA replication begins classically with the formation a bidirectional replication fork at the origin of replication (*ori*) producing daughter circular dsDNAs, which in turn are used as the template to produce more viral dsDNA molecules. During the latter stages of infection, the rolling circle DNA replication mechanism is employed, producing linear multimeric concatemers. The concatemers are eventually cleaved at *cos* sites into individual linear λ phage genomes through the function of λ terminase holoenzyme, fueled by ATP (adenosine triphosphate) hydrolysis.

Other early gene transcripts include cII and cIII, whose gene products are important in the determination of whether the λ phage DNA proceeds toward lytic infection versus lysogenic infection. Stable cII structure favors the lysogenic pathway, whereas if cII is cleaved by the membrane-bound *E. coli* protease, FtsH, lytic infection occurs. FtsH is an ATP-dependent zinc metalloprotease complex that plays a role in providing membrane protein quality control in bacterial cells. The stability of cII is regulated by cIII, which acts as a competitive inhibitor of FtsH. Factors that influence this process in favor of lysogeny include low temperature, lack of nutrients for the cell, and high dose of infection.

The major capsid protein (MCP) encoded by λ late gene E is used to form the procapsid, which is an empty precursor head structure. The tail is formed from the major tail protein gpV that polymerizes to form a long tube. The length of this tube is controlled by λ protein gpH. Lambda heads and tails are assembled independently, then are later joined before the linear dsDNA genomic molecule is translocated into the preformed procapsid structure. Genome translocation is mediated by 2 subunits of λ terminase. The gpNu1 subunit of λ terminase is both necessary and sufficient for DNA packaging to regulate the assembly and stability of the packaging machinery. It possesses a DNA-binding domain that recognizes the *cos* site specifically. The gpA subunit of λ terminase, on the other hand, is responsible for initiating the translocation of the viral genome. GpA possesses endonuclease activity that introduces staggered nicks at the *cos* site that are required for DNA packaging. This subunit also possesses helicase activity that separates the nicked strands to generate a 5′ overhang at each end of the viral genome. Last, gpA possesses DNA translocase and ATPase activities that stimulate the loading of the linear dsDNA into the procapsid.

Cell lysis of the *E. coli* host is promoted by 5 lambda proteins: S105, S107, R, Rz, and Rz1. λ R protein is an endolysin that is responsible for the cleavage of the peptidoglycan cell wall. Once produced in the cytoplasm, λ S105 and S107 proteins regulate the release of λ R protein into the cell's periplasm, which is the space between the inner and outer cell membrane. Rz and Rz1 are membrane lipoproteins that span the entire periplasm, forming the "spanin" complex. The spanin complex disrupts the outer membrane, allowing λ R proteins to reach and degrade the cell wall.

Lysogenic Cycle

Lambda phage cI protein is required for the stimulation of the lysogenic cycle by directly repressing the 2 lytic promoters pL and pR. While repressing all lytic gene expression, cI activates its own transcription from the autoregulated pRM promoter. Furthermore, the presence of cI can cause immunity to superinfection (or secondary and tertiary infection) by other lambda phages by blocking the promoters of immediate early genes. CI has also been shown to downregulate host *pckA* gene, which encodes phosphoenolpyruvate carboxykinase required for growth on succinate and other carbon sources, although the function of this activity is still unclear. At this point, cIII blocks FtsH activity, thus maintaining a high level of stable cII molecules to stimulate lysogeny.

Insertion of the prophage into the *E. coli* host chromosome is regulated by the tetramer complex intasome composed of λ integrase (IN), integration host factor (IHF), bacteriophage λ excisionase (Xis), and factor for inversion stimulation (Fis). This complex is responsible for not only site-specific recombination to integrate the prophage into the cellular chromosome, but also prophage excision removal during the lysogeny to lytic cycle progression. IN is a member of the tyrosine recombinase family that can cleave and religate DNA. The many functions of IHF include its involvement in recombination and gene expression. IHF binds to recombination site in the λ genome to facilitate efficient recombination. IHF can regulate gene expression of both phage and host bacterial cell by altering or bending DNA, forming protein-protein contacts with RNA polymerase, or inhibiting the use of competing promoters. IHF also plays a role in transposition and the initiation of DNA replication. Last, IHF can modulate cII and cIII expression levels.

Lysogeny to Lytic Cycle Switch

Environmental stressors can stimulate the switch from lysogeny to the lytic cycle. These conditions may include lack of nutrients (or starvation), exposure to toxic chemicals, extreme changes in temperature and pH, and the binary fission process. This switch is controlled by the balance in the expression and concentration of cI and Cro proteins. The lysogenic cycle is achieved and maintained when the λ cI concentration is higher. However, a higher concentration of Cro protein drives lysogeny into the lytic cycle. *E. coli* SOS response, which is activated in response to DNA damage, is upregulated when the cell is under certain stress that can cause DNA damage. SOS (named after the naval distress signal "Save Our Souls") response is a bacterial DNA damage mechanism. This DNA repair system can in turn eliminate λ cI proteins by SOS-induced cleavage. The lytic promoters, pL and pR, are now de-repressed, allowing Cro proteins to be produced once again to repress cI expression.

Emesvirus Zinderi 2 (MS2) Phage

Infection Cycle of MS2 Phage

Emesvirus zinderi 2 (MS2) phage belongs to the Fiersviridae family (formerly named Leviviridae) and genus *Emesvirus*. Closely related bacteriophages include bacteriophage f2, Qβ, R17, and GA. MS2 phage is a non-enveloped viral particle with a diameter of 27 nm. Its linear plus-strand RNA genome of 3569 bases is encapsidated inside an isocahedral capsid (Figure 12.3). The maturation protein is inserted through 1 of the vertices of the capsid. The MS2 phage (+) ssRNA genome serves as messenger RNA (mRNA) for the translation of viral proteins immediately after entry into the host cell. MS2 phage is one of the smallest known viruses with a simple viral genome encoding only 4 proteins:

- Coat protein (CP) is the MCP.
- Replicase (RP) subunit is the catalytic subunit of RNA-dependent RNA polymerase (RdRP).
- Maturation protein (AP), or "A" protein named for well-studied ssRNA viruses of genus *Allolevivirus* in the Levivirus family, is required for attachment of the virion to the pilus structures of the host bacterial cell.
- Lysis protein (LP) causes host cell lysis.

The MS2 phage infects enteric gram-negative bacteria that carry the fertility (F) factor, which is a plasmid that can be replicated and translocated from the donor to recipient cell during the process of bacterial conjugation (Figure 12.4). The F plasmid contains genes that encode for proteins that form the F pilus. One of these bacterial proteins is

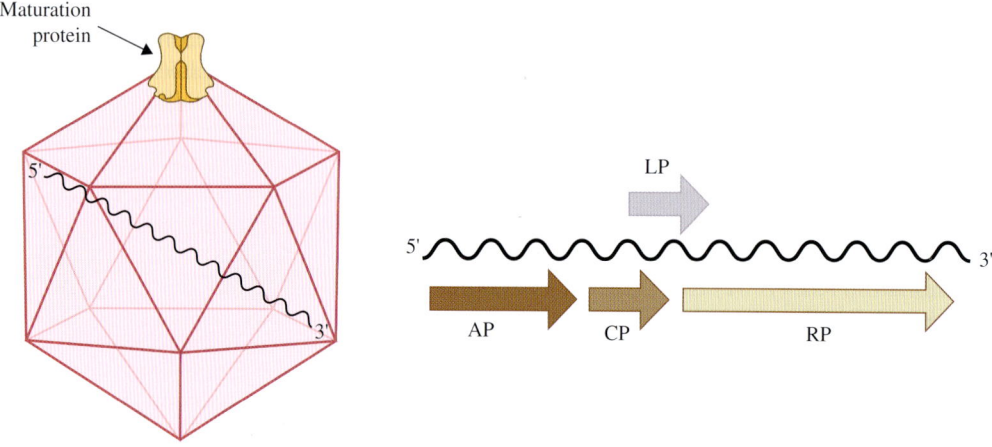

FIGURE 12.3 MS2 phage structure and genome. MS2 phage is a non-enveloped virion with its linear plus-strand RNA genome surrounded by an icosahedral capsid. Its genome encodes 4 proteins: maturation protein (AP), replicase subunit (RP), coat protein (CP), and lysis protein (LP).
Credit: Figure partially created in BioRender.

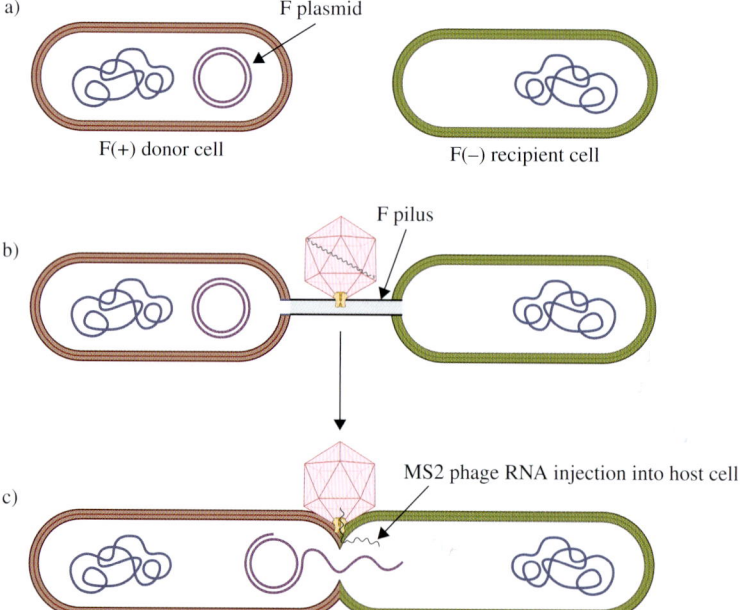

FIGURE 12.4 MS2 phage entry into host cell. (a) The donor F(+) bacterial cell is on the left, and the recipient cell is on the right. (b) MS2 phage binds to the F-pilus that was formed during bacterial conjugation. (c) MS2 phage genome is injected into the recipient cell as the newly synthesized F-plasmid is being transferred from the donor cell to the recipient cell.
Credit: Figure partially created in BioRender.

F-pilin, which acts as the host surface receptor for the MS2 phage. Attachment of the MS2 phage occurs through the interaction between AP and F-pilin on the F pilus of cells that are undergoing bacterial conjugation. The viral ssRNA genome enters the target cell via the conjugative pore that is intended for transferring the F plasmid.

Once inside the host cell, the MS2 (+) ssRNA genome serves directly as the template for protein synthesis. At this point, only the CP gene is accessible to ribosomes; therefore, the CP gene is expressed first, followed by the synthesis of RP. RP is only the catalytic subunit; therefore, other cellular translation subunits from the host cell are sequestered to form a fully functional RdRP. These subunits are the elongation factor thermal unstable (EF-Tu), elongation factor thermal stable (EF-Ts), and ribosomal protein S1. During translation, EF-Tu delivers an aminoacyl (or charged)-tRNA to the small ribosomal subunit, while EF-Ts is a guanine nulceotide exchange factor whose function is to return EF-Tu into its active form. Once assembled, replication of the viral genome can now begin. A conformational change at the 5' end of the (+) ssRNA occurs during genome replication, which in turn allows AP to be expressed at a very slow rate. Ribosomes and RdRP compete for the same 5' end of the (+) ssRNA to regulate translation and replication, respectively. During the early stage of infection, the binding of RdRP ceases all translation events. This allows time for the amplification of the viral genome. As the infection develops, there are now many more genomic RNAs to be bound by both ribosomes and RdRP for their use.

Viral particle assembly occurs once enough genomic RNA, CP, and AP molecules have accumulated in the cytosol. Host cell lysis is induced by LP synthesis. LP molecules are inserted into the host cell membrane, which creates channels that allow nonspecific diffusion of ions. This high amount of ion movement disrupts the electrostatic potential of the cell membrane, which in turn activates cellular autolysins to cause cell lysis.

Viruses That Infect the Archaeal Host

In 1974, Hanna Oksanen and Tatiana Demina discovered the first archaeal virus, Hs1, which infects halophilic archaea species of the genus *Halobacterium*. Since then, over 100 archaeal viruses have been identified worldwide in extreme environments, such as acidic hot springs, saline bodies of water, and the ocean floor. They have also been isolated from the human body. Archaeal viruses infect species in the domain Archaea. Archaeal viruses are unique and very different from viruses that infect the other 2 domains of living organisms. These viruses do not form the typical helical or icosahedral capsid seen in those infecting the other 2 domains of organisms. Their unusual virion morphologies include shapes like spindles, lemons, rods, bottles, droplets, and coils. Some are enveloped, and others are non-enveloped. Their unique structures are tough to withstand the harsh environment where they exist since these are the very same environments where their susceptible and permissible cells thrive. Harsh environments can include extreme temperature, extreme pH, and extreme salt concentration.

All known archaeal viruses have genomes that are either ssDNA or dsDNA. So far, no RNA archaeal virus has been discovered. Archaeal viral DNAs contain higher GC contents to maintain genome stability under extreme conditions. Most archaeal viruses infect hyperthermophilic archaea and thus exhibit characteristics that enable them to survive in these environments. Like their extremophile host cells, archaeal viruses possess very stable biomolecules to help them accommodate and survive in the extreme environment where they reside. The viral lipid envelope and surface proteins are most affected by extreme environments. In order to survive and remain functional, these lipid envelopes and surface proteins adopt conformations that are very different from those found in relatively temperate environments where eubacteria and eukaryotes exist. This aspect may explain the formation of unique viral structures constructed from capsid proteins.

Archaeal lipid envelopes lack a peptidoglycan cell wall, and their lipid composition and conformation are different from other membranes. The envelope is composed of ether-linked membrane lipids with a glycerol-1-phosphate backbone instead of ester-linked lipids with a glycerol-3-phosphate backbone, like bacterial and eukaryotic cells. In some cases, ether linkages of phospholipids allow the formation of a monolayer membrane instead of a bilayer membrane. Phospholipid molecules are packed more tightly in a monolayer. A monolayer structure is bendable (or has high curvature) and more stable than the bilayer structure. Their characteristics provide better resistance to high temperature and allow less influx of extracellular materials. Envelope surface proteins are highly glycosylated. Glycosylation not only is important for adsorption to target cells, but also helps the virus cope with changing environments. The presence of carbohydrate residues is also important for protein folding, stability, and function.

Proteins are components that are most affected by extreme environments. In order to survive and remain functional, archaeal proteins adopt conformations that are very different from those found in environments where bacterial and eukaryotic cells exist. This aspect may explain the formation of unique viral structures constructed from capsid proteins. The following are types of proteins with distinguishing characteristics:

- Thermophilic proteins are stable at temperatures between 41°C and 122°C. They have a prominent hydrophobic core with high intramolecular electrostatic interactions to maintain activities at high temperature.

- Psychrophilic (cryophilic) proteins are stable at temperatures between −20°C and 10°C. They have a core that has low hydrophobicity. The protein surface is mostly nonpolar to maintain flexibility and remain active under low temperature.

- Halophilic proteins are stable under high salt concentrations. They have a core that has low hydrophobicity. These proteins are rich in acidic amino acid and therefore have high negative charges on the surface, along with peptide insertions to compensate for extreme ionic conditions.

- Acidophilic proteins are stable in an acidic environment. Their characteristics are similar to those of thermophilic proteins.

- Alkaliphilic proteins are stable in high pH. Their characteristics are similar to those of halophilic proteins.

- Haloalkaliphilic proteins are stable in both high salt and acidic environment. There is a high proportion of acidic residues on the protein surface.

- Piezophilic proteins are stable in a high-pressure environment. These proteins have a compact, dense hydrophobic core with reduced hydrogen bonding but increased multimerization. Their characteristics are similar to those of thermophilic proteins.

In August 2022, the ICTV updated the classification system for bacterial and archaeal virus. As of 2024, the ICTV had identified over 45 classified families and 135 species. The following is an incomplete list of families with some distinguishing features:

- Ampullaviridae: enveloped, bottle shaped, linear dsDNA
- Bicaudaviridae: non-enveloped, spindle/lemon shaped, 2-tailed, circular dsDNA
- Clavaviridae: non-enveloped, spindle/lemon shaped, circular dsDNA
- Fuselloviridae: non-enveloped, spindle/lemon shaped, circular dsDNA, (eg, *Sulfolobus* spindle-shaped virus 1)
- Guttaviridae: non-enveloped, ovoid/droplet shaped, circular dsDNA
- Halspiviridae: non-enveloped, spindle/lemon shaped, linear dsDNA
- Globuloviridae: enveloped, spherical, linear dsDNA
- Lipothrixviridae: enveloped, filamentous/rod shaped, linear dsDNA (eg, *Acidianus* filamentous virus 1)
- Ovaliviridae: enveloped, ellipsoidal, linear dsDNA
- Pleolipoviridae: enveloped, pleomorphic, circular ssDNA, circular dsDNA, or linear dsDNA
- Spiraviridae: non-enveloped, coil shaped, linear dsDNA
- Rudiviridae, non-enveloped, filamentous/rod shaped, linear dsDNA (eg, *Sulfolobus islandicus* rod-shaped virus 1)
- Thaspiviridae: non-enveloped, spindle/lemon shaped, linear dsDNA
- Tristromaviridae, enveloped, filamentous/rod shaped, linear dsDNA

Acidianus Filamentous Virus 1

Infection Cycle of *Acidianus filamentous* Virus 1

Acidianus filamentous virus 1 (AFV1) belongs to the Lipothrixviridae family and genus *Captovirus*. The structure of the virion consists of an enveloped, filamentous- (or rod-) shaped capsid 900 nm long and 24 nm wide in dimension (Figure 12.5). There is a claw-like structure at each end of the virion, which is connected to the main body by narrow appendages with collars. The lipid monolayer envelope of AFV1 is 2 nm thick, which is built to survive in acidophilic and hyperthermophilic environments. The viral envelope is acquired from the host membrane, which adopts a U-shaped, horseshoe conformation created by bending long, flexible lipid molecules. These lipid molecules are glycerol dibiphytanyl glycerol tetraether lipid, which lacks cyclopentane rings. This U-shaped structure allows for the high curvature that wraps around the long, thin, filamentous capsid. AFV1 is one of the first enveloped filamentous viruses with a linear dsDNA genome to be discovered. Its linear dsDNA genome of 20,869 base pairs contains 40 ORFs. This virus was isolated from an acidic hot spring in Yellowstone National Park in Wyoming, where the temperature is between 75°C and 85°C and the pH is between 1.5 and 3.

The study of the archaeal virus life cycle is still in its infancy, and a complete life cycle of any individual species is not yet available. The next material concerns some of what is known about AFV1 so far.

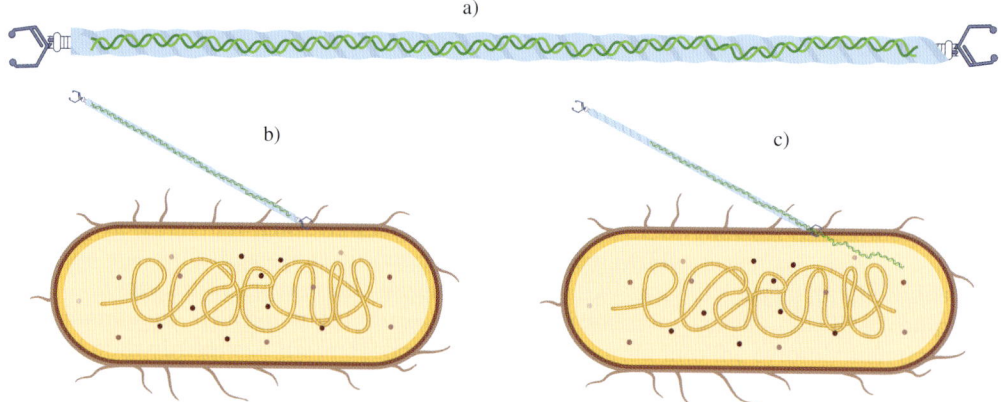

FIGURE 12.5   AFV1 virion structure and genome. (a) Schematic of AFV1 with a claw-like structure at each end of the virion. (b) The claw-like structure of AFV1 attaches to a host cell filament. (c) AFV1 genome is injected through the claw-like structure into the cytosol. *Credit:* Figure partially created in BioRender.

AFV1 infects the hyperthermophilic archaea *Acidianus hospitalis*. Attachment occurs when the claw-like terminal structures of AFV1 adhere to host cell filaments. There are different types of filaments that are formed on the surface of archaeal cells to be used for cell motility. These filaments include archaellum (or archaeal flagellum), type IV adhesive pili, type IV ultraviolet-inducible pili, cannulas, hami, fimbriae, threads, and protein sheaths. Once attached, irreversible conformational changes to the viral particle occur, possibly membrane fusion and uncoating. The dsDNA genome is then injected or released into the host cell cytosol, possibly through the claw-like structure. It is proposed that, to propagate, AFV1 uses some of the host transcription, translation, and DNA replication machineries, while providing some essential viral components encoded in its genome. After virion assembly, a possible mechanism for progeny viral particle release from the host cell is via budding using the endogenous endosomal sorting complexes required for transport, which is assumed to be used by enveloped archaeal viruses. The induction of host cell lysis is another mechanism for viral particle egress. These mechanisms include virus-associated pyramids (VAPs) and VAP-independent disruption of the cell membrane.

Viruses That Infect the Plant Host

Transmission of Virus to Plant Host

Of all plant viruses, 93% have an RNA genome. In fact, 65% of plant viruses have a plus-strand ssRNA genome, while 10% have a minus-strand ssRNA genome and 5% have a dsRNA genome. Another 3% of viruses possess a plus-strand ssRNA genome that expresses a reverse transcriptase. Most of the remaining 17% of viruses have a ssDNA genome, and very few have a dsDNA genome. This is in contrast with other viruses mentioned in this textbook, for which roughly 25% of animal viruses and 75% of bacteriophages have a dsDNA genome.

Plant viruses have varied host ranges, from a few species such as the *Citrus tristeza* virus to over 1000 species, such as the *Cucumber mosaic* virus. Regardless, they have been found to be able to infect virtually all species of cultivated and wild plants. Their life cycle begins with the entry into the host plant cell. Unlike animal cells, all plant cells are fortified by an impenetrable cell wall in addition to the outermost plant cuticle. Being unequipped to bore through these defensive layers, plant viruses can only enter the cell, and thus the plant itself, passively through areas of the plant that have been mechanically damaged, caused by, for example, injury to the plant, damage by insects, and grafting. Thus, transmission occurs from wounds or damaged areas of one plant to wounds or damaged areas of the next host plant.

There are 3 modes of transmission for plant viruses: vectors, seeds, or direct contact. Most plant viruses are actively transmitted through vectors, including living organisms such as plant-feeding arthropods (insects), nematodes, plant-parasitic fungi insects, mites, soil microorganisms, and other animals. Aphids and whiteflies are the major vectors that are capable of transmitting the largest number of virus species. Humans can also act as a vector by transmitting viruses through cultivation practices using contaminated tools. Farmers may also deliver viral particles that adhere to clothing and footwear used in the field into mechanical wounds on the plant. Viruses can be transmitted through living infected plant seeds and pollen. Direct contact between a healthy and an infected plant is a very efficient and rapid mode of transmission. Direct contact can include not only the physical contact between diseased and healthy plants in the field, but also through tubers and grafting. The weather, such as the wind, is another likely mechanical means for transmission.

Once the virus has managed to replicate in the initial infected host cell, viral particles must disseminate throughout the plant. Cell-to-cell dissemination occurs through the diffusion of viral particles from one cell to a neighboring still healthy cell via plasmodesmata. These are small channels that form connections between cells for communication and transport of necessary vital components such as ions. Viral particles may also be released through damaged cells; however, once released, the virus must again find another damaged host cell to infect. Cell-to-cell dissemination is a slow process; therefore, in order carry out a rapid systemic spread of the virus to colonize the entire plant successfully, the virus needs to enter the plant's vascular system. Viral particles can be released into the phloem, then infect surrounding cells where reproduction and continued dissemination occur.

Symptoms of a viral infection can be exhibited in multiple parts of the plant, including the leaves, flowers, and fruits. The most common symptoms exhibited by leaves may include mosaic patterns, striping or streaking, yellowing, leaf rolling and curling, vein clearing and banding, and small necrotic or chlorotic spots known as local lesions. Common symptoms exhibited by flowers may include deformation of flower shape and changes in color, including color breaking, which is the appearance of dramatic color mosaics. Typical symptoms exhibited by fruits and vegetables may include stunting, discoloration, malformation, chlorotic ring spots, and mosaic patterns. In response to a viral infection, stems of plants may develop stem pitting and grooving or even tumors.

To resist infection, plants possess the intracellular machinery to carry out RNA silencing to target viral mRNAs as a general means for antivirus defense. Some plants also possess active and passive mechanisms against a viral infection. Passive defenses include the obstruction of expressing host factors that are required by the virus for replication and dissemination. Active defenses include the expression of antivirus resistance genes to target and eliminate specific viruses.

Plant viruses are classified into 49 families and 73 genera of cultivated plants.

Circular dsDNA viruses include Caulimoviridae family divided into 6 genera (eg, genus *Badnavirus*, species commelina yellow mottle virus). Circular ssDNA viruses include the Geminiviridae family, divided into 3 genera (eg, genus *Mastrevirus*, species maize streak virus). Plus-strand ssRNA viruses include Bromoviridae family divided into 4 genera (eg, genus *Alfamovirus*, species alfalfa mosaic virus); Closteroviridae family divided into 2 genera (eg, genus *Closterovirus*, species beet yellows virus); Comoviridae family divided into 3 genera (eg, genus *Comovirus*, species cowpea mosaic virus); Potyviridae family divided into 3 genera (eg, genus *Potyvirus*, species potato virus Y); Sequiviridae family divided into 2 genera (eg, genus *Sequivirua*, species parsnip yellow fleck virus); Tombusviridae family divided into 5 genera (eg, genus *Carmovirus*, species carnation mottle virus); and 16 genera not assigned to families. Minus-strand ssRNA viruses include the Rhabdoviridae family divided into 2 genera (eg, genus *Cytorhabdovirus*, species lettuce necrotic yellows virus), and Bunyaviridae family with 1 genus (eg, genus *Tospovirus*, species tomato spotted wilt virus). Double-stranded RNA viruses include the Partitiviridae family, divided into 2 genera (eg, genus *Alphacryptovirus*, species white clover cryptic virus 1); Reoviridae family divided into 3 genera (eg, genus *Fijivirus*, species Fiji disease virus); and 4 genera not assigned to families.

Tobacco Mosaic Virus

Infection Cycle of Tobacco Mosaic Virus (TMV)

Tobacco mosaic virus (TMV) is a plant virus that belongs to the Virgaviridae family and genus *Tobamovirus*. This virus infects the leaves of tobacco, tomato, bean, and pepper plants. Its structure is a non-enveloped virion of 300 nm long and 18 nm wide, with a hollow central cavity that measures 4 nm in diameter (Figure 12.6). The plus-strand

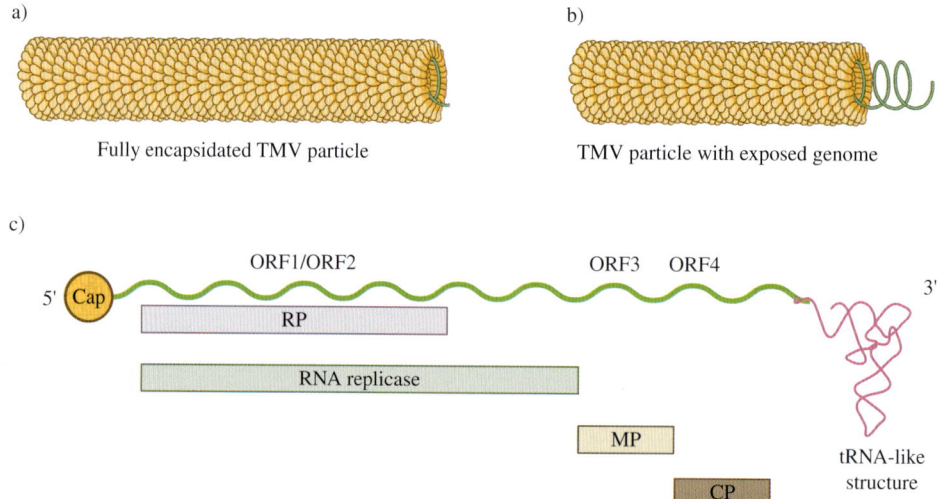

a)

Fully encapsidated TMV particle

b)

TMV particle with exposed genome

c)

FIGURE 12.6   TMV virion structure and genome. (a) Fully encapsidated TMV particle. (b) Encapsidated TMV particle exposing RNA genome. (c) TMV genome contains 4 ORFs encoding for a replication-associate protein (RP), RNA replicase, movement protein, and coat protein. *Credit:* Figure partially created in BioRender.

ssRNA genome of 6395 bases containing 4 ORFs is encapsidated by a helical, rigid, rod-like capsid. The ORFs encode for a 126-kDa replication-associate protein (RP) that has methyl transferase and RNA helicase activities, 183-kDa RP that is an RNA replicase (that has methyl transferase and RNA helicase and polymerase activities), movement protein (MP) to mediate cell-to-cell movement of the viral RNA in the plant tissues, and structural capsid or coat protein (CP). There is a 7-methyl guanosine cap at the 5'-terminus and a tRNA-like structure at the 3'-terminus of the ssRNA molecule. Like the genome for Poliovirus, TMV (+) ssRNA genome acts as a viral mRNA immediately after entry into the host cell.

TMV enters the plant cell through the damaged cell wall where mechanical wounds in the plant exist (Figure 12.7). Entry may be either through the transiently opening in the plasma membrane or via pinocytosis. Uncoating of the virion to expose the viral genome may involve the lower Ca^{2+} concentration and slightly more acidic cytosolic condition compared to the external environment. Uncoating may also involve a process called cotranslational disassembly, by which ribosomes remove capsid proteins while catalyzing translation simultaneously.

The (+) ssRNA serves directly as an mRNA for the translation of the 183-kDa RP and 126-kDa proteins. The (+) ssRNA also serves as the template for the synthesis of the full-length, complementary, minus-strand ssRNA catalyzed by both RPs. Even though the

TMV life cycle

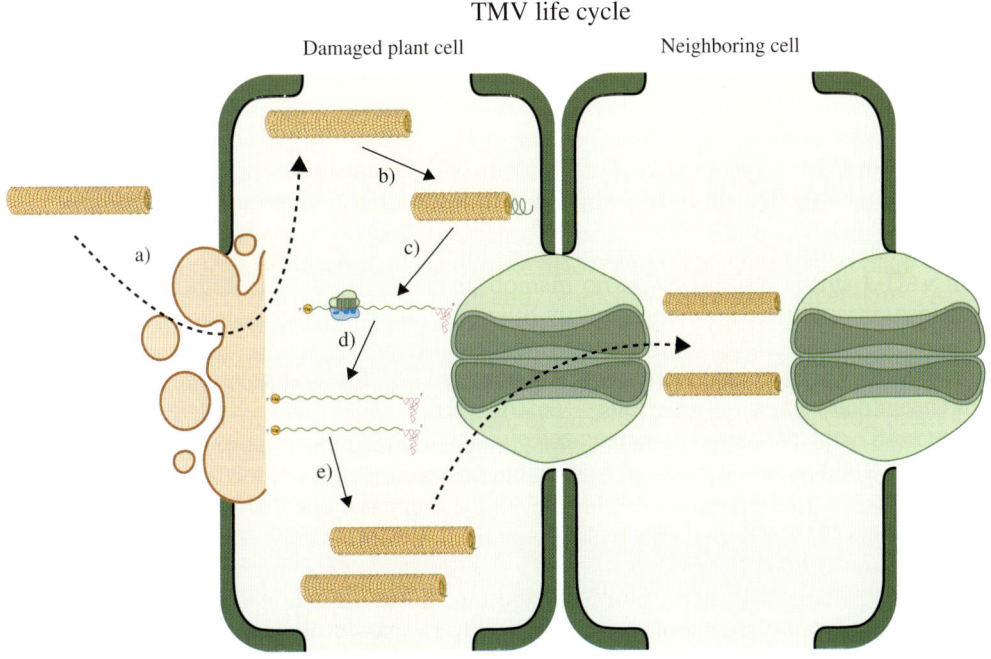

FIGURE 12.7 TMV life cycle. (a) TMV enters damaged part of plant cell. (b) Uncoating occurs in the cytosol. (c) The viral genome is immediately bound by ribosomes for protein synthesis. (d) Viral RNA replicase regulates viral genome replication. (e) Assembly of TMV virions occurs in the cytosol. (f) Completely assembled virions migrate through plasmodesmata to neighboring cells. *Credit:* Figure partially created in BioRender.

183-kDa replicase subunit contains RdRP activity and is thus sufficient to replicate the viral genome, the 126-kDa subunit provides support for maximum progeny RNA synthesis. This minus strand in turn serves as a template to produce full-length (+) ssRNA molecules known as subgenomic RNA (sgRNA). The host ribosome then translates the sgRNA to produce molecules of MP and coat/capsid protein CP.

Synthesis of CP is coupled with RNA replication, leading to assembly. The virion assembly is initiated when CPs interact with a site at the 3' end of the sgRNAs, termed the origin-of-assembly sequence (OAS). TMV CPs interact with the hairpin structure of OAS, then recruit more subunits to wrap the (+) ssRNA molecule tightly into a helical structure in a bidirectional manner. Once formed, mature viral particles are stable and can be released from the plant when cells are damaged or leaves dried up to transmit to other plants. The sgRNAs may also be bound and wrapped up by MPs, in which case this complex can disseminate to a neighboring cell and/or throughout the plant. Virus reproduction and virion assembly are localized at the ER membrane. TMV can disseminate to neighboring cells through plasmodesmata and cause systemic infection of the plant via the phloem.

TMV is one of the few plant viruses that can survive for long periods of time in the external environment, up to 50 years in storage at 4°C. Passive mechanical transmission can occur when an infected leaf rubs against the leaves of a healthy plant or via human-related direct contact. Plants germinating from contaminated seeds can also become infected.

Further Readings

Casjens S, Hendrix R. Bacteriophage lambda: Early pioneer and still relevant. *Virology*. 2015;479–480:310–330.

Ge X, Wang J. Structural mechanism of bacteriophage lambda tail's interaction with the bacterial receptor. *Nature Commun*. 2024;15:4185.

Krupovic M, Cvirkaite-Krupovic V, Iranzo J, Prangishvili D, Koonin E. Viruses of archaea: Structural, functional, environmental and evolutionary genomics. *Virus Res*. 2018;244:181–193.

Kuiper B, Schontag A, Oksanen H, Daum B, Quax T. Archaeal virus entry and egress. *Microlife*. 2024;5:uqad048.

Lee S, Lewis D, Adhya S. The developmental switch in bacteriophage λ: A critical role of the Cro protein. *J Mol Biol*. 2018;430(1):58–68.

Liu C, Nelson R. The cell biology of *Tobacco mosaic* virus replication and movement. *Front Plant Sci*. 2013;4:12.

Saunders K, Thuenemann E, Peyret H, Lomonossoff G. The tobacco mosaic virus origin of assembly sequence is dispensable for specific viral RNA encapsidation but necessary for initiating assembly at a single site. *J Mol Biol*. 2022;434(24):167873.

Tars K. ssRNA phages: Life cycle, structure and applications. *Biocommun Phages*. 2020 Jun 30:261–292.

Zhu Y, Shang J, Peng C, Sun Y. Phage family classification under *Caudoviricetes*: A review of current tools using the latest ICTV classification framework. *Front Microbiol*. 2022;13:1032186.

Viruses in Biotechnology and Vaccine Development

Application of Viruses in Biotechnology

Historical Perspective

Biotechnology, as defined by the American Chemical Society, is the application of biological organisms, systems, or processes to learn about the science of life, solve problems, and make useful products for industrial and other purposes. Most people today think of biotechnology as a very new science discipline that involves the integration of the natural sciences and engineering sciences. Far from the truth, the processes of making wine or producing a new breed of dog involves as much biotechnology as the technique for polymerase chain reaction that is used to solve crimes and determine family lineages. In fact, biotechnology covers a wide range of techniques to manipulate living organisms for human purposes. We can look as far back as the ancient Egyptians to learn about the biotechnology of making leavened bread using yeast. Fermentation techniques used to brew alcoholic beverages and make cheese were developed by the ancient Chinese thousands of years ago. The Aztecs used spirulina algae to make cakes. Both the domestication of animals and the cultivation of plants require biotechnology.

What people think of as biotechnology today is mostly about the use of genetic engineering for the manipulation of DNA (also known as recombinant DNA). Present-day or modern biotechnology is based on different areas of biological sciences, including biochemistry, cell biology, embryology, genetics, microbiology, and molecular biology, in addition to chemical engineering and digital computing. Biotechnology is used in 4 major areas: health care, agriculture, environmental uses, and industrial uses. Applications for health care biotechnology include gene therapy, molecular diagnostics, regenerative medicine, personalized medicine, and drug development. Applications for agricultural biotechnology include controlling insect pests, managing weeds, protecting crops from disease, and increasing the nutritional value of plants, animals, and even microbes. Applications for environmental biotechnology include cleaning up and reducing pollution and restoring damaged ecosystems. Applications for industrial biotechnology include producing nonedible products like biofuels, vegetable oil, and biodegradable plastics.

For most of the applications mentioned above, genetic engineering is key. But genetic engineering was not possible until the discovery of restriction enzymes. The phenomenon of restriction digestion was actually observed by S. E. Luria and Mary Human in 1952 and G. Bertani and J. J. Weigle in 1953. In 1970, H. O. Smith and K. W. Wilcox purified an enzyme they called endonuclease R that was able to cleave bacteriophage T7 DNA into specific fragments. Kathleen Danna and Daniel Nathans reported the utility of endonuclease R in 1971. They were able to separate the DNA fragments using a technique that was, at the time, recently developed in 1964 by Ulrich Loening; the technique was called polyacrylamide gel electrophoresis. In 1972, John Morrow and Paul Berg introduced methods for preparing recombinant DNAs.

Genetic engineering or recombinant DNA requires 3 types of biological tools: enzymes like restriction endonuclease, vector or vehicle DNA, and passenger DNA or DNA fragments to be manipulated. In this chapter, we focus on the discussion of vehicle DNA. Vehicle DNA (also referred to as a cloning vector or gene carrier) houses and transfers a fragment of foreign DNA into a suitable host. Examples of cloning vectors are plasmid, phage DNA, eukaryotic viral DNA, and cosmid, which are special hybrid plasmids that contain a λ phage cos sequence.

Viruses in Bacteriophage Biotechnology

Their ability to infect bacteria and take over cellular machinery to express genes found in their genome, whether native or foreign, makes bacteriophages incredibly useful for biotechnology. They are widely used in medical and industrial research to produce proteins that improve human and nonhuman animal conditions and agricultural crop production. Bacteriophage contributions to biotechnology stem from their ability to infect specific pathogens and integrate into the host genome. This ability allows them to utilize cellular machinery to express messenger RNAs (mRNAs) and proteins from desired genes without killing the host cell, although they are capable of destroying the cell wall to cause bacterial cell lysis, thus releasing virions to spread the infection.

The viral receptor-binding protein (RBP), located on the surface of the virion, determines host cell specificity. RBPs are called spikes or fibers, and in tailed phages, they are called tail fiber proteins or spike proteins. Adsorption occurs when these tail fiber proteins interact with either polysaccharides (eg, lipopolysaccharide, capsular polysaccharide, or teichoic acid) or proteins (eg, OmpC and FhuA) on the surface of host bacterial cells. OmpC is protein porin and FhuA is a ferrichrome transporter. Both are bacterial surface proteins that are involved in phage attachment and infection. Bacteriophages can alter their RBPs over generations of propagation, which in turn enables them to alter their host range. In other words, phages can be induced to produce different RBPs in response to the changing host's range in the environment to ensure their ability to reproduce in specific bacterial species available. In the laboratory, phages can be manipulated to produce modified RBPs that target a customized host range. Thus, a particular phage can deliver a desired gene to an expanded host range, depending on which version of RBP it expresses.

Lambda, T4, ΦX174, and M13 are among the most commonly used bacteriophages for biotechnology. Phages have been used in tissue engineering and cloning, where they transfer DNA fragments into cells in culture. Bacteriophages can be genetically engineered to act as antibacterial agents by serving as cloning vectors to deliver antibiotic genes to their host cell. This technology is better because phages can target their host cell specifically, while the use of antibiotics lacks specificity. Phages can act as types of "natural" biocontrol agents to treat bacterial contamination to ensure food safety by preventing biofilm formation in food-related environments or be used in industries such as agriculture, water treatment, and petroleum. Phages also encode viral proteins, such as endolysins that degrade peptidoglycan and holins that degrade the peptidoglycan layer and break the cell membrane, leading to the physical destruction of bacterial cells.

Clinical applications involving the use of phages as a delivery vehicle is called phage therapy. Phages can be engineered to act as vehicles to deliver vaccines by carrying vaccine antigens directly on their surface or carrying antigen-encoding genes whose gene products can be displayed on the surface of viral infected cells or even cancer cells. These proteins tag the infected or cancer cells to be recognized and attacked by the host's immune system. The same technology can be used to deliver drug-encoding genes, like chemotherapeutic agents, to target host cells. Phages can also be used as biosensors to identify bacteria, detect phage particles, and recognize specific molecules in patients.

Bacteriophages can affect eukaryotic cells and host immune systems. Phages like T4 have been shown to interact with mucin glycoproteins in the mucous layer of the gastrointestinal tract. This localization of phages likely increases the probability of their encounter with host bacteria. To control the colonization of pathogens and help shape the gastrointestinal tract's microbiome. T4 phages have also been shown to interact with other organ systems, including the respiratory system, central nervous system, and urinary tract. Other phages, like ES2, can induce the expression of surface proteins such as cluster of differentiation 86 (CD86), CD40, and major histocompatibility complex class II (MHCII) in immune cells. Pattern recognition receptors (PRRs) on the dendritic cell surface can also be induced by certain phages to activate nuclear factor kappa B signaling and the production of pro-inflammatory cytokines interleukin (IL) 6, IL-1α, IL-1β, and tumor necrosis factor alpha α.

Viruses in Eukaryotic Cell Biotechnology

Viral genomes can be genetically engineered to contain target genes for gene therapy. The viruses themselves can in turn deliver these genes to target host cells for a multitude of applications. In medical treatment, viruses are used in the development of vaccines and carry genes that encode substances to prevent and treat diseases of specific cells and tissues, like infectious diseases and cancers.

For instance, a live-attenuated poliovirus vaccine was developed by Albert Sabin to eradicate paralytic poliomyelitis. Other live-attenuated virus vaccines include those that are used to immunize against yellow fever, typhoid fever, mumps, and shigella. A live vaccinia virus has been developed to eliminate smallpox.

For agricultural applications, viruses can deliver genes that improve plant and crop growth and production. These desirable features include resistance to stresses, resistance to pests, resistance to drought, and improvement of plant breeding.

Viruses are used in environmental biotechnology, such as for controlling the aquatic microbial communities to stimulate algal growth and reducing the amounts of carbon dioxide in the deep ocean.

Frequently used viruses as vectors for all such applications include adenovirus (AdV), herpes simplex viruses (HSVs), lentivirus, and retrovirus. Virus-like particles (VLPs) are used in nanotechnology, for example, as nanovaccines and drug nanocarriers. VLPs are protein complexes that are made up of 1 or more different viral structural proteins. These complexes mimic the structure and size of viral particles; however, they do not contain viral genetic material. This means that they can be recognized but are incapable of infecting the host cell. VLPs can be produced and undergo self-assembly in living or cell-free expression systems.

Each virus species is initially propagated in its natural host cell and isolated for research and other applications. For instance, human AdV is produced efficiently in the human embryonic kidney 293 (HEK293) cell line and canine adenovirus type 2 (CAV-2) vectors are replicated in dog kidney cells.

Cell lines that are used to propagate retroviruses include National Institutes of Health (NIH) 3T3 mouse embryonic fibroblast cell line, the human cell lines (HT1080, TE671, HEK293), and CEM (T-lymphoblast cell line). Lentiviral vectors can be produced in COS (fibroblast-like cell lines derived from monkey kidney tissue) and HEK293T cell lines.

HSVs can be propagated in a variety of cell lines, including HeLa (Henrietta Lacks [cells]; cervical adenocarcinoma cell line), Vero (African green monkey kidney cell line), BHK-21 (baby hamster kidney cell line), McCoy (from human synovial fluid), HaCaT (keratinocyte cell line), Hep-2 (laryngeal squamous carcinoma cell line), A549 (lung carcinoma epithelial cell line), and fetal lung cell lines (MRC-5 and WI38), FRhK (fetal rhesus monkey kidney [cell line]), HEL (human embryonic lung [cell line]), and human fetal foreskin cell line.

Adenovirus Uses in Biotechnology

Adenoviruses are commonly used vectors for gene therapy to treat cancer and other diseases. More than 400 gene therapy trials have been conducted using human AdVs as vectors. Human and non-human AdVs are small DNA tumor viruses that can induce carcinogenesis in experimental systems, such as hamsters, mice, rats, and even baboons. However, there is no definitive evidence that AdVs stimulate human cancers. The first human AdV was isolated from human adenoids (lymphoid organs in the nasal cavity) in 1953. There are 7 species of AdVs that are further separated into at least 57 human serotypes.

AdVs are favorable vectors for gene therapy because they can infect a wide range of vertebrates, including bats, birds, dogs, horses, humans, mice, ruminants, and swine. AdVs also infect a wide range of tropism due to their ability to recognize and bind to a variety of host receptors, such as Coxsackie and adenovirus receptor, CD46, desmoglein 2, glycans GD1a, and polysialic acid. AdVs can infect both dividing and nondividing cells and are maintained as an episome after entry into the host cell. AdVs have been shown to be safe vectors, have therapeutic influence, and are well tolerated by the host cell and organism.

AdVs can moderately activate the innate immune responses by interacting with surface and intracellular host proteins such as integrins, PRRs, and various toll-like receptors (TLRs) (like TLR-2, TLR-4, and TLR-9). The adaptive immune responses can also be activated by using transgene viral products without excessive release of pro-inflammatory cytokines. The ability to stimulate both immune systems makes AdVs very effective for their intended role in biotechnology, especially in vaccine development.

Two types of genetically modified AdV vectors are used for gene therapy and vaccine development: replication defective (RD) and replication competent (RC).

The RCs are oncolytic vectors that can replicate in and destroy cancer cells preferentially by progressing through their natural lytic infection.

ONYX-015 is an RC (oncolytic) AdV vector that has been examined in clinical trials. This vector lacks the gene that encodes the protein E1B-55K, which is required for AdV replication in noncancerous cells. However, many infected cancer cell lines do not have this requirement. Therefore, ONYX-015 can only cause the lysis of cancer cells but leave uninvolved cells undamaged. Ad5-cytosine deaminase/TKrep (TK, tyrosine kinase) is another version of the E1B-55K-deleted vector that expresses a cytosine deaminase–thymidine kinase (TK from HSV) fusion protein. This vector is used in combination with 5-fluorocytosine plus ganciclovir, followed by radiation therapy. This combination of activities inhibits the replication of cellular DNA as well as the vector, causing the tumor cells to be sensitive to radiation therapy.

CV706 is another RC AdV vector engineered to have the AdV E1A promoter-enhancer replaced with the prostate-specific antigen (PSA) promoter-enhancer. CV706, therefore, replicates preferentially in PSA-positive cancer cells.

RDs contain a strong promoter, like the cytomegalovirus (CMV) immediate early promoter, to regulate high expression of the desired transgene.

For cancer gene therapy, 2 types of RD AdVs vectors carrying the transgene p53, Advexin® and Gendicine® driven by the CMV and Rous sarcoma virus promoters, respectively, have been developed and used in clinical trials.

The E1 genes of these vectors is deleted. E1 gene products are involved in viral replication. Without these genes, these versions of AdV cannot propagate, but can still express the desired transgene. The rationale is to increase the level of p53 since 50% of cancer cells lack a functional p53 gene, and almost all cancer cells have a defect in the p53 tumor suppressor pathway. A high p53 expression in turn can cause these cells to undergo cell cycle arrest and eventually activate the intrinsic apoptotic pathway, thus killing the cancer cells.

Sitimagene ceradenovec is another RD AdV with E1 gene deletion that carries the TK gene driven by the CMV promoter. Sitimagene ceradenovec is used synergistically with ganciclovir, which is an antiviral agent used to treat symptoms of CMV infection. TK activity, supplemented by ganciclovir, can cause DNA chain termination and eventually promote cellular apoptosis. Additionally, neighboring tumor cells can be affected by the combined activities of TK and ganciclovir that were initiated in the original cell infected by RD AdV.

For regenerative medicine, RD Ad-BMP2 (bone morphogenetic protein 2) is used in gene therapy for bone regeneration. This vector contains the gene that encodes osteo-inductive BMP2, which plays an important role in osteogenesis. RD Ad-VEGF121 (adenovirus containing gene encoding vascular endothelial growth factor) is used in gene therapy to reduce wall thickening and improve myocardial perfusion. VEGF is a signaling molecule that regulates angiogenic functions.

For vaccine development, several AdV-based vaccines were developed during the COVID-19 pandemic. ChAdOX1 nCoV-1, Sputnik V, and Ad5-nCOV are adenoviral vectors that express a full-length spike protein of SARS-CoV-2 (severe acute respiratory syndrome coronavirus 2). AdV-based vaccines were also developed against other viral infections. Ad26.ZEBOV expresses glycoproteins from Ebola, Sudan, Marburg, and Tai Forest viruses. Ad26.ZIKV.001 expresses neutralizing antibodies against Zika virus). ChAdOx1 expresses influenza virus antigens nucleoprotein and matrix protein 1 (M1).

Herpes Simplex Virus Uses in Biotechnology

Herpes simplex virus vectors have many applications in biotechnology, including gene therapy, vaccine development, and comparative genomics. Three types of genetically modified HSV vectors are used: replication-competent attenuated vector, replication-incompetent recombinant vector, and amplicon (defective helper-dependent vector). The use of HSV in biotechnology is still under investigation and development.

Replication-competent attenuated vectors contain an HSV genome that lacks some nonessential viral genes, such as ones involved in virus replication, virulence, and immune evasion. Note that these vectors can still reproduce in vitro. The major genes that have been identified and studied are ones that encode for TK, ribonucleotide reductase (RR), the virion-host shutoff (vhs) and the ICP (infected cell protein) 34.5 proteins. In addition to causing DNA chain termination and promoting cellular apoptosis, TK is necessary for efficient virus replication in neurons. RR is required for new viral DNA synthesis during virus replication. Vhs reduces host protein synthesis by destabilizing host RNAs and inducing translational arrest. ICP34.5 is a neurovirulence factor that is essential for HSV pathogenicity by blocking host antiviral response.

All of these attenuated vectors can reproduce in vitro but not in vivo. This ability is important for HSV to be used for gene therapy because wild-type virus is highly toxic and pathogenic in the host organism. Limiting the ability of this virus to replicate in the host organism makes using it as a vector much safer. A major advantage of the replication-competent attenuated vectors over replication-incompetent recombinant vector is the fact that the attenuated systems are more efficient and prolific for gene transfer.

Replication-incompetent recombinant vectors lack essential genes for virus replication; therefore, they can only propagate in established transformed cell lines. Immediate early genes from these vectors have been deleted or mutated in different combinations. ICP0, ICP4, and ICP27 regulate the expression of early and late genes. Additionally, ICP27 is involved in RNA splicing and polyadenylation and providing stability to the newly synthesized viral mRNAs. ICP22 sequesters cellular DNA polymerase to provide optimal expression of the ICP0 protein. ICP47 provides a mechanism for the virus to evade host immune surveillance by inhibiting antigen presentation by MHC class I.

Amplicon vectors are modified HSV-1 particles that have the same external structure and immunogenicity. The genome includes bacterial plasmids that can accommodate 1 or 2 transgenes along with HSV-1 genes providing helper functions for replication. These "helper function" genes include ICP8/UL29 (HSV DNA binding protein that is involved in viral DNA replication, recombination, and the formation of replication compartments), UL5/8/52 (helicase-primase gene), and UL30/42 (DNA polymerase gene).

The circular amplicon is replicated into a concatemeric form of a plasmid DNA (called amplicon plasmid), then packaged into HSV-1 particles. Amplicon vectors are produced by transfecting the amplicon into a cell that is superinfected (or co-infected) by helper HSV-1. These vectors can accommodate a large transgene, up to 40–59 kb, forming a genome of 152 kb in total size. Another advantage of amplicons is their lack of genes necessary for virus reproduction; therefore, there's limited toxicity imposed on the host cell.

HSV Vector in Gene Therapy

HSV can deliver human genes to be expressed in epithelial cells of the skin and mucous membranes, especially tissues of the nervous system. HSV-1 has the ability to cross the blood-brain barrier under certain conditions and therefore can be used for treatment by delivering neurotransmitters.

Similar to RC AdV, HSVs are oncolytic vectors that can also selectively target and destroy cancer cells. Last, they can deliver prophylaxis and therapeutics against HSV infection or other infections targeting specific tissues and organs.

Advantages of HSV vectors include their ability to accommodate a large transgene up to 40–59 kb and their ability to persist in a latent state. Once in the host cell, the genome does not integrate into the host genome because it's maintained as a circular amplicon, thus avoiding the potential of insertional mutagenesis.

HSV vectors can be genetically modified to carry the gene encoding erythropoietin, which is controlled by a regulatable (inducible or repressible) promoter such as the human CMV immediate early gene promoter plus the tetracycline response element. Erythropoietin provides neuroprotection to midbrain dopaminergic neurons, which tend to be progressively depleted in individuals with Parkinson disease. The depletion of these dopamine-producing neurons leads to motor defects and cognitive impairment. The ability not only to express therapeutics, but also to regulate the timing and levels of expression is valuable for treatment.

Another example is the insertion of the dystrophin gene, controlled by muscle-specific promoter, into the ICP0 locus of the HSV-1 genome. This vector induces an increase of dystrophin in dystrophin-deficient muscle cells in culture.

An HSV-1 amplicon was engineered to synthesize the RNA complementary strand to brain-derived neurotrophic factor (BDNF), which is a protein implicated in epilepsy. The RNA complementary strands have been shown to knock down BDNF mRNAs in the cytoplasm of amplicon-infected cells.

HSV Vector in Cancer Gene Therapy

rQNestin34.5v.2 is a genetically modified HSV-1 that lacks ICP6 but contains the ICP34.5 gene. ICP6 allows HSV to replicate in quiescent cells. ICP34.5 is normally required for HSV virulence in noncancerous cells; however, in this vector construct, ICP34.5 is controlled by the nestin promoter, which is only activated in gliomas. Therefore, only gliomas experience cell lysis when the nestin promoter is upregulated.

Talimogene laherparepvec (IMLYGIC®) is an attenuated HSV-1 prescription medication that is used for treating inoperable metastatic melanoma. HSV1716 has been shown to be effective in treating highly invasive cancers, like pediatric high-grade glioma and diffuse intrinsic pontine glioma. HF10 (or high-frequency spinal cord stimulation) in combination with erlotinib and gemcitabine has been shown to be effective in treating inoperable pancreatic cancer. HF10 is a minimally invasive treatment approved by the Food and Drug Administration (FDA) that uses electrical pulses to help reduce chronic pain. The combination of erlotinib and gemcitabine is FDA approved to treat advanced and metastatic pancreatic carcinoma. The chemotherapeutics erlotinib is a tyrosine kinase inhibitor targeting epidermal growth factor receptor, and gemcitabine is a nucleoside analogue.

HSV Vector in Vaccine Development

HSV vectors carry transgenes that encode for antigens to stimulate protective immunity against HSV infection and other infectious diseases. HSV vectors have been used in vaccines for Ebola virus, foot and mouth disease virus, and human immunodeficiency viruses (HIV).

HSV Vector in Comparative Genomics

Studying herpesviruses is a useful tool to examine the evolution of humans because they co-evolve as host and pathogen. HSVs have existed in human populations as early as 5000 years ago, during the Bronze Age, but existed in nonhuman species as far as millions of years before. Since these viruses can remain latent in their host, their genomes can be detected in specimens of archeological excavations.

Beside humans, the primate host range of HSVs includes chimpanzees, baboons, macaques, and spider monkeys. Only humans, however, are infected with more than 1 species of HSV, HSV-1 and HSV-2. These 2 viral species share 87% genomic identity;

however, they have different origins. It's been proposed that HSV-1 arose through virus-host codivergence, while HSV-2 arose from cross-species transmission. Another interesting observation is the fact that HSV-2 genomic sequence is more closely related to the chimpanzee herpesvirus (ChHV) than that of HSV-1. Comparing the genomics of HSV-1 and HSV-2 has provided insights on the evolution of both the viruses and their host organisms.

Here's what we know so far about host-pathogen evolution of this particular pair (human and HSV, that is). Ancient HSV appeared roughly 45 million years ago. About 6 million years ago, HSV-1 and ChHV diverged. Around 1.6 million years ago, ChHV appeared to be transmitted from *Paranthropus boisei*, a common ancestor of today's chimpanzees, to *Homo habilis*, which is an extinct species that preceded modern humans. *Paranthropus boisei* and *Homo habilis* coexisted near Lake Turkana in Kenya. They presumably shared water sources and common tools for hunting and scavenging, which are likely modes of transmission.

HSV-1 has been used as a marker for studying human migration patterns due to co-evolution with their host through millions of years. Analysis of their genomic sequence suggests that HSV-1 originated from Africa, then was carried by host populations that migrated into Europe and Asia. This pattern of migration is consistent with the "out of Africa" model of the origin of the modern human. Analysis of their genomic sequence also suggests that HSV-2 first co-evolved with hosts in Africa. At some point, HSV-2 containing HSV-1 recombinant fragments was carried by *Homo sapiens* out of Africa.

Lentivirus Uses in Immunodeficiency Treatments

The current third-generation lentiviral vectors are derived from HIV and are used to deliver genes into mammalian tissues. Lentivirus vectors have been used to deliver desired genes to the liver, retina, skeletal muscle, and central nervous system safely and effectively in clinical trials. The vector can integrate the transgene, in the form of complementary DNA, into the genome of both dividing and nondividing cells to maintain stable expression of the desired gene. The advantage of infecting nondividing cells is the reduction of oncogenic potential. They can deliver large amounts of desired genetic material, up to 8.5 kb. They are nonpathogenic and do not stimulate an inflammatory response in the host organism.

The latest version is called the third-generation, self-inactivating (SIN) lentiviral vector is very safe, and there is practically no chance that replication-competent lentiviruses (RCLs) can be produced through recombination events. This vector lacks the genes that are critical for lentivirus in vivo replication and pathogenesis, including *tat, vif, vpr, vpu,* and *nef*. Moreover, the *gag* and *pol* genes, whose functions are described in the Classification and Structure of Human Immunodeficiency Virus section of chapter 8, are modified to mitigate the synthesis of RCLs with the *pol* gene modification to prevent insertional mutagenesis.

Insertional mutagenesis has been shown to lead to genotoxicity when a lentiviral vector is inserted into or near genes that confer possible replication and survival of viral particles. Furthermore, SIN lentiviral vectors can be transduced efficiently to non-proliferating or slowly proliferating cells, for example, CD34⁺ hematopoietic stem cells (HSCs). This feature makes these vectors very useful for clinical applications.

SIN lentiviral vectors have also been modified to carry genes to generate immunity to cancer when transduced into mature T lymphocytes. These transgenes include chimeric antigen receptors, T-cell receptors (TCRs), and cloned TCRs. T lymphocytes transduced with these vectors can recognize antigens that are not naturally recognized by their endogenous TCRs. Treatment using this vector has been shown to be successful on individuals with advance B-cell malignancies.

Applications of SIN Lentiviral Vector

Conditionally, RCLs have been modified to encode a complementary (antisense) RNA strand to HIV envelope mRNA. This vector can silence envelope protein expression and thus prevent the production of mature HIV virions. Treatment of natural HIV infection of mature peripheral blood T lymphocytes was shown to be effective using this vector. RCLs encoding different antisense RNA strands, short-hairpin RNA or microRNA, have also been constructed and transferred into $CD34^+$ HSCs to treat other genetic conditions, such as β-thalassemia, metachromatic leukodystrophy, Wiskott-Aldrich syndrome and X-linked adrenoleukodystrophy.

Genetically modified lentivirus containing the gene encoding platelet-derived growth factor B (PDGF-B) was used to treat diabetic wounds in mice. PDGF-B promotes angiogenesis and collagen organization in connective tissue during wound healing.

Lentivirus can even be modified structurally when the HIV envelope protein gp120 is replaced by a different protein to switch host cell targets. This significantly broadens the tropism by allowing the desired gene to be transferred to a variety of cell types. This vector is often referred to as a pseudovirus. For instance, gp120 is required for binding to $CD4^+$ T-helper cells and macrophages. However, by exchanging this envelope protein with the vesicular stomatitis virus G protein, the lentiviral vector can now bind to human low-density lipoprotein receptor and attach to different cell types, such as bone marrow-derived macrophages, lymphocytes, human monocyte macrophages, and fibroblasts.

Glycoprotein gp120 can be replaced by SARS-CoV-2 spike proteins. This version of pseudovirus can attach to cells that exhibit the angiotensin-converting enzyme 2 (ACE2) on their surface and thus can be used as therapy to treat SARS-CoV-2 infection.

"Chimeric antigen receptor" (CAR) T-cell therapies using lentiviral vectors encoding antisense RNA to the B-cell marker CD19 have been used to treat individuals with B-cell acute lymphoblastic leukemia and non-Hodgkin lymphoma. This technology enables T-cell receptors (TCRs) of T lymphocytes to detect specific target antigens. Through clinical trials, this treatment demonstrated high efficacy, especially for patients with relapsed (or refractory) disease.

Retrovirus Vector Applications in Biotechnology

The retrovirus genera (alpha to gamma) and lentivirus genus are very similar in most aspects of their life cycle; however, the types of cells that they can infect can be different. Species of retrovirus genera can only infect dividing cells, whereas species of the genus *Lentivirus* can infect and integrate into the genome of both dividing and nondividing cells. This feature renders using retroviruses as delivery vectors more limiting. Nevertheless, retroviruses are equally useful tools for biotechnology. Like lentiviral vectors, retroviruses have the ability to integrate their genome into cellular chromosomes of their host cell with efficacy. These vectors can also be stably replicated and disseminated throughout the host organism.

An advantage of using retroviral vectors is the fact that all viral coding regions have been removed in these vectors since the regulation of viral protein synthesis is not required for the early phase of retroviral life cycle. This means that only genes of interest are packaged in these vectors. Commonly used retroviral vectors are gamma-retroviral vectors that are derived from the Moloney murine leukemia virus MoMLV or MMLV), whereas lentiviral vectors are derived from the HIV genome. The following are popular gamma-retroviral transfer vectors used in research labs:

- pBABE-puro is an MMLV-based vector used for cloning and expressing genes of interest with puromycin selection. A part of the acronym pBABE stands for "Baceterial Gene Expression" containing a puromycin-resistant gene.

- pMKO.1 puro is an MuLV-based vector that expresses small interfering RNAs to direct the cellular RNAi machinery to silence target genes. pMKO.1 puro refers to the pLKO.1 lentiviral vector containing a puromycin-resistant gene.

- CAG-GFP-IRES-CRE (CAG promoter, Green Fluorescent Protein, Internal Ribosome Entry Site, and Cre recombinase) is used to induce site-specific recombination events and is widely used for mouse transgenics.

- pdCas9 (CRISPR-associated protein 9)-humanized expresses human codon-optimized Cas9 under the control of MuLV promoter used for mammalian gene knockdown.

Baculovirus Expression Vector System (BEVS) in Biotechnology

The appearance of baculoviruses is linked to the development of the silk industry in China over 5000 years ago. This industry spread to Korea and Japan, then Europe over 2000 years ago. Silk production finally spread to Mexico by the 1500s. With the intensification and abundance of silk production, infectious diseases arose in silk worm populations. High interest in identifying and studying baculoviruses is due to the threat of potential harm to the silk industry.

Baculoviruses infect larvae of a broad range of about 600 host insect species, including immature lepidopteran species, sawflies, and mosquitoes. Their infection causes the host organism to liquify to release newly formed viral particles into the environment for transmission to new hosts. Humans have known about baculoviruses for over 2000 years, with the earliest recorded infection occurring in the domestic silk moth *Bombyx mori*, causing silkworm "jaundice." These viruses have been studied and used as biopesticides in crop fields since the 1940s.

Max Summer developed the first recombinant baculovirus for the expression of human beta interferon in insect cells in 1983. The baculovirus expression vector system (BEVS) is a system used to express proteins in eukaryotic cells. This system can accommodate a large DNA insert that yields relatively high expression levels. Protein synthesis in this system includes translational modification similar to those of mammalian cells.

Traditional BEVSs use cultured insect cells to allow a shuttle vector carrying the gene of interest to recombine at low frequency with the circular double-stranded DNA (dsDNA) genome of *Autographa californica* nuclear polyhedrosis virus. This virus species belongs to a subfamily of baculoviruses that produce characteristic large proteinaceous occlusion bodies (OBs) during the very late stages of infection. These OBs are highly stable in typical ambient conditions, which allows them to remain infectious indefinitely. Birds may be a vector for baculovirus transmission and dispersal.

Baculoviruses possess circular, covalently closed dsDNA genomes of 60–160 kb. Interestingly, their genomes encode for their own RNA polymerase, used for the expression of their late and very late genes; however, the expression of viral early genes requires the use of cellular transcription machinery. The value of BEVSs is their ability to express very high levels of intracellular recombinant proteins. These proteins can then be released into the culture medium following virus-induced host cell lysis. High amounts of protein can be easily purified.

Uses of Baculovirus Expression Vector System

The BEVSs have been used in many applications, including vaccine development, biological control such as pest insect control, gene therapy, and production of bulk amounts of recombinant proteins.

Vaccine development using BEVSs has been shown to be rapid and inherently safe while producing flexible product designs in large scale. A BEVS-based vaccine was developed in 2007 against human papillomavirus (HPV) infection to protect against cervical cancer. To date, a number of BEVS-derived products, both vaccines and therapeutic agents, have been approved for human and veterinary use.

For human use, there are the following:

- Cervarix® is a vaccine used to prevent cancers related to HPV infection, including premalignant anogenital lesions of the cervix, vulva, vagina, and anus, as well as cervical and anal cancers. This treatment is used for individuals from 9 years and older.

- Provenge® (sipuleucel-T) was the first cancer vaccine to receive FDA approval in April 2010. It is used to treat asymptomatic or minimally symptomatic metastatic, advanced prostate cancer. This therapeutic vaccine uses a patient's own immune system to identify and target prostate cancer cells.

- Glybera® (alipogene tiparvovec) was approved to treat adult patients diagnosed with familial lipoprotein lipase deficiency who are affected by severe or multiple pancreatitis attacks despite dietary fat restrictions. Currently, this therapeutic agent is no longer available in the market due to its cost.

- Flublok® provides immunization against diseases caused by influenza A and B and is recommended for individuals 65 years and older.

For veterinary use, there are the following:

- Porcilis® (*Porcilis ileitis* swine vaccine) provides immunization to pigs from 3 days of age against diseases associated with porcine circovirus 2 (PCV2) and ileitis (inflammation of the ileum) caused by *Lawsonia intracellularis* (an obligate intracellular bacterium).

- BAYOVAC CSF E2® provides immunization against clinical classic swine fever (CSF).

- Circumvent® PCV is used to prevent viremia and reduce virus shedding caused by PCV2 for swine 3 weeks and older.

- Ingelvac CircoFLEX® is a vaccine used for pigs 2 weeks and older to prevent viremia, lymphoid depletion, inflammation, and colonization of lymphoid tissue caused by PCV2.

BEVSs are biopesticides to protect crops. They act as effective pathogenic viruses to suppress a variety of insects without causing serious harm to environmental conditions. The use of BEVSs is a much safer alternative to protect the health of humans and other living organisms as well as the environment compared to conventional insecticides.

BEVSs provide insect pest management, reduce insecticide resistance, and target specific host organisms.

Vaccine Development

Historical Perspective

Dr. Edward Jenner was credited for creating the world's first successful vaccine; however, the history of immunization did not start with his discovery. In fact, the vaccination procedure known as "inoculation" or "variolation" was likely invented by Taoist or Buddhist monks or nuns around AD 1000. Vaccination (or inoculation or variolation) is the act of introducing a vaccine into an individual. Immunization is the process by which one's body becomes immune to a disease. This version of history was not from written records but rather transmitted verbally. Earliest trials of variolation were documented in China and India during the 16th century. Written records of smallpox dated back to China in the fourth century. Smallpox-like rashes were found on Egyptian mummies that are as old as 3000 years.

We first need to distinguish between the term vaccination and immunization. Vaccination is the act of introducing a vaccine into the body to elicit immune responses against a pathogen and hence build immunity to a specific disease. The physical mechanism of introducing vaccine is called an inoculation. Inoculation via an injection can be subcutaneous, intramuscular, or intravascular. Variolation is the same process as inoculation but refers specifically to smallpox. There are also other methods of inoculation, such as introducing the vaccine orally. Immunization, on the other hand, is the process by which a person becomes protected or resistant against a pathogen and therefore gains immunity from a disease. This immunity can be generated by means of a vaccine or by being previously exposed to the actual pathogen. Vaccination and immunization are often used interchangeably.

Inoculation was practiced in the Ottoman Empire in the 1500s, reached Constantinople by 1650, then reached Europe and Northern Africa. This practice migrated to North America in the Massachusetts colony through an enslaved man named Onesimus. Inoculation was used to fight against the 1721 smallpox epidemic in Boston as well as Britain early in the 18th century. Benjamin Franklin promoted inoculation, which he referred to as variolation after the death of his son in 1736 from smallpox. General George Washington enacted a mandate that required all troops to be vaccinated. This was the first medical mandate in US history. Divisions of troops were inoculated in 5-day intervals to offset the temporary lack of soldiers that were recovering from the vaccination.

Edward Jenner observed that those, like milkmaids, who were previously infected with cowpox were immune to smallpox. He discovered that a person inoculated first with cowpox, usually showing signs of mild reaction, showed no reaction or disease symptoms to smallpox inoculation. This method involved inoculating a sample taken from a blister of someone infected into another person's skin. The technique was called arm-to-arm inoculation. In May 1796, an 8-year-old boy was successfully vaccinated with a sample collected from a cowpox sore on the hand of a milkmaid. Even though he did not develop this technique, Edward Jenner was credited for his scientific approach to prove that vaccination works. The term *vaccination* comes from *vacca*, meaning "cow" in Latin.

Inoculation with cowpox was the only vaccine available against smallpox for almost 80 years. In the late 1800s, the French biochemist Louis Pasteur proved that rabies was a communicable disease. In July 1885, he and his colleagues injected doses of rabbit spinal cord suspensions into a 9-year-old boy who was bitten by a rabid dog 2 days previously. Louis Pasteur was credited for developing an extremely successful live-attenuated (or weakened) vaccine against rabies in humans. He later developed another vaccine against cholera in humans.

In the early 1900s, toxoid vaccines against tetanus and diphtheria were introduced. Around the same time, vaccines against influenza and yellow fever were also developed. The Polish virologist Hilary Koprowski developed the first effective polio vaccine in 1950. The *Haemophilus influenzae* type b (Hib) vaccine, also known as the Hib vaccine, was introduced in the United States in 1985. *Haemophilus influenzae* is a small, facultatively anaerobic bacteria that causes diseases and infections such as pneumonia (lung infection), ear infection or otitis media (infection of space behind the eardrum), bronchitis (inflammation of the bronchioles), cellulitis (skin infection from open wound), epiglottitis (swelling of the throat), infectious arthritis or septic arthritis (swelling of the joint), meningitis (swelling of the lining of the brain and spinal cord), and septicemia (infection of the bloodstream).

In 1992, the first hepatitis A vaccine was developed. The first vaccine against HPV was available in 2006. The first Ebola vaccine was approved in 2019. In December 2020, the first vaccine against SARS-CoV-2, which causes COVID-19, was approved.

What Is the Function of Vaccination?

The first and most important aspect about vaccination is to recognize the fact that a vaccine does not prevent an infection. Again, an infection can still occur after an individual has been vaccinated. Vaccination simply provides the body the ability to mount a more robust defense against a pathogen more quickly and efficiently. These robust immune responses greatly reduce the chance of the individual acquiring the disease.

Vaccines contain a component or components (also known as antigens) of a particular pathogen to mimic an infection. The body's immunological responses to these antigens are identical to ones that are activated in response to an actual pathogen. The goal of vaccination is to prepare the body's immune system to defend the host against diseases caused by pathogenic infections.

Vaccination can be thought of as a primary infection, but without the chance of causing a disease. Therefore, when the person experiences an actual infection of the same pathogen after immunization, mounting secondary immune responses would be many times more effective. The chance of a pathogen establishing a disease is greatly decreased when the immune system can clear the pathogen more quickly during secondary immune responses.

The end result of vaccination is immunization. Let's walk through how vaccination leads to immunization. The inoculated vaccine activates the mucosal and/or innate immune system, depending on the route of inoculation. The innate immune system responds to the vaccine like it does with any other foreign entities involving leukocytes, such as macrophages, dendritic cells, and naïve, mature B lymphocytes. The inflammatory response of innate immunity causes the vaccinated individual to have symptoms and "feel sick." However, since the vaccine cannot induce the propagation of the pathogenic antigens, this individual will not acquire the disease.

Vaccination also involves the activation of the adaptive immune system since dendritic cells and naïve, mature B lymphocytes can also encounter the vaccine antigens. The result of the adaptive immune response is the production of memory B and T cells that recognize the specific antigens that were presented in the vaccine. Memory B cells express antibodies that have high affinity for the antigens, whereas, memory T cells express TCRs that also have high affinity for the antigens.

If this individual is later infected by the actual pathogen, the body can mount an adaptive immune response more quickly and with higher affinity to pathogenic components. Additionally, memory B cells that are reactivated can produce not only higher affinity antibodies, but also at higher amounts. Increased affinity and vigor of activities leads to an increase in the speed of pathogen clearance, which in turn greatly reduces the chance of pathogen replication.

After vaccination, a person may experience symptoms of the actual pathogenic infection. This is because the vaccine tricks the body into thinking that it's been invaded by the virulent pathogen itself, but in fact the antigen or antigens found in the vaccine are incapable of replication and thus cannot mount a full-blown infection and certainly cannot cause a disease. The symptoms that are experienced after vaccination reflect the physiological manifestation of the immune responses.

Considerations When Designing a Vaccine

When designing and creating a vaccine, 3 considerations are noted by the developers: (1) The vaccine must be safe, should not cause the disease, and presents minimal side effects. (2) The vaccine must be effective in protecting the individual from illness caused by the particular pathogenic infection through the appropriate immune responses and confer long-term immunity. (3) The vaccine must be stable in storage and affordable for those who are most susceptible to the infection.

Ethical issues such as beneficence, justice, nonmaleficence, and autonomy must also be considered in the process of development, distribution, and implementation. Key ethical issues include mandates, research and humane testing, informed consent, and access disparities.

In the United States, state policies can mandate certain vaccinations to groups of individuals, such as children entering the public school system (Table 13.1). This mandate was first enacted in the 1850s to prevent the spread of smallpox in the school. However, there are those who have objections to vaccination due to individuals' beliefs, including ideological and religious concerns.

Research and testing through clinical trials must account for diverse contributions from many experts, including public health, epidemiology, immunology, statistics, and pharmaceutical companies willing to mass produce the vaccine. The above-mentioned considerations of safety, effectiveness, and affordability when developing a vaccine are guiding principles.

There are specific informed consent laws created by lawmakers. To be clear, only a handful of individuals in the community, no matter how big or small, is tasked with making these laws. These lawmakers, however, should have been provided ample information from experts to consider before deliberating such laws. Not everyone would agree with these laws and lawmakers; therefore, there are allowances for potential vaccine recipients, or their parents or legal representatives, clear and concise information via the Vaccine Information Statement (VIS). The VIS provides the recipient and/or relevant parties with the appropriate information about vaccine risks and benefits.

Vaccine	1940's	1950's	1960's	1970's	1985–1994	1994–1995	2000	2005	2010	2020	2023
Smallpox	+	+	+	+							
Diphtheria	+*	+*	+*	+*	+*	+*	+*	+*	+*	+*	+*
Tetanus	+*	+*	+*	+*	+*	+*	+*	+*	+*	+*	+*
Pertussis	+*	+*	+*	+*	+*	+*	+*	+*	+*	+*	+*
Polio (OPV)		+	+	+	+	+					
Measles			+	+**	+**	+**	+**	+**	+**	+**	+**
Mumps			+	+**	+**	+**	+**	+**	+**	+**	+**
Rubella			+	+**	+**	+**	+**	+**	+**	+**	+**
Hib					+	+	+	+	+	+	+
Hepatitis B						+	+	+	+	+	+
Varicella							+	+	+	+	+
Hepatitis A							+	+	+	+	+
Polio (IPV)							+	+	+	+	+
Pneumococcal								+	+	+	+
Influenza								+	+	+	+
Rotavirus									+	+	+
COVID-19											+

*Given in combination as DTP

**Given in combination as MMR

Combination vaccine available in 2020	Diphtheria, tetanus, acellular pertussis
	Diphtheria, tetanus, acellular pertussis, inactivated polio
	Diphtheria, tetanus, acellular pertussis, inactivated polio, hepatitis B
	Diphtheria, tetanus, acellular pertussis, inactivated polio, Hib
	Diphtheria, tetanus, acellular pertussis, inactivated polio, Hib, hepatitis B
	Measles, mumps, rubella
	Measles, mumps, rubella, varicella
	Hepatitis A, hepatitis B (only for 18 years of age and older)

These are vaccines recommended in the United States through the decades as compiled by the Children's Hospital of Philadelphia.

TABLE 13.1 Recommended Vaccines

285

Types of Vaccines

There are 4 major types of vaccines: live-attenuated vaccines, inactivated (or killed) vaccines, toxoid vaccines, and subunit and polysaccharide vaccines (Table 13.2). Each type of vaccine has its strengths and weaknesses. In general, live-attenuated vaccines offer long-lasting, even lifetime, protection, while the protection provided by nonlive vaccines fades over time and can only be restored with booster shots. No vaccine, however, contains the full-strength, virulent pathogen that causes human disease. Each type of vaccine is designed to train the body's immune system against specific antigens of specific pathogens. The greater the speed and strength of immune responses, the lower the risk of acquiring the disease conferred by the pathogen. At the cellular level, if the body's immune system can prevent or impede the replication of the pathogen, then the pathogen cannot cause a disease.

Live-Attenuated Vaccines

Live-attenuated (nonvirulent) vaccines are created by using a weakened or attenuated form of the pathogen that causes a disease. The major component of the vaccine is the pathogen itself. There are 2 types of viral particles used. (1) The virion can no longer infect a human host cell because of the removal of pathogenic and/or virulent factors. (2) The virion can still infect but cannot replicate in the human host cell, which is the primary objective of any pathogen. The fact that these vaccines are so similar to the natural virus enables them to stimulate a resilient and long-term immune response that may last the person's lifetime. The most common technique for producing viruses that are attenuated is by growing the disease-causing virus through a series of nonhuman cell cultures or animal embryos. This passage may require upward of 200 cell cultures or embryos in a series. The aim is to create a virus that can still be recognized by human cells. However, this virus had accumulated genetic mutations to prevent it from replicating in human cells. Furthermore, the virus can still replicate in nonhuman cells to produce more viral particles.

The development of such vaccines is complex, difficult, and more expensive than other methods. They are also heat labile and unstable. Although weakened or attenuated, they are still virulent. Being virulent means that they can still stimulate a strong immune response, which makes them inappropriate for use in individuals with weakened immune systems or long-term health problems or those who have had organ transplants. Examples of live-attenuated vaccines include measles, mumps, and rubella (MMR); varicella (chicken pox); influenza (flu); and yellow fever.

Inactivated (or Killed) Vaccines

Inactivated or killed vaccines are created by inactivating or killing the pathogen with high heat or chemicals such as formaldehyde and formalin. The structure of the virus, however, is still intact, allowing the retention of the antigenicity of the pathogen. These treatments eradicate the pathogen's ability to replicate in any cell type, thus eliminating infectivity. Inactivated vaccines are very stable and safe and are much less expensive to produce than live-attenuated vaccines. Last, inactivated vaccines are safer to use in individuals with a weakened immune system.

One disadvantage of inactivated vaccines is the fact that they do not elicit a full cellular immune response and/or long-term immunity, which is especially true with T-cell–mediated response. Therefore, repeated doses of inoculation or booster shots may be required to create long-term immunity. Inactivated vaccines are often administered with an adjuvant to boost their immunogenicity. An adjuvant is a chemical that can stimulate both an inflammatory response and the cellular immune system.

Vaccine	Antigen	First Developer	Route of Inoculation and Effects
Live-attenuated vaccines:			
Measles*	Edmonston-Enders (or Moraten) strain	John Enders in 1963	Subcutaneous injection
Mumps*	Jeryl Lynn strain	Maurice Hilleman in 1966	Subcutaneous injection
Rubella*	RA 27/3 strain	Stanley Plotkin and Leonard Hayflick in 1971	Subcutaneous injection
Influenza (LAIV)	Strains vary by year; first flu vaccine was monovalent (influenza A)	Jonas Salk and Thomas Francis in 1945	Intranasal (nasal spray)
Yellow fever	YF 17D strain	Max Theiler in 1930's	Subcutaneous injection
Varicella (Chickenpox)	Oka strain	Maurice Hilleman in early 1970's	Subcutaneous injection
Polio (OPV)	Poliovirus types 1, 2, and 3	Albert Sabin in 1962	Intramuscular or subcutaneous
Hepatitis A	H2 or L-A-1 strain	China in 1992	Subcutaneous
Rabies		Louis Pasteur and Emile Roux in 1885	Intramuscular or subcutaneous
Rotavirus (RotaTeq)	Rotavirus serotypes G1, G2, G3, G4, and G9	Merck & Co. in 2006	Oral
* Given in combination as MMR			
Inactivated vaccines:			
Polio (IPV)	Poliovirus types 1, 2, and 3	Jonas Salk in 1955	Intramuscular or subcutaneous
Hepatitis A	HAV	Maurice Hilleman in 1992	Intramuscular
Influenza (Flu vaccine)	First vaccine used PR8 strain	Jonas Salk and Thomas Francis in 1945	Intramuscular
Pertussis** (Whooping cough)	Toxin gene is encoded by Lambdoid prophage	Leila Denmark and Pearl Kendrick in 1930	Intramuscular
Rabies (HDCV and PCEC)		Lieutenant-Colonel Sir David Semple in 1911	Intramuscular or subcutaneous

TABLE 13.2 Types of Vaccines

Vaccine	Antigen	First Developer	Route of Inoculation and Effects
Toxoid vaccines:			
Diphtheria**	Toxin gene is encoded by prophage Corynephage β	Gaston Ramon in 1923	Intramuscular
Tetanus**	Toxin gene is encoded by prophage	P. Descombey in 1924	Intramuscular
Pertussis** (Whooping cough)	Toxin gene is encoded by Lambdoid prophage	Leila Denmark and Pearl Kendrick in 1930	Intramuscular
Cholera	Toxin gene is encoded by Cholera toxin phage (CTXφ)	Jaime Ferran in 1885	Intramuscular
**Given in combination as DTaP			
Subunit vaccines:			
Hepatitis B (purified)	HBsAg surface antigen	Baruch Blumberg and Irving Millman in 1969	Intramuscular or subcutaneous
HPV (Recombinant vaccine)	L1 major capsid protein	Jian Zhou and Ian Frazer in 1991	Intramuscular
Shingles (Herpes Zoster or Shingrix)	Varicella-Zoster virus glycoprotein E	Merck & Co. in 2006	Intramuscular or subcutaneous
COVID-19 (mRNA vaccine)	mRNA encoding SARS-CoV-2 spike glycoprotein from B.1.351 virus	Drew Weissman and Katalin Kariko developed mRNA technique used for Pfizer-BioNTech and Moderna's vaccines in 2021	Intramuscular

The first column shows examples of the 3 major types of vaccines. The second column indicates the type of antigen used to create each vaccine. The third column identifies the first developer of each vaccine and the date. The fourth column indicates the route of inoculation or administration of vaccine.

TABLE 13.2 Types of Vaccines (*Continued*)

Toxoid Vaccines

Toxoid vaccines are produced by using a toxin that causes a disease. This toxin is expressed by the pathogen or virus itself. In this case, host immunity is cultivated against the portion of the pathogen that causes the disease instead of the entire viral structure. Toxoid vaccines may also require multiple doses or booster injections to provide ongoing protection against diseases.

Subunit Vaccines

Subunit vaccines are created using components of the virus such as protein, carbohydrate, and nucleic acid. Viral proteins or protein complexes are most commonly used and can be purified directly from viral particles. Viral surface carbohydrates can also be isolated and used for subunit vaccines. Subunit vaccines can elicit a very strong immune response that is highly specific with high affinity. Like inactivated vaccines, subunit vaccines are safe to use in individuals with a weakened immune system and those with long-term health problems.

One disadvantage of subunit vaccines is their ineffectiveness in inducing T-cell–mediated immunity, although they are safe and effective in inducing humoral (or B-cell–mediated immunity). Also, multiple doses of inoculation or booster shots are often required to provide ongoing protection against diseases.

To create subunit vaccines, other techniques have been employed to produce the viral component other than purification directly from the viral particles. Conjugate vaccines, recombinant vector vaccines, and DNA/RNA-based vaccines have also been used.

Conjugate vaccines are created using epitopes of viral antigens. Targeting the purified epitope can stimulate the production of antibodies that have higher specificity and affinity to the antigen when compared to other techniques to produce subunit vaccines. The disadvantage of this system is the fact that these epitopes may not be naturally and readily exposed in a live pathogen. This inaccessibility may cause these vaccines to be ineffective and inefficient. Another difficulty with this technique is the fact that epitopes are too small in size to be able to stimulate the immune system by themselves. These epitopes need to be attached to an inert protein, such as serum albumin or keyhole limpet hemocyanin, to provide the structural bulk to stimulate an immune response.

DNA or RNA vaccines are created using segments of nucleic acid, either DNA or RNA, that encode specific viral proteins. The nucleic acids are taken up by host cells, where they are used to express viral proteins or peptides. These viral proteins or peptides can be presented on the surface of host cells to be recognized as nonself molecules by the adaptive immune system to elicit an immune response.

There are many advantages to DNA/RNA vaccines. There is no chance of instigating an infection since this technique does not involve infectious agents. The DNA can be integrated into the cellular chromosome, which means that viral proteins or peptides can be expressed long term and provide long-term immunity. Viral proteins and peptides on the host cell surface can elicit a strong T-cell–mediated immune response. This technique is inexpensive once the nucleic acid is cloned. Last, no adjuvant is required.

Recombinant vector vaccines (also known as recombinant virus vaccines or platform-based vaccines) are created using a recombinant virus, which is a nonvirulent viral vector that contains the gene of interest. This recombinant virus can infect the host without causing the disease. However, during the infection, the gene of interest can be expressed to stimulate both B-cell–mediated and T-cell–mediated immunity. In essence, these vaccines act like natural infections. These vaccines are similar to live-attenuated vaccines; therefore, they have similar disadvantages in terms of cost and should not be used in individuals with weakened immune systems and health problems.

Further Readings

Abril A, Carrera M, Notario V, Sanchez-Perez A, Villa T. The use of bacteriophages in biotechnology and recent insights into proteomics. *Antibiotics (Basel)*. 2022;11(5):653.

Boylston A. The origins of inoculation. *J R Soc Med*. 2012;105(7):309–313.

Cockrell A, Kafri T. Gene delivery by lentivirus vectors. *Mol Biotechnol*. 2007;36:184–204.

Crystal R. Adenovirus: The first effective in vivo gene delivery vector. *Hum Gene Ther*. 2014;25(1):3–11.

Danna K, Nathans D. Specific cleavage of simian virus 40 DNA by restriction endonuclease of *Hemophilus influenzae*. *Proc Natl Acad Sci U S A*. 1971;68(12):2913–2917.

Elois M, Da Silva R, Pilati G, Rodriguez-Lazaro D, Fongaro G. Bacteriophages as biotechnological tools. *Viruses*. 2023;15(2):349.

Kost T, Condreay J, Jarvis D. Baculovirus as versatile vectors for protein expression in insect and mammalian cells. *Nat Biotechnol*. 2005;23:567–575.

Marconi P, Argnani R, Epstein AL, Manservigi R. HSV as a vector in vaccine development and gene therapy. In: Madame Curie Bioscience Database. Landes Bioscience; 2000–2013: https://www.ncbi.nlm.nih.gov/books/NBK7024/

Milone M, O'Doherty U. Clinical use of lentiviral vectors. *Leukemia*. 2018;32(7):1529–1541.

Mody P, Pathak S, Hanson L, Spencer J. Herpes simplex virus: A versatile tool for insights into evolution, gene delivery, and tumor immunotherapy. *Virology (Auckl)*. 2020;11:1178122X20913274.

Roberts R. How restriction enzymes became the workhorses of molecular biology. *Proc Natl Acad Sci U S A*. 2005;102(17):5905–5908.

Rohrmann G. Chapter 1, Introduction to the baculoviruses, their taxonomy, and evolution. In *Baculovirus Molecular Biology*. 4th ed. National Center for Biotechnology Information (US); 2019. https://www.ncbi.nlm.nih.gov/books/NBK543452/

Roldao A, Silva A, Mellado M. Viruses and virus-like particles in biotechnology. *Compr Biotechnol*. 2011;2011:625–649.

Saleh A, Qamar S, Tekin A, Singh R, Kashyap R. Vaccine development throughout history. *Cureus.* 2021;13(7):e16635.

Scarsella L, Ehrke-Schulz E, Paulussen M, Thal S, Ehrhardt A, Aydin M. Advances of recombinant Adenoviral vectors in preclinical and clinical applications. *Viruses*. 2024;16(3):377.

Wold W, Toth K. Adenovirus vectors for gene therapy, vaccination and cancer gene therapy. *Curr Gene Ther*. 2013;13(6):421–433.

Classification of Viruses Based on Epidemiological Criteria

Arbovirus

Historical Perspective

The Cuban physician and scientist Carlos Finlay first proposed the idea that the *Aedes aegypti* mosquito was a carrier (or vector) for the transmission of yellow fever in 1881. In 1901, the US Army physician Walter Reed confirmed that yellow fever is transmitted by a mosquito and not by direct contact. Decades later, in 1932, Fred Soper identified monkeys as a reservoir for the original yellow fever virus.

In 1906, Thomas Bancroft suggested that the same *A. aegypti* mosquito could also transmit dengue fever. The following year, in 1907, Ashburn and Craig proved that dengue was caused by a virus. The first dengue virus (DENV-1) was discovered in 1943 by Ren Kimura and Susumu Hotta from blood samples of patients in Nagasaki, Japan. In 1945, Albert Sabin independently identified what he called DENV-2 from patients in Hawaii. Both DENV-1 and DENV-2 are now referred to as DENV-1.

The disease tick-borne encephalitis was initially reported by Lev Zilber and colleagues in 1937 in Arkhangelsk, Northern Russia. The tick-borne encephalitis virus was later isolated in 1939 from both ticks and human patients in the region of Belarus. The West Nile virus (WNV) was first isolated from a febrile patient in the West Nile district of Northern Uganda in 1937. The first WNV epidemic was recorded in Haifa, Israel, in 1951. Several large outbreaks ensued in Egypt between 1951 (these began in the upper Nile Delta region) and 1954. The largest WNV epidemic was recorded in 1974 in the Karoo and Northern Cape provinces of South Africa.

In 1947, the Zika virus was first identified by Alexander Haddow in a captive rhesus macaque monkey in the Zika forest. Haddow was a member of a research team that routinely surveilled for yellow fever infection in primates at the time. Zika virus was identified in the *A. aegypti* mosquitoes in the same forest in 1948. In 1952, this virus was isolated in humans in Nigeria and Uganda. The latest Zika virus outbreak occurred in 2015 in Brazil, where approximately 274,000 cases were reported.

An outbreak of chikungunya virus (CHIKV) was recognized in 1952 in the southern region of the United Republic of Tanzania. CHIKV was isolated in 1953 from human patients' blood as well as from the *A. aegypti* mosquito. Outbreaks of CHIKV have been recorded in Thailand in 1967 and in India in the 1970s. In recent years, CHIKV have become more frequent and widespread and have been identified in more than 100 countries in Asia, Africa, Europe, and the Americas.

Common Features of Arboviruses

Arboviral diseases are caused by infections caused by a group of viruses that are transmitted to humans by arthropods or insects and are referred to as arthropod-borne virus or arbovirus. Mosquitoes and ticks are the most common vectors for their transmission. Transmission leading to infections occur most often during the warmer time of the year when arthropods are more active. In other words, transmission of arboviruses is dependent on the life cycle of their cognate arthropod vector.

Families of arbovirus include the Bunyaviridae, Togaviridae, Flaviviridae, and Reoviridae (Table 14.1). The most prominent common feature of all arboviruses is the fact that they are small RNA viruses. They can be transmitted by insect bites, blood transfusions, organ transplants, sexual contact, or transplacentally. Arboviruses are also pathogenic agents for emerging and re-emerging infections. As such, they quickly rise to the level of a public health emergency of international concern (PHEIC), which is a formal declaration by the World Health Organization (WHO) to indicate that these infections pose a worldwide health risk through international spreading of the disease. This risk therefore requires a coordinated international response.

Some arbovirus-infected individuals are asymptomatic. Symptoms of arbovirus that do occur can range from mild to severe. Mild symptoms include fever, headache, muscle pain or joint pain, and skin rash. These are symptoms of non–neuro-invasive arbovirus infections and are related to a febrile illness. Other more serious symptoms may occur, such as diarrhea, vomiting, stiff neck, excessive sleepiness, unconsciousness, and bleeding. Examples of nonneurotropic arboviruses are dengue virus, Zika virus, and CHIKV.

Severe symptoms such as rapid onset, high fever, confusion, tremors, seizures, paralysis, coma, and even death are related to neurological syndromes caused by neuro-invasive arbovirus infections. Examples of neurotropic arboviruses are WNV, Japanese encephalitis virus, and St. Louis encephalitis virus.

Blood-Borne Virus

Historical Perspective

Historical events of transmission of blood-borne viruses (BBVs) that were recorded include the 1812 outbreak of hepatitis in Norfolk, Virginia, and hepatitis A virus (HAV) infection that affected troops during World War I and World War II. However, Paul Beeson was the first to implicate hepatitis as a blood-borne disease.

In 1943, Paul Beeson reported the outbreak of an illness resembling catarrhal jaundice in individuals who had received blood transfusion. The incubation period varied between 4 and 30 weeks. This was the first observation of the transmission of hepatitis B virus (HBV) via blood transfusion, hence the identification of a blood-borne disease. During World War II, contaminated syringes were blamed for the transmission of the blood-borne disease to more than 100 British soldiers on the battlefield.

Virus family	Virus	Vector	Disease
Bunyaviridae	California encephalitis virus	*Aedes triseriatus* mosquito	Viral encephalitis and sub-acute encephalomyelitis
	Jamestown Canyon virus	*Aedes triseriatus, Culex pipiens linnaeus*, and *Coquillettidia perturbans* mosquitoes	Jamestown Canyon encephalitis
	La Crosse virus	*Aedes triseriatus* mosquito	La Crosse encephalitis
Flaviviridae	Dengue fever virus	*Aedes aegypti* mosquito	Dengue fever (or also known as breakbone fever), Dengue hemorrhagic fever, and Dengue shock syndrome
	Japanese encephalitis virus	*Culex pipiens Linnaeus, Culex tritaeniorhynchus*, and *Culex vishnui* mosquitoes	Japanese encephalitis can lead to febrile illness and neurological disease
	Murray valley encephalitis virus	*Culex annulirostris* mosquito	Murray valley encephalitis that can lead to encephalitis, cognitive impairment, seizures, hypotonia, and quadriplegia; potentially fatal
	Powassan virus	*Ixodes scapularis, Ixodes cookei*, and *Ixodes marxi* ticks	Encephalitis and meningitis
	Tick-borne encephalitis virus (TBEV)	*Ixodes ricinus* and *Ixodes persulcatus* ticks	Tick-borne encephalitis (also known as Russian spring–summer encephalitis and Far Eastern encephalitis)
	West Nile virus	*Culex pipiens Linnaeus, Culex quinquefasciatus*, and *Culex tarsalis* mosquitoes	Encephalitis, meningitis, and meningoencephalitis
	Yellow fever virus	*Aedes aegypti* mosquito	Yellow fever (also known as tropical hemorrhagic fever)
	Zika virus	*Aedes aegypti* mosquito	Zika fever, Guillain-Barré syndrome, encephalitis (brain swelling), meningitis, myelitis, and Congenital Zika syndrome (microcephaly) in newborn

TABLE 14.1   Human Arboviruses and Their Diseases

Virus family	Virus	Vector	Disease
Reoviridae	Banna virus	*Culex pipiens Linnaeus* mosquito	Viral encephalitis
	California hare coltivirus, Colorado tick fever virus, and Salmon River virus are serotypes	Rocky Mountain wood tick (*Dermacentor andersoni*)	Flu-like syndromes, meningitis, and encephalitis
	Eyach virus	*Ixodes ricinus* and *Ixodes ventalloi* ticks	Febrile illnesses; neurologic syndromes such as Tick-borne encephalitis, polyradiculoneuritis, and meningopolyneuritis
	Kemerovo virus	*Ixodes persulcatus* tick	Kemerovo tickborne viral fever that can lead to meningitis
	Tribeč virus	*Ixodes persulcatus* tick	Febrile meningoencephalitis
Togaviridae	Chikungunya virus	*Aedes aegypti* and *Aedes albopictus* mosquitoes	Chikungunya fever that can lead to meningoencephalitis
	Eastern equine virus	*Culiseta melanura*, *Aedes aegypti*, *Culex pipiens linnaeus*, and *Coquillettidia perturbans* mosquitoes	Eastern equine encephalitis can lead to meningitis and encephalitis
	Ross River virus	*Aedes triseriatus* and *Culex pipiens Linnaeus* mosquitoes	Ross River fever (also known as epidemic polyarthritis)
	Venezuelan equine encephalitis virus	*Aedes aegypti*, *Culex* vomerifer, and *Culex pedroi* mosquitoes	Venezuelan equine encephalitis
	Western equine virus	*Aedes triseriatus* and *Culex pipiens Linnaeus*, and *Culiseta melanura* mosquitoes	Western equine encephalitis that can lead to encephalitis, myelitis, and meningitis

These are examples of species in each of the four virus families. Arthropods that transmit these viruses are overwhelmingly mosquitoes and ticks. They cause mainly febrile illnesses and neurological syndromes.

TABLE 14.1   Human Arboviruses and Their Diseases (*Continued*)

A significant milestone in dentistry occurred in the 1970s and 1980s when over 100 cases of HBV transmission from oral health care providers to their patients were reported between 1969 and 1985 in California, Connecticut, Georgia, Indiana, Maryland, and Pennsylvania.

A number of transfusion recipients were later diagnosed with acquired immuno-deficiency syndrome (AIDS). These patients were either not in known high-risk groups or in individuals with hemophilia who had received antihemophilic factor concentrate. The first 3 cases of *Pneumocystis carinii* pneumonia (PCP) in patients with hemophilia were reported in 1982, which implies that immune dysfunction symptoms were transmitted through blood and blood products. These data provided the Centers for Disease Control and Prevention (CDC) the evidence to conclude in 1983 that AIDS can be transmitted via blood and blood products. Six cases of human immunodeficiency virus (HIV) transmission from oral health care providers to their patients were also reported in Florida between 1987 and 1989.

Based on standards set by the CDC, the Occupational Safety and Health Admin-istration published the "Occupational Exposure to Bloodborne Pathogens standard" in 1991 outlining practices for handling bodily fluids to reduce risks associated with exposure to viruses and other microorganisms that cause blood-borne diseases.

Baruch Blumberg discovered the Australia antigen (now referred to as HBV sur-face antigen or HBsAg) in 1963, which was the first specific marker for viral hepati-tis. He and his colleagues later identified HBV in 1969 in samples from patients with transfusion-associated jaundice. Blumberg and his colleagues further developed a blood test to detect HBV and later the first HBV vaccine in 1969. These accomplish-ments earned him the Nobel Prize in Medicine in 1976.

In 1989, a virus that caused liver damage was discovered by Harvey Alter, Michael Houghton, and Charles Rice. These 3 scientists were awarded the 2020 Nobel Prize in Physiology or Medicine for the discovery of hepatitis c virus (HCV). The most recent discovery of a BBV is human hepegivirus 1 (HHpgV-1). HHpgV-1 was isolated from serum samples of patients with hemophilia and blood transfusion recipients. HHpgV-1 infection appears to be associated with an increased risk of lymphoma and sporadic encephalitis.

Common Features of Blood-Borne Viruses

Blood-borne viruses are viruses that are transmitted through contact with blood or other bodily fluids and passed from one person to another, not with just blood, as the name may suggest. In addition to blood, the source of contaminated bodily fluids includes amniotic fluid, breast milk, cerebrospinal fluid, pericardial fluid, peritoneal fluid, pleu-ral fluid, saliva (through dental procedures), semen, synovial fluid, and vaginal secre-tions. BBVs can also be found on fomites.

Common modes of transmission of these viruses include

- contact of contaminated substance with damaged skin or open wound;
- contact of contaminated substance with mucous membrane;
- puncture from contaminated needles, broken glass, or other sharps;
- transfusion with blood or blood product;
- transplacental transmission from mother to baby before birth;

- perinatal transmission;
  - o microtransfusion as the placenta detaches;
  - o or through mucosal contact with maternal blood during birth;
- treatment using nonsterile equipment;
- body piercing and tattooing using nonsterile equipment;
- sexual activities (oral, vaginal, or anal).

Depending on the mode of transmission, a number of portals of entry into the host organism are possible. These portals of entry include

- direct interaction with interstitial or soft tissues and blood vessels of open wounds;
- directly into the bloodstream, leading to passive viremia;
- oral and nasal cavities;
- urogenital cavities.

The initial infection of some BBVs may yield an asymptomatic infection, whereas others can produce an initial severe illness. However, any BBV can cause a lifelong or chronic infection. Viremia, which is the presence of virus in the bloodstream, can induce virus species-specific symptoms, such as chills, cough, diarrhea, dizziness, fatigue, fever, headache, muscle and joint pain, nausea and vomiting, poor appetite, runny or congested nose, and sore throat. Swollen lymph glands and rash also sometimes are seen.

Human immunodeficiency virus 1 (HIV-1) and HIV-2, HBV, and HCV are the most prevalent BBVs (Table 14.2). Incidentally, HAV causes inflammation of the liver; however, HAV is transmitted via the fecal-oral route instead. Other BBVs include Ebola virus, Lassa virus, erythrovirus B19, Epstein-Barr virus (EBV), and Zika virus.

These common BBVs pose enhanced risks to health care workers, patients, and family members alike. Risk and management of blood-borne infections are especially important for health care workers. It is critical that protocols are in place in clinical settings that outline risks of occupational transmission, methods of preventing exposure, assessment and management of occupational exposures, detection and diagnosis methods, and available treatments.

To diagnose BBV infections, blood tests can be used by detecting viral DNA or antibodies against viral components in the individual's blood sample. Serology tests and nucleic acid testing (NAT) are two commonly used methods to diagnose HBV. For patients who are known to have been exposed to HBV, the viral load can be detected by serology tests that identify and quantify the levels of HBsAg, antibodies against HB surface proteins (anti-HBs), and antibodies against HB core proteins. In addition, NAT can determine the rate of virus replication, identify specific virus variants, and reveal virus reservoirs. Techniques that are most commonly used to detect HBV are enzyme-linked immunosorbent assays, radioimmunoassay, enzyme immunoassay (EIA), polymerase chain reaction, and recently developed techniques such as microparticle enzyme immunoassay, electrochemiluminescence immunoassay, and chemiluminescence microparticle immunoassay (CMIA).

Virus	Tropism	Common Mode of Transmission	Disease
Cytomegalovirus	Endothelial cells, leukocytes, dendritic cell, lymphocytes	Blood transfusion; contact with infected body fluids (like saliva, urine, tears, semen, and breast milk)	Mononucleosis, pneumonitis, congenital infections, urinary infections
Dengue virus	Monocyte, macrophage, B and T lymphocytes, immature dendritic cell, hepatocyte, Kupffer cell, neuron, microglia, endothelial cell	Perinatal transmission, breastfeeding, blood transfusions, bone marrow transplants, organ transplants, and sexual contact	Dengue fever (also known as break-bone fever or 7-day fever)
Ebola virus	Monocyte, lymphocytes, hematopoietic lineage cells, hepatocytes, Kupffer cells, endothelial cells	Direct contact with blood, other bodily fluids, or fomites	Ebola hemorrhagic fever (also known as Ebola virus disease)
Epstein-Barr virus	B lymphocyte	Blood transfusion and organ transplants; contact with blood and semen during sexual contact; contact with saliva and contaminated semen.	Infectious mononucleosis (also known as mono, glandular fever, or kissing disease)
Erythrovirus B19 virus	Human erythroid progenitor cell and myocardial endothelial cell	Blood transfusion, contact with blood or blood product, transplacental transmission	Erythema infectiosum, transient aplastic crisis, hydrops fetalis, arthritis
Hepatitis B virus	Hepatocyte	Blood transfusion, transplacental or perinatal transmission, sexual contact, use of contaminated needles and equipment (tattooing and body piercing), invasive medical procedure	Hepatitis, Hepatocellular carcinoma, liver cirrhosis

TABLE 14.2   Blood-Borne Viruses and Their Diseases

Virus	Tropism	Common Mode of Transmission	Disease
Hepatitis C virus	Hepatocyte, monocytes, macrophage, B and T lymphocytes, dendritic cell, peripheral blood mononuclear cells (PBMC), and Kupffer cell	Blood transfusion, perinatal transmission, use of contaminated needles and equipment (tattooing and body piercing), sexual contact	Hepatitis, Hepatocellular carcinoma, liver cirrhosis
Human Immunodeficiency virus-1	CD4 T lymphocyte (with co-receptor CCR5),	Sexual contact, use of contaminated needles and equipment (like for drug use),	Acquired immunodeficiency syndrome
	monocyte, macrophage, dendritic cell	transplacental transmission, breast-feeding	
Human Immunodeficiency virus-2	CD4 T lymphocyte (with co-receptor CCR5 or CXCR4), monocyte, macrophage, dendritic cell	Rare in U.S., usually through sexual contact	Acquired immunodeficiency syndrome
Lassa virus	Macrophage, dendritic cell	Contact of broken skin with infected bodily fluids (like blood, urine, saliva, or semen), sexual contact	Lassa fever (hemorrhagic illness), long-term sequela includes myocarditis
Rubella (German measles) virus	Microglia	Transplacental transmission, direct contact with the nasal or throat secretions	German measles, transplacental transmission can lead to birth defects, miscarriage, or stillbirth
Rubeola (measles) virus	Endothelial cell, immature lymphocytes, T and B lymphocytes, activated monocyte macrophage, and mature dendritic cells	Exposure to body fluids (like sputum, respiratory and nasal secretions, and saliva)	Measles that can lead to complications such as conjunctivitis, bronchitis, laryngitis, croup, diarrhea, pneumonia, and encephalitis
West Nile virus	Brain microvascular endothelial cells, neuron, and glial cells	Blood transfusions and organ transplants	Encephalitis, meningitis, and meningoencephalitis
Yellow fever virus	Dendritic cell, B lymphocyte, Natural Killer cell, endothelial cell	Blood transfusion, perinatal transmission	Yellow fever or liver disease

TABLE **14.2**   Blood-Borne Viruses and Their Diseases (*Continued*)

Virus	Tropism	Common Mode of Transmission	Disease
Zika virus	Neural progenitor cell, glioblastoma stem cell, monocyte, dendritic cell, and microglia	Transplacental transmission, blood transfusion, contact with blood products, and organ transplant	Zika fever, Guillain-Barré syndrome, encephalitis, meningitis, myelitis, and Congenital Zika syndrome (microcephaly in newborn)

Blood-borne viruses use different route of entry into the host organism. During an infection, they have different tropism and cause a variety of diseases.

TABLE 14.2   Blood-Borne Viruses and Their Diseases (*Continued*)

The United States National Institutes of Health have clinical guidelines detailing the screening and diagnosis of BBVs, including HCV. Like testing for HBV, similar techniques are used to detect antibodies against HCV components and HCV RNA. EIA and CMIA are commonly used for HCV antibody testing.

For screening and diagnosis, NAT, antibody tests, and antigen-antibody tests are used to detect and quantify HIV as described above. The Mayo Clinic recommends additional tests to be performed to identify other infections and complications that might cause AIDS. Tests for cervical and anal cancer, cytomegalovirus (CMV), HBV, HCV, liver or kidney damage, sexually transmitted infections, toxoplasmosis, tuberculosis, and urinary tract infections are included.

Emerging Viruses

Historical Perspective

In 1948, the US Secretary of State George Marshall, in support of the science community, famously declared that scientists have the means to eradicate infectious diseases globally. The epidemiologist Aidan Cockburn, who was an advisor to the WHO in 1962, believed that it is entirely practical and probable that infectious diseases can be eradicated. Optimism was high as reputable scientist after scientist suggested that it is, in fact, possible to "close the book on infectious diseases," as declared by US Surgeon General William Steward in 1969.

The HIV pandemic in the early 1980s, however, caused virologists and epidemiologists to reevaluate their prediction that infectious diseases can be eradicated in the near future. AIDS was deemed a new disease in 1981, but the identification of HIV as the etiologic pathogen was not identified until 1983. AIDS became one of the worst pandemics in recorded history. In 1988, US Surgeon General C. Everett Koop produced a pamphlet called "Understanding AIDS" to increase awareness and concern about AIDS. This pamphlet underlines the importance of lessons learned during this pandemic, including performing formative research, achieving a consensus on scientific knowledge, trusting and using communications experts, centralizing the final decision-making function, and building a base of support among constituency groups.

Even with the development of effective vaccines against many virulent viruses, recurrence of numerous outbreaks had occurred through the past century. We see recurrences of the 1889–1892 Russian flu outbreak as the 1918–1920 Spanish flu pandemic, as the "Asian flu" outbreak in 1957, and as the 1968–1970 "Hong Kong flu" outbreak. We see recurrences of the 1976 Ebola virus outbreak as the 2014 outbreak that was initiated in Nigeria and the 2022 outbreak that started in Uganda. We see recurrences of coronavirus outbreaks in the 21st century in the form of an outbreak of severe acute respiratory syndrome coronavirus (SARS-CoV) that started in China in late 2002, the Middle East respiratory syndrome coronavirus outbreak in 2012, and the latest COVID-19 (coronavirus 19) pandemic that started in 2019.

Joshua Lederberg, the 1953 Nobel Prize winner for Medicine, coined the term *emerging and reemerging diseases* in the 1980s to describe diseases of infectious origin that have been increasing and threatens to increase in the near future. He presented this concept in his Institute of Medicine report in 1992 and again in 2003. However, David Sencer actually described the concept of emerging disease two decades earlier in his 1971 article, "Emerging Diseases of Man and Animal," where he defined emerging diseases as "infectious diseases of man and animals currently emerging as public health problems."

Common Features of Emerging Viruses

Emerging (or emergent) virus and re-emerging virus are viruses that adapted to an environment that is normally not associated with the original species of virus and emerge as new pathogenic strains that had not before been identified. Different virus strains belong to the same species; however, they have differences in stable and heritable biological, serological, and molecular features. These new strains, in turn, cause emerging infectious diseases (EIDs) that have not been observed previously within a population or geographic location. Influenza virus, coronavirus, HIV, and Ebola virus are some of the most recent examples of emerging virus outbreaks (Table 14.3).

The changes that occur to form the emerging virus and allow the virus to survive in different environments are driven by the accumulation of genomic mutations. Factors that contribute to the emergence or re-emergence of viruses include human-induced activities such as urbanization, globalization, international mobility, deforestation, and reforestation that lead to increased contact between human and wild animals; environmental and ecological factors like weather and climate change; and alteration in the genetic composition of the virus via either antigenic drift or antigenic shift. Most emerging viruses are zoonoses, especially emerging viruses that belong to the group arbovirus.

Emerging viruses as a group tend to possess genomes that are less stable, most often RNA genomes and particularly minus-strand RNA viruses. Nucleic acid synthesis is rapid, which yields a high rate of damages and mutations. There are a number of DNA repair mechanisms in the host cell that can reduce the retention of damages and mutations in the DNA molecule. Since RNA repair mechanisms are limited, especially in host eukaryotic cells, RNA mutations accumulate through many generations of replication creating new virus strains. These constantly evolving virus species cause evolving EIDs in response to rapid changes in the relationship between pathogen and host.

Virus Family	Virus	Outbreaks and Reservoirs	Disease
Bunyaviridae	Hantavirus	1951–1953 in wild mice (endemic in Korea)	Hantavirus pulmonary syndrome (HPS)
	Sin Nombre virus	1993 in wild mice (sporadic in U.S.)	Hantavirus pulmonary syndrome (HPS)
Coronaviridae	SARS-CoV	2002–2003 in bats (sporadic in China)	Severe acute respiratory syndrome (SARS)
	MERS-CoV	2012–2015 in camels (pandemic)	SARS-like
	SARS-CoV-2	2019–2020 most likely in bats (pandemic)	Severe acute respiratory syndrome (COVID-19)
Filoviridae	Ebola virus	1976 in bats (epidemic in Africa)	Hemorrhagic fever
		2013–2016 (epidemic in Western Africa)	Hemorrhagic fever
Flaviridae	Dengue fever virus	1953 in monkeys (endemic in Asia, Africa, South America)	Hemorrhagic fever
	Japanese encephalitis virus (JEV)	1935 in birds and bats (endemic in Asia)	Encephalitis
	West Nile virus (WNV)	1937 in birds (endemic in Uganda)	Encephalitis
		2012 in birds (endemic in U.S.)	Encephalitis
	Zika virus	2015 in non-human primate (endemic in Brazil)	Microcephaly, Guillain-Barre syndrome, neuropathy, myelitis
Orthomyxoviridae	H5N1 influenza virus (Avian flu)	1987 in wild birds (pandemic)	Respiratory disease influenza (or flu)
		2003 in wild birds (sporadic in Asia)	Respiratory disease influenza (or flu)
	H1N1 influenza virus (Swine flu)	2009 in pig (pandemic)	Respiratory disease influenza (or flu)
	H7N9 influenza virus (Avian flu)	2013 in wild birds (sporadic in China)	Respiratory disease influenza (or flu)
Orthopoxviridae	Variola virus	1771, 1752, 1764, 1775 in humans (epidemic in U.S.)	Smallpox
		1864 in humans (epidemic in Angola, Central Africa)	Smallpox
		1870–1871 in humans (epidemic during Franco-Prussian War)	Smallpox
		1975 in humans (endemic in Bangladesh)	Smallpox

TABLE 14.3 Emerging and Re-emerging Viruses and Emerging Infectious Diseases

Virus Family	Virus	Outbreaks and Reservoirs	Disease
Paramyxoviridae	Hendra virus	1994 in bats (sporadic in Australia)	Hemorrhage in Lung
	Measles virus	2008 in humans (endemic in the United Kingdom)	Measles or Rubeola
	Nipah virus	1999 in bats (sporadic in Malaysia)	Encephalitis
Phenuiviridae	Thrombocytopenia syndrome virus (SFTSV)	2003 in *Haemaphysalis longicornis* tick (endemic and epidemic in Asia)	Thrombocytopenia syndrome
Retroviridae	Human Immunodeficiency virus-1 (HIV-1), HIV-2	1981 in humans (pandemic)	Acquired Immunodeficiency syndrome (AIDS)
Rhabdoviridae	Rabies virus	1800's in wild animals (endemic in Europe)	Paralysis and hydrophobia
		2022 in wild animals (endemic in Asia)	Paralysis and hydrophobia
Togaviridae	Chikungunya virus (CHIKV)	2004–2006 in humans (epidemic in Kenya, Indian Ocean islands, India)	Chikungunya fever
		2013–2017 in humans (epidemic in Caribbean islands, Brazil)	Chikungunya fever
		2022–2023 in humans (epidemic in Paraguay)	Chikungunya fever

This is a list of emerging and re-emerging viruses. Most known emerging viruses are zoonotic. A few that cause human diseases have only been isolated in humans. Diseases caused by these emerging viruses are varied.
- Endemic: small to large local outbreak that last a long time (>1 year) with persistency
- Epidemic: small to large local outbreak that lasts a short time (<1 year) with no persistency
- Pandemic: large outbreak in multiple continents that lasts a short time (<1 year) with no persistency
- Sporadic: small outbreak in multiple locations that lasts a short time (<1 year) with no persistency

TABLE 14.3 Emerging and Re-emerging Viruses and Emerging Infectious Diseases (*Continued*)

Most emerging viruses are known to be zoonotic. They gain flexibility in adaptation as they are transmitted from one host species to another, with the host organisms themselves surviving in different environments. Emerging viruses are also known to be highly virulent, causing outbreaks as they appear in the host populations. As a group, these viruses are resilient, having survived throughout human history, and reintroduce themselves (or re-emerge) numerous times in the form of endemics, epidemics, and pandemics.

Gastroenteritis Virus

Historical Perspective

Acute infectious gastroenteritis, or diarrheal disease, is the most common food-borne disease seen around the world. These nonbacterial diseases can cause significant morbidity and even mortality, especially among infants and young children in low- and middle-income countries (or "developing countries"). Even in developed or high-income countries, gastroenteritis is a major cause of morbidity. However, the etiology of these illnesses remained unknown until the Norwalk virus was discovered in 1972.

The identification of the Norwalk virus was possible through the use of immune electron microscopy (IEM). IEM is a technique that permits the direct visualization of antigen-antibody interaction. Antibodies coated viral particles found in stool filtrate from samples collected at an elementary school in Norwalk, Ohio, in 1968. This technique led to the discovery of related gastroenteritis viruses, belonging to the genus *Norovirus*, such as sapovirus in Japan in 1977. Other noroviruses include Montgomery County virus, Hawaii virus, and Snow Mountain virus, which were identified from stool samples collected during epidemics in these regions.

In 1973, rotavirus was identified as the etiologic cause of severe gastroenteritis in infants and young children. Rotavirus are different from norovirus. Rotavirus is more likely to cause infantile diarrhea, while norovirus can affect people of all ages. Enteric adenovirus and astrovirus were discovered in the stool of children with acute diarrhea in 1975. Both virus genera infect mostly infants and young children; however, they've also been isolated in adults. Other viruses that have been linked to gastroenteritis include Aichi virus, bocaviruses, coronaviruses, CMV, enteroviruses, parvovirus B19, picobirnaviruses, poliovirus, sapovirus, and toroviruses have often been associated with gastroenteritis as well.

Common Features of Gastroenteritis Viruses

Viral gastroenteritis, also known as the "stomach flu," that causes intestinal symptoms and diseases is a very common global illness. The pathogens that cause this illness are diverse (Table 14.4). Since they share common clinical presentations and modes of transmission, treatments of these viral infections are similar.

Symptoms of viral gastroenteritis include nausea, vomiting, watery diarrhea, abdominal pain, fever, headache, muscle aches, anorexia, and weight loss. Loss of appetite, bloating, and bloody or pus-filled stools have also been reported. Dehydration can be caused by fluid loss through vomiting and diarrhea and is a notable outcome of this illness and a major concern. Decreased urine output, dark-colored urine, dry skin, thirst, and dizziness are indications of dehydration. If not treated, severe dehydration may be fatal.

Gastrointestinal infection outbreaks are most common in close communities, such as in day care centers, nursing facilities, and crowded workplaces and on cruise ships, although ingesting contaminated food or water, direct contact with contaminated surfaces, and sexual contact with infected individuals can occur in isolated cases as well. In the United States and other industrialized countries, these infections are mild and self-limiting in most individuals. There is a significant morbidity rate of more severe acute viral gastroenteritis in young children, elderly patients, and individuals who are immunocompromised, but the mortality rate is low (0.01% of those with symptoms). In countries with lower socioeconomic opportunities, however, viral diarrheal diseases

Major Group of Gastroenteritis Virus	Cause of Damage	Molecular Function	Physiological Effect
Adenovirus	Heat-labile enterotoxin (LT)	Inhibit GTPase activity, rendering adenylate cyclase constitutively active	Cause loss of fluid and electrolytes into the intestinal lumen, thus disrupting balance of electrolytes; Promote leukocyte migration into gastric mucus through inflammatory lesions; Cause diarrhea
Astrovirus	Capsid protein	Disrupt tight junctions	Damage intestinal epithelium leading to diarrhea
Norovirus (Norwalk virus)	Not fully elucidated, may secrete enterotoxin similar to NSP4	Act on enterocytes to blunt villi	Cause epithelium dysfunction and malabsorption; Cause diarrhea
Rotavirus	Non-structural protein 4 (SP4)	Increase intracellular calcium	Alter epithelial cell integrity and tight junction structure and formation; Cause chloride secretion and diarrhea

Column 1 is a list of the 4 major gastroenteritis viruses. Column 2 describes the mechanism used by each virus to cause tissue damage. Column 3 indicates the molecular activity leading to tissue damage. Column 4 describes the physiological effect of tissue damage.

TABLE 14.4 Gastroenteritis Viruses and Their Effects

result in a high mortality rate, especially in infants, accounting for over 200,000 deaths of children per year globally according to the CDC.

The 4 major viruses that are known to cause gastroenteritis are enteric adenovirus, astrovirus, norovirus (formerly known as Norwalk virus), and rotavirus. These viruses cause gastrointestinal problems through the activity of their secreted enterotoxin. Viral infections cause the destruction of the intestinal epithelium, which leads to malabsorption that can cause diarrhea and vomiting.

Most gastroenteritis viruses are transmitted via the fecal-oral route. Outbreaks often occur as the result of individuals in a particular community ingesting the same source of contaminated food and water. Therefore, effective measures to control these infections depend on proper food handling, hand hygiene, and clean water. Vaccination is the best means to protect infants and young children from this illness.

Gastroenteritis viruses infect mainly the upper gastrointestinal tract, but different virus species target different cell types. They cause rapid acute onset of diarrhea, whose diagnosis allows clinicians to differentiate viral infections from those mediated by bacteria and protozoa. Symptoms such as nausea, vomiting, fever, and abdominal pain often accompany diarrhea. The infection itself is mild and self-limiting, usually resolving within 1 to 3 days.

Since the symptoms of viral gastroenteritis are common and the diagnosis of the illness is fairly generic, further nucleic acid and diagnostic testing must be performed to distinguish among the different infecting viruses. For most patients with mild symptoms, medical therapy is not necessary. Hydration and electrolyte replenishment are sufficient treatments to maintain comfort while the viral infection resolves itself. The use of oral rehydration solution has greatly reduced the mortality rate from severe diarrheal disease in low-income nations. Antisecretory or antidiarrheal drugs such as bismuth subsalicylate (or Pepto-Bismol®) and crofelemer (Mytesi®) may be used to control diarrhea. Loperamide (Imodium®) and diphenoxylate/atropine (Lomotil®) inhibit segmental contractions of the intestine to slow down the movement of intraluminal fluid and allow increased fluid absorption.

Respiratory Virus

Historical Perspective

The description of a very contagious disease with flu-like symptoms that occurred in the northern region of Greece was recorded around 410 BCE in Hippocrates' "Book of Epidemics." Throughout the Middle Ages (ADE 476 to 1450), epidemics that most likely stemmed from respiratory infections were described as a "plague" or "pest."

The term *influenza*, meaning "influence" in Italian, was coined to describe the 1357 outbreak in Florence. At the time, people believed that contagious diseases were caused by the influence of celestial bodies, the weather, or climates. However, respiratory infections or illnesses were not considered contagious, despite the fact that certain diseases were known to be contagious centuries before.

Several iterations of the "flu" pandemic occurred in 1889–1892, 1918–1920, 1957–1958, and 1968. Several coronavirus pandemics occurred in the 21st century in 2002–2003, 2012, and 2019–2020. Viruses that can cause respiratory infections in addition to influenza viruses and coronaviruses include adenoviruses, parainfluenza, parvovirus B19, respiratory syncytial virus (RSV), metapneumovirus, measles, mumps, and rhinoviruses.

Common Features of Respiratory Viruses

Respiratory viruses can cause problems in both the upper and lower respiratory tracts. Organs and tissues that are affected include the nasal cavity, sinuses, larynx, pharynx, trachea, bronchus, bronchiole, and lungs. Common symptoms of an upper respiratory tract infection may include coughing, fatigue, fever, runny or stuffy nose, sneezing, and sore throat. Common symptoms of a lower respiratory tract infection may include breathlessness, rapid breathing, tight chest, and wheezing. Additional symptoms of a respiratory tract infection may include chills, decreased appetite, diarrhea, headache, loss of taste or smell, muscle or body aches, vomiting, and weakness. A severe lower respiratory tract infection can lead to pneumonia, whose symptoms may include short and shallow breathing, struggling to breathe, and poor feeding in infants.

Depending on which section of the respiratory tract is infected and whether or not major tissue damage has occurred, various physiological conditions and diseases can be exhibited. Conditions and diseases that may arise as the result of upper respiratory tract infections include the common cold, flu, sinusitis, and croup (also known as acute laryngotracheobronchitis). Conditions and diseases that arise from lower respiratory tract infections include bronchitis and bronchiolitis. Once the lungs are infected, pneumonia is very likely to occur. Pneumonia is the most common viral infection in individuals diagnosed with AIDS. Other conditions that may result from a lung infection are bronchiectasis, chronic obstructive pulmonary disease, lung leiomyoma, and lung leiomyosarcoma. Finally, pharyngitis and tonsillitis can also be associated conditions.

Respiratory viruses are quite contagious. The degree of the contagiousness of a virus is largely correlated with transmissibility. There are 4 major modes of transmission for respiratory viruses: direct contact, indirect contact, through large respiratory droplets, and through small respiratory aerosols.

Direct contact occurs when viral particles are transferred directly from an infected individual and the uninfected individual. When they make physical contact, viral particles that remain on the skin of the infected individual can attach to the mucous membrane of the nose, mouth, or even eyes.

Indirect contact occurs when the uninfected individual interacts with fomites. Viral particles can be transferred directly from the surface of the fomite to the nose, mouth, or eyes of this individual. Viral particles can also be transferred to another part of this individual (ie, the hands). This individual in turn touches the nose, mouth, or eyes. This is why one should wash hands thoroughly and often and always refrain from touching one's nose, mouth, or eyes.

The mechanics of transmission via large respiratory droplets and small respiratory aerosols are similar. Transmission through these modes is most commonly accomplished through coughing and sneezing by the infected individual without covering the mouth and nose. There are many factors that dictate the projection of viral particles via these 2 modes, such as gravitational settling rate, force of transport, dispersion in a turbulent air jet, environmental factors, and viral load. Larger viral particles are generally transmitted by large droplets. Large droplets do not usually travel very far due to gravity. However, large droplets are more resistant to evaporation, whereas aerosols can not only travel farther but also evaporate more quickly.

During the fall and winter of the 2023–2024 season, the 3 most common respiratory viruses in the United States were SARS-CoV-2 of the Coronaviridae family; influenza (A, B, and C) virus of the Orthomyxoviridae family; and RSV, not the Rous sarcoma virus, of the Paramyxoviridae family.

Other common respiratory viruses that season included human adenovirus A–G (HAdV A–G) of the Adenoviridae family; bocavirus of the Parvoviridae family; 229E, NL63, HKU1, and OC43 of the Coronaviridae family; enterovirus (A–E) of the Picornaviridae family; human parainfluenza viruses of the Paramyxoviridae family; human metapneumovirus of the Paramyxoviridae family; and rhinovirus of the Picornaviridae family.

In past years, prevalent respiratory viruses included erythrovirus B19 of the Parvoviridae family; measles virus and mumps viruses of the Paramyxoviridae family; and varicella zoster or varicella virus of the Herpesviridae family. Last, CMV (or human herpes virus [HHV] 5), EBV (herpes simplex virus [HSV] 4), HSV-1, and roseolovirus (HSV-6), also of the Herpesviridae family, have also been associated with respiratory disease in individuals who are immunocompromised.

Tumor Virus

Historical Perspective

Frances Peyton Rous is considered the founder of the scientific field of tumor virology when he discovered the Rous sarcoma virus in 1911, although he was not actually the first scientist to discover a tumor virus. In 1907, Giuseppe Ciuffo showed that a virus causes the formation of human warts. Seventy-six years later in 1983, Harald zur Hausen reported the link between the human papillomavirus (HPV) and genital warts and cervical cancer. The conventional wisdom at the time was that these conditions were caused by the sexually transmitted HSV2.

In 1908, Vilhelm Ellermann and Olaf Bang reported that chicken leukemia can be transmitted by a cell-free filtrate, which was identified 40 years later as the avian leukemia virus. In the 1930s to 1960s, several mammalian viruses were discovered including cottontail rabbit papillomavirus, mouse mammary tumor virus, mouse/murine leukemia virus, mouse polyomavirus, and simian virus.

Until the link between HPV and cervical cancer was made in 1983, the EBV was considered the first human tumor virus isolated and identified. EBV was identified in 1964 by Anthony Epstein, Yvonne Barr, and Burt Achong in a cell line derived from Burkitt lymphoma. EBV is also strongly associated with other cancers, such as nasopharyngeal carcinoma and Hodgkin lymphomas. As with identifying Rous sarcoma virus as the first tumor virus discovered instead of HPV, John Trentin identified HAdV in 1962 in experimentally infected animals but not in the natural environment.

Soon after, in 1967, HBV was discovered by Baruch Blumberg, Alfred Prince, Kazuo Okochi, and Seishi Murakami in blood samples from patients with serum hepatitis. Chronic HBV infection was later found to be associated with a 100-fold increased risk for hepatocellular carcinoma. Maurice Hilleman developed an effective HBV vaccine in 1976, which is credited as the first vaccine that is capable of preventing the development of a human cancer. Since the link between HPV and cervical cancer was made by Harald zur Hausen in 1983, efforts were invested in the development of a safe HPV vaccine. In 2005, large-scale clinical trials were finally conducted for HPV/cervical cancer vaccines.

Additional human tumor viruses were identified including human T-cell leukemia virus (HTLV) in 1977 linking to T-cell leukemia, HCV linking to hepatocellular carcinoma, and Kaposi sarcoma herpes virus (KSHV) in 1994 linking to Kaposi sarcoma. The discovery of oncogenes like src, erbB, raf, and myc as well as tumor-suppressor genes like p53 and retinoblastoma were made through studying tumor viruses. All of these oncogenes and tumor-suppressor genes play important roles in the cell cycle and cell proliferation that can contribute to carcinogenesis.

Common Features of Tumor Viruses

Tumor viruses represent a group of different sizes, shapes, chemical compositions, and genetic materials; however, they all share the common ability to induce carcinogenesis (or cancer development) and tumorigenesis (or tumor formation). They are classified as either small DNA tumor virus, large DNA tumor virus, retrovirus, or RNA tumor virus.

The genome of small DNA tumor virus is too small to encode their own gene products to carry out virus replication. The genome is ultimately integrated into the cellular chromosome, then get replicated during the host cell cycle. This group of tumor viruses includes adenoviruses, papillomaviruses, and polyomaviruses. Of these, the HPVs and

Tumor Virus	Viral Oncoproteins	Cellular Binding Partners	Target Pathways and Processes
Epstein-Barr virus (EBV)	LMP1	p53, Mdm2, pRb, p300, Chk2, c-Myc, HDAC1, SUMO-1, SUMO-3, Cyclin A, E and D1, TRAFs, TRADD, JAK	NFκB, MAPK, PI3K/Akt, JAK/STAT, TNF, cellular transcription, cell cycle, metastasis, ub-proteasome, apoptosis, inflammation, chromatin remodeling, cellular signaling, autophagy
	LMP2	TNFR associated factors, RAS, JAK	Apoptosis, metastasis, MAPK, PI3K/Akt, JAK/STAT, TNF, BCR signaling
	EBNA2	RBP-Jκ, PU.1, AUF1, DDX20, SMN	Notch, cellular transcription, metastasis
	EBMA3C	p53, Mdm2, pRb, p300, RBP-Jκ, Chk2, Nm23-H1, c-Myc, HDAC1, SUMO-1, SUMO-3, SCF^{Skp2}-complex, DDX20, SMN, CtBP, Cyclin A, E and D1	Cell-cycle, Notch, ub-proteasome, metastasis, chromatin remodeling, cellular transcription, apoptosis, inflammation
Hepatitis B virus (HBV)	HBx	NFκB, p53, c-jun, c-fos, PKC, c-myc	Wnt/β-catenin, NFκB, TGFβ, JAK/STAT, cell cycle, apoptosis, cellular transcription, cellular signaling, metastasis
Hepatitis C virus (HCV)	NS3	p53, PKA, H2B, H4, Arginine methyltransferase 1	PKC, inflammation
	NS5A	p53, Bax, IFN-induced dsRNA activated protein kinase (PKR), growth factor receptor-binding protein 2 (Grb2), TRADD, CDK1, TRAF2, TBP PI3K p85 subunit,	Cell-cycle, apoptosis, Ras-Erk MAPK pathway, PI3K, NFκB
Human papillomavirus (HPV)	E6	p53, p73, c-Myc, pRb, p21CIP1, p27KIP1	Cell cycle, ub-proteasome
	E7	IRF-1, cyclin A and E	Cell cycle, ub-proteasome
Human T-cell lymphotropic virus (HTLV-1)	Tax	Cyclic AMP, p300/CBP, MAD-1, MAD-2, cyclin D1, Chk1 and 2	Cell-cycle, apoptosis, cellular transcription, NFκB, PI3K/AKT, chromatin remodeling

TABLE 14.5   Oncoproteins and Their Effects

Tumor Virus	Viral Oncoproteins	Cellular Binding Partners	Target Pathways and Processes
Kaposi's sarcoma-associated herpesvirus (KSHV or HHV-8)	vGPCR, vIL-6, vBcl2, vCyclin, LANA and vFLIP	p53, pRb, c-Myc, GSK3β, FADD, core histones, TRAF2, MAPK2, Transcriptional activators (Brd2, Brd4, Sp1, AP-1, CBP), and transcriptional inhibitors (HP1, Dnmt3, mSin3)	Notch, Wnt/β-catenin, NFκB, JNK/AP1, cellular transcription, cell cycle, apoptosis, ub-proteasome, chromatin remodeling, cellular signaling
Merkel cell polyomavirus (MCPyV)	LT	p53, pRb	Cell cycle

Oncoproteins expressed by specific tumor viruses. These oncoproteins associate with specific binding partners to modulate specific cellular functions that affect carcinogenesis.

TABLE 14.5 Oncoproteins and Their Effects (*Continued*)

Merkel cell polyomavirus (MCPyV) are the only ones that are considered true human tumor viruses. HPVs have been linked to cervical, anogenital, and oropharyngeal cancers. MCPyV has been linked to Merkel cell carcinoma, a lethal form of skin cancer. A definitive etiological link between human adenoviruses and human cancers has not been established. HAdVs have only been shown to induce cancers in laboratory animals.

The genome of large DNA tumor virus does encode the viral DNA polymerase; however, some host cellular components are required for viral DNA replication. Because they require the host DNA replication machinery, permissive host cells for both small and large DNA tumor viruses must enter and progress through the synthesis (S) phase of the cell cycle. Increased cell proliferation activities greatly increase the chance of cancer formation. This group of tumor viruses includes EBV, KSHV (or HHV-8), and HBV. EBV has been linked to Burkitt lymphoma and nasopharyngeal carcinoma. KSHV has been linked to Kaposi sarcoma. HBV has been linked to hepatocellular carcinoma.

Retroviruses can convert their RNA genome into proviral DNA using their own genome-encoded, RNA-dependent DNA polymerase. Retroviruses induce carcinogenesis by expressing oncoproteins (from oncogenes) to upregulate cell cycle activities. This group of tumor viruses includes HTLV-1 and HBV since it expresses a reverse transcriptase even though it possesses a large DNA genome. HTLV-1 has been linked to T-cell leukemia.

HCV is the only known RNA tumor virus that is not a retrovirus. HCV has been linked to hepatocellular carcinoma and B-cell, non-Hodgkin lymphomas. HCV's role in cancer development includes immune-mediated inflammation leading to chronic tissue damage, HCV replication leading to impairment of cellular DNA damage response, and promoting host cell proliferation.

There are 5 biologic criteria that associate a viral infection with cancer development and tumor formation. These biologic criteria are the following:

1. Is the virus found in tumor tissues?
2. Is the virus present before disease onset?
3. Does the virus (or viral components) persist or remain present and active in the body's tissues following the acute infection?
4. Is the virus located in the appropriate site in the body?
5. Can preventing viral infection block oncogenesis?

Tumor viruses induce carcinogenesis and tumor formation mostly by manipulating various cellular pathways, which eventually lead to the transformation and immortalization of the host cell. Transformation is defined as the process by which normal, uninvolved cells acquire the properties of cancer cells. Immortalization is the process by which a population of cells can reproducing indefinitely while avoiding the process of replicative cell senescence. Examples of cellular pathways that can be upregulated by tumor viruses include the phosphoinositide 3-kinase/protein kinase B (or Akt) and mitogen-activated protein kinase pathways. Both of these pathways are important for canonical cell proliferation.

In addition to modulating pathways involved in cell cycle regulation and cell proliferation, other cellular processes are can be activated may include increased oxidative stress, glucose transport, and glycolytic metabolism. Tumor viruses may actually inhibit certain cellular pathways, such as cell cycle arrest and DNA damage responses.

Tumor viruses modulate cellular pathways and processes through the expression of their oncogenes. Oncogene products, or oncoproteins, bind to important cellular binding partners to dysregulate signaling pathways. Examples of oncogenes include E6 and E7, which are expressed by HPV16 and HPV18; and EBV nuclear antigen 2 (EBNA2), EBNA3C, latent membrane protein 1 (LMP1), and LMP2, which are expressed by EBV (Table 14.5). These oncoproteins bind to target cellular components to either upregulate or downregulate cellular activities that are beneficial to the survival of the virus and promote virus replication.

To survive and replicate, tumor viruses require the host cell proliferation machineries. These viruses either require a permissive host cell that is already going through this process or they must stimulate and increase the chance of the host cell to enter the cell cycle after the infection. There is also a requirement that the host cell remains in this cycle for a long time while disregarding the methodical steps to progress through the different cell cycle stages. The altered pattern of cell proliferation in turn increases the chance of carcinogenesis in the particular tissue targeted by the virus.

Further Readings

Beckham J, Tyler K. Arbovirus infections. *Continuum (Minneap Minn)*. 2015;2015:1599–1611.
Beeson P. Jaundice occurring one to four months after transfusion of blood or plasma: report of seven cases. *JAMA*. 1943;121:1332–1334.
Beltrami E, Williams I, Shapiro C, Chamberland ME. Risk and management of bloodborne infections in health care workers. *Clin Microbiol Rev*. 2000;13(3):385–407.
Javier R, Butel J. The history of tumor virology. *Cancer Res*. 2012;68(19):7693–7706.

Kapikian A. Overview of viral gastroenteritis. *Arch Virol*. 1996;12:7–19.

Kumar M, Pahuja S, Khare P, Kumar A. Current challenges and future perspectives of diagnosis of hepatitis B virus. *Diagnostics (Basel)*. 2023;13(3):368.

Leung N. Transmissibility and transmission of respiratory viruses. *Nat Rev Microbiol*. 2021;19:528–545.

Mackie P. The classification of viruses infecting the respiratory tract. *Paed Resp Rev*. 2003;2003:84–90.

Noor R, Maniha S. A brief outline of respiratory viral disease outbreaks: 1889–till date on the public health perspectives. *Virusdisease*. 2020;31(4):441–449.

Orenstein R. Gastroenteritis, viral. *Encyclopedia of Gastroenterology*. 2020:652–657.

Purushothaman P, Uppal T, Verma S. Human DNA tumor viruses and oncogenesis. *Animal Biotech*. 2020;2020:131–151.

Ryu W-S. New Emerging viruses. *Mol Vir Hum Path Viruses*. 2017;2017:289–302.

Saha A, Kaul R, Murakami M, Robertson E. Tumor viruses and cancer biology. *Cancer Biol Ther*. 2010;10(10):961–978.

Sencer D. Emerging diseases of man and animals. *Ann Rev Microbiol*. 1971;25:465–486.

Snowden F. Emerging and reemerging diseases: A historical perspective. *Immunol Rev*. 2008;225(1):9–26.

Spurgeon M. Small DNA tumor viruses and human cancer: Preclinical models of virus infection and disease. *Tumour Virus Res*. 2022;14:200239.

Weston S, Frieman. Respiratory viruses. *Encyclopedia of Microbiology*. 2019:85–101.

Index

Note: Page numbers followed by *f* refer to figures; those followed by *t* refer to tables.

A

Abacavir (Ziagen), for HIV, 191
Acanthamoeba castellanii Mamavirus *(ACMV)*, 243, 250
Acanthamoeba polyphaga Mimivirus *(APMV)*, 243–249
  classification and structure of, 245–247, 246*f*
  current status of, 249
  infection cycle in humans, 248–249
  infection cycle in protists, 247–248, 247*f*
ACE2 receptor, SARS-CoV-2 and, 66–67, 67*f*
Ace Lake mavirus, 253
Achong, Burt, 307
Acidianus filamentous virus (AFV1), 265–266, 266*f*
Acidophilic proteins, 264
Ackermannviridae, 256
Acquired (adaptive) immune system, 16, 33–42, 35*f*
  affinity maturation and isotope switching and, 41–42
  B-lymphocyte maturation and, 34–37, 35*f*
  naïve
    mature B-lymphocyte activation and, 39–41, 40*f*
    mature T-lymphocyte activation and, 37–39, 37*f*
  T-lymphocyte maturation and, 36–37
Acute-phase proteins (APPs), hepatocytes and, 31, 32*f*
Acyclovir, for herpes simplex virus, 160
Ad26.ZEBOV (Zabdeno), 99
Adaptive (acquired) immune system, 16, 33–42, 35*f*
  affinity maturation and isotope switching and, 41–42

Adaptive (acquired) immune system *(Cont.)*:
  B-lymphocyte maturation and, 34–37, 35*f*
  naïve, mature B-lymphocyte activation and, 39–41, 40*f*
  naïve, mature T-lymphocyte activation and, 37–39, 37*f*
  T-lymphocyte maturation and, 36–37
Adenoviridae, 140, 306
Adenovirus-based vaccines, 275
Adenoviruses (AdVs)
  biotechnology uses of, 274–275
  gastroenteritis caused by, 304, 304*t*
Affinity maturation, 41
Alfonso XIII, 76
Alkaliphilic proteins, 264
Alloherpesviridae, 140
Allomimiviridae, 245
Alper, Tikvah, 223, 226
Alphacoronavirus, 63
Alphainfluenzavirus, 76
Alphapolyomavirus, 163
Alpharetrovirus, 179, 193
Alter, Harvey, 295
Amantadine, for influenza A virus, 86
Amdoparvovirus, 122
Amenamevir, for herpes simplex virus, 161
Ampullaviridae, 140, 265
Anelloviridae, 122
Ansuvimab (Ebanga), for Ebola virus, 99
Antibody test, for HIV, 190
Antiretroviral therapy, for HIV, 178–179, 191
Aphthoviruses, 52
Aplastic crisis, transient, B19V and, 129
Apple dimple fruit viroid, 230
Apple fruit crinkle viroid, 230

Apple scar skin viroid, 230
Apscaviroid, 231
Aquambidensovirus, 122
Aquareovirus, 104
Arber, Werner, 45, 46
Arbovirus, 291–292
   common features of, 292, 293t–294t
   historical background of, 291–292
Archael viruses, 263–266
Arthropathy, B19V and, 129
Artiparvovirus, 122
Ascoviridae, 140, 245
Aseptic meningitis, poliovirus and, 57–58
Asfarviridae, 140
Asfaviridae, 245
Asian flu, 76
Astrovirus, 304, 304t
Atazanavir (Reyataz), for HIV, 191
Atkinsviridae, 256
Attachment, 8
Australian grapevine viroid, 230
Autographiviridae, 256, 257
Autolykiviridae, 256
Aveparvovirus, 122
Avocado sun blotch viroid, 231
Avsunviroid, 231
Avsunviroidae, 231

B

Bacteria, viruses infecting, 255–257
Bacteriophages, 272–273
   bacterial defense against, 45–49
      CRISPR and, 47–49
      restriction digestion and, 45–47, 46f
Baculoviridae, 140
Baculovirus expression vector system (BEVS),
   biotechnology applications of, 280–282
Baloxavir (Xofluza), for influenza A virus, 86
Baltimore, David, 2–3, 193
Bancroft, Thomas, 291
Bang, Olaf, 307
Banna virus, 294t
Barr, Yvonne, 307
Barré-Sinoussi, Françoise, 177–178
BAYOVAC CSF E2, 281
B-cell receptors (BCRs), 36
Beeson, Paul, 292
Behring, Emil, 15
Beijerinck, Martinus, 1, 223
Berg, Paul, 272
Bertani, G., 272

Bertani, Joe, 45
Betacoronavirus, 63
Betainfluenzavirus, 76
Betapolyomavirus, 163
Betaretrovirus, 179
Bicaudaviridae, 140, 265
Bictegravir sodium-emtricitabine-tenofovir
    alafenamide fumarate (Biktarvy),
    for HIV, 191
Bidnaviridae, 122
Biotechnology, 271–282
   adenovirus uses in, 274–275
   bacteriophage, 272–273
   baculovirus expression vector system in,
    280–281
   definition of, 271
   eukaryotic cell, 273–274
   genetic engineering and, 271
   herpes simplex virus uses in, 275–278
    for cancer gene therapy, 277
    for comparative genomics, 277–278
    for gene therapy, 276–277
    for vaccine development, 277
   lentivirus uses in immunodeficiency treatments
    and, 278–279
   retrovirus vector applications in, 279–280
Birnaviridae, 104, 116
Bishop, Michael, 193
Blattambidensovirus, 122
Blood-borne viruses (BBVs), 292, 295–299
   common features of, 295–296, 297t–299t, 299
   historical background of, 292, 295
Blumberg, Baruch, 199, 223, 295, 307
Blumeviridae, 256
B-lymphocytes
   maturation of, 34–37, 37f
   naïve, mature B-lymphocyte activation and,
    39–41, 40f
Bocaparvovirus, 122
Bocavirus, 306
Bone marrow stromal cells, 34
Bortezomib, for herpes simplex virus, 162
Bovine spongiform encephalopathy (BSE), 224
Bovispumavirus, 179
"Brazilian flu," 75–76
Brevihamaparvovirus, 122
Brivudine, for herpes simplex virus, 160
Bromoviridae, 268
Brugge, Joan, 193
Bulbar poliomyelitis, 58
Bunyaviridae, 76, 268, 292, 293t, 301t
Burnet, Frank, 51

C

Cafeteriavirus-dependent mavirus, 250–251
CAG-GFP-IRES-CRE, 280
California encephalitis virus, 293*t*
California hare coltivirus, 294*t*
Canyon, 54
Capsids, 4–5, 5*f*
Cardioviruses, 52
Cardoreovirus, 104
Casirivimab, for SARS-CoV-2, 73
Cauliviridae, 200, 268
CD8+T lymphocytes, 37
CD155, poliovirus susceptible cells and, 54
Cedratvirus lena, 249
Cell adhesion molecules (CAMs), inflammatory
        response and, 28
Cervarix, 281
Chandler, R. L., 223
Chaphamaparvovirus, 122
Charpentier, Emanuelle, 48
Chaseviridae, 256
Chelle, P. L., 223
Chesebro, Bruce, 224
Chick-embryo system, 91
Chikungunya virus, 291–292, 294*t*
Chimeric antigen receptor (CAR) T-cell
        therapies, 279
Chlorella virus virophage SW01 (CVv-SW01), 251
Cholera vaccine, 288*t*
Chrysanthemum chlorotic mottle viroid, 231
Chrysanthemum stunt viroid, 230
Chrysoviridae, 104
Cidofovir, for human polyomavirus, 172
Circoviridae, 122, 132
Circumvent PCV, 281
Citrus bark cracking viroid, 230
Citrus bent leaf viroid, 230
Citrus dwarfing viroid, 230
Citrus exocortis viroid (ceVd), 230
Citrus tristeza virus, 267
Ciuffo, Guiseppe, 307
Clavaviridae, 256, 265
Clinical trials, for vaccines, 284
Closteroviridae, 268
Clustered regularly interspaced short
        palindromic repeats (CRISPR), bacterial
        defense against bacteriophages and, 47–49
CMX001 (brincidofovir), for human
        polyomavirus, 172
Cocadviroid, 231
Cockburn Aiden, 299

Coconut cadang-cadang viroid, 230
Coconut tinangaja viroid, 230
Cold sores, 159
Coleus blumei 1 and 3 viroids, 230
Coleviroid, 231
Colorado tick fever virus, 294*t*
Coltivirus, 104
Columnea latent viroid, 230
Comoviridae, 268
Comparative genomics, herpes simplex virus
        vector in, 277–278
Complement cascade
    acute-phase proteins and, 31–33
    host cell protection against damage from,
        33, 34*f*
Conjugate vaccines, 289
Convalescent plasma, for SARS-CoV-2 treatment,
        62, 71
Copiparvovirus, 122
Coronaviridae, 52, 63, 301*t*, 306
Corticoviridae, 140
Cossart, Yvonne, 121
COVID-19 vaccines, 71, 288*t*
    Moderna, 62–63
    Pfizer-BioNTech, 62–63
    recommendations for, 285*t*
Creutzfeldt, Hans, 223
Creutzfeldt-Jakob disease (CJD), 223, 224
CRISPR, bacterial defense against bacteriophages
        and, 47–49
Cucumber mosaic virus, 267
Cuevavirus, 92
Cuille, J., 223
CV706, 275
Cypovirus, 104
Cystoviridae, 104, 256
Cytokines, inflammatory response and, 26
Cytomegalovirus (CMV), 140, 297*t*, 306

D

Daazvirus, 216
Dagavirus, 216
Daletvirus, 216
Dalvirus, 216
Danna, Kathleen, 46, 272
Darunavir (Prezista), for HIV, 191
Deevirus, 216
Deltacoronavirus, 63
Deltainfluenzavirus, 76
Deltapolyomavirus, 163
Deltaretrovirus, 179
Deltavirus, 216

Demerecviridae, 256
Demina, Tatiana, 263
Dendritic cells, 37–39, 37f
Dengue virus, 293t, 297t
  first (DENV-1), 291
Dependoparvovirus, 122
D'Herelle, Félix, 255
Dianlovirus, 92
Diciambidensovirus, 122
Diener, Theodor O., 229
Dinovernavirus, 104
Diphtheria vaccine, 285t, 288t
Direct transmission, 7–8
Disease
  ability to cause, factors affecting, 11–12
  causation of, 10
Dissemination strategy, 9–10
DNA vaccines, 289
DNA with reverse transcriptase. *See* Hepatitis B
    virus (HBV); Hepatitis delta virus (HDV)
Dobrovirus, 216
n-Docosanol, for herpes simplex virus, 161
Doravirine (Pifeltro), for HIV, 191
Double-stranded DNA (dsDNA). *See* Herpes
    simplex virus (HSV); Human polyomavirus
Double-stranded RNA (dsRNA).
    See Picobirnavirus; Rotavirus
Doudna, Jennifer, 48
Drexlerviridae, 256
Duinviridae, 256
Dushi Lake virophages, 253

E

Eastern equine virus, 294t
Ebola vaccines, 99, 100
Ebola virus, 91–101, 297t
  classification and structure of, 92–93, 93f
  current status of, 100
  diagnosis, treatment, and prevention of, 99–100
  dissemination in host body, 97
  historical background of, 91–92
  host cell infection. viral replication, and tissue
      damage and, 94–97, 95f–97f
  host range, transmission, tropism, and suscep-
      tible host cell and, 93–94
  pathogenesis and clinical manifestations of,
      98–99
  scientific significance and discoveries and, 101
Eczema herpeticum, 159
Efavirenz, for HIV, 191
Eggplant latent viroid, 231
Ehrlich, Paul, 15

Ellermann, Vilhelm, 307
Emerging viruses, 299–302
  common features of, 300, 301t–302t, 302
  historical background of, 299–300
Emesvirus zinderi 2 (MS2) phage, 261–263
  infection cycle of, 261, 262f, 263
Emtricitabine (Emtriva), for HIV, 191
Emtricitabine-tenofovir alafenamide fumarate
    (Descovy), for HIV, 191
Emtricitabine-tenofovir disoproxil fumarate
    (Truvada), for HIV, 191
Enterobacteria phage λ, 257–261
  lysogenic cycle and, 260–261
  lysogeny to lytic cycle switch and, 261
  lytic infection and, 257–261
Enteroviruses, 52, 306
Epidemics, designation of, by CDC, 12
Epsilonpolyomavirus, 163
Epsilonretrovirus, 179
Epstein, Anthony, 307
Epstein-Barr virus (EBV), 297t, 306, 307, 308t
Equispumavirus, 179
Erikson, Raymond, 193
Erythema infectiosum, 128–129
Erythroparvovirus, 122
Erythrovirus B19, 121–131, 297t, 306
  classification and structure of, 122–124, 123f
  current status of, 130
  diagnosis, treatment, and prevention of, 129–130
  dissemination in host body, 128
  historical background of, 121–122
  host infection, viral particle replication, and
      tissue damage and, 124–128, 125f–127f
  host range, transmission, tropism, and
      susceptible host cell and, 124
  pathogenesis and clinical manifestations of,
      128–129
  scientific significance and discoveries and, 131
Escherichia virus lambda, 257–261
  lysogenic function and, 260–261
  lysogeny to lytic cycle switch and, 261
  lytic cycle and, 258–260
Ethical issues, in vaccine development, 284
Eukaryotic cell biotechnology, 273–274
Evangelina Circumflex, 281
Everolimus, for human polyomavirus, 172
Exit portals, 7
Eyach virus, 294t

F

Famciclovir, for herpes simplex virus, 160
Fatal familial insomnia (FFI), 224

Felispumavirus, 179
Fevrier, Benoit, 223
Field, Francis, 51
Fiersviridae, 256, 261
"Fifth disease," 128
Fijivirus, 104
Filoviridae, 76, 92, 301*t*
Finlay, Carlos, 291
Finnlakeviridae, 256
Flaviviridae, 52, 292, 293*t*, 301*t*
Flublok, 281
Foscarnet, for herpes simplex virus, 161
Fostemsavir (Rukobia), for HIV, 191
Francis, Thomas, 51, 91
"French flu," 75–76
Function laesa, 27
Fuselloviridae, 140, 265

G

Gajdusek, Daniel Carleton, 199, 223, 226
Gallic acid-based small compounds, for human
    polyomavirus, 172
Gallo, Robert, 178
Gammacoronavirus, 63
Gammainfluenzavirus, 76
Gammapolyomavirus, 163
Gammaretrovirus, 179
Ganciclovir, for herpes simplex virus,
    160, 162
Gastroenteritis viruses, 303–305
    common features of, 303–305, 304*t*
    historical background of, 303
Geminiviridae, 122, 268
Gene expression, regulation of, signal transduction
    pathways leading to, 23–25, 24*f*
Gene rearrangement, 35
Gene therapy
    adenovirus vectors for, 274–275
    herpes simplex viruses for, 276–277
Genetic engineering, 271, 273–274
Genital herpes, primary, 159
Genome replication, 9
Genomics, comparative, herpes simplex virus
    vector in, 277–278
Genomic sequences, 5
Genomoviridae, 122
"German flu," 75–76
Gerstmann-Sträussler-Scheinker disease (GSS)
    syndrome, 224
Giant viruses, 243–245, 244*f. See also Acanthamoeba
    polyphaga* Mimivirus *(APMV)*
Gibbs, C. J., 226

Globuloviridae, 140, 256, 265
Glutamine, for herpes simplex virus, 162
Glybera (alipogene tiparvovec), 281
Goblet cells, 16
Goodpasture, Ernest William, 91
Grapevine yellow speckle 1 and 2 viroids, 230
"Great Influenza Epidemic," 75–76
Griffith, J. S., 223
Gross, Hans, 230
Gross, Ludwik, 162
Guarani virophage, 253
Guenliviridae, 256
Guttaviridae, 140, 265

H

H1N1 influenza A virus. *See* Influenza A virus
H2N2 influenza A virus, 76
H3N2 influenza A virus, 76
Haddow, Alexander, 291
Haloalkaliphilic proteins, 264
Halophilic proteins, 264
Halspiviridae, 265
Hanbury, Ernest, 255
HBV vaccines, 200, 211*t*–213*t*
Hematopoietic stem cells (HSCs), 33–34, 35*f*
Hemiambidensovirus, 122
Hepadnaviridae, 200
Hepanhamaparvovirus, 122
Heparin sulfate proteoglycans (HSPGs), 192
Hepatitis A vaccine, 285*t*, 287*t*
Hepatitis A virus (HAV), 292
Hepatitis B vaccine, 285*t*, 288*t*, 295
Hepatitis B virus (HBV), 199–215, 292, 296, 297*t*,
    307, 308*t*
    classification and structure of, 200–202,
        201*f*, 202*f*
    current status of, 210, 213
    diagnosis, treatment, and prevention of,
        209–210, 211*t*–213*t*
    dissemination in host body, 208–209
    historical background of, 199–200
    host infection, viral replication, and tissue
        damage and, 203–208, 203*f*, 204*f*,
        207*f*, 208*f*
    host range, transmission, tropism, and
        susceptible host cell and, 202–203
    pathogenesis and clinical manifestations of, 209
    scientific significance and discoveries and,
        213–215
Hepatitis C virus (HCV), 295, 296, 298*t*, 307,
    308, 308*t*

Hepatitis delta virus (HDV), 215–222
  classification and structure of, 216–217, 217*f*
  current status of, 221–222
  diagnosis, treatment, and prevention of, 221
  dissemination in host body, 220
  historical background of, 216
  host infection, viral replication, and tissue
    damage and, 218–220, 218*f*, 219*f*
  host range, transmission, tropism, and
    susceptible host cell and, 217
  pathogenesis and clinical manifestations of,
    220–221
Hepatocytes, acute-phase proteins and, 31, 32*f*
Hepatoviruses, 52
Herelleviridae, 256
Herpes encephalitis, 159
Herpes gladiatorum, 159
Herpes labialis, 159
Herpes simplex keratitis (keratoconjunctivitis), 159
Herpes simplex virus (HSV), 139–162
  biotechnology uses of, 275–276
    for cancer gene therapy, 277
    for comparative genomics, 277–278
    for gene therapy, 276–277
    for vaccine development, 277
  classification and structure of, 140–142, 141*f*,
    143*t*–146*t*, 147*f*
  current status of, 161
  diagnosis, treatment, and prevention of,
    160–161
  dissemination in host body, 157–158
  historical background of, 139–140
  host infection, viral replication, and tissue
    damage and, 148–157, 148*f*, 149*f*
    latent infection and, 154–156, 155*f*
    lytic infection and, 150–153, 151*f*,
      153*f*, 154*f*
    reactivation and, 156–157, 157*f*
  host range, transmission, tropism, and
    susceptible host cell and, 142, 146–148
  pathogenesis and clinical manifestations of,
    158–160
  scientific significance and discoveries and,
    161–162
Herpes simplex virus (HSV-6) (roseolovirus),
  140, 306
Herpesviridae, 140, 306
Herpetic gingivostomatitis, acute, 158
Herpetic pharyngostomatitis, acute, 159
Herpetic whitlow, 159
Hib vaccine, recommendations for, 285*t*
Highly active antiretroviral therapy (HAART), 178

Hilleman, Maurice, 307
Hippocrates, 140
HIV vaccine, 192
HKU1, 306
Hong Kong flu, 76
Hop latent viroid, 230
Hop stunt viroid (HSVd), 230
Host cell
  entry into, 8–9
  exit from, 9
Host cell susceptibility. *See under specific viruses*
Host immune responses, 15–50
Host range, 7. *See also specific viruses*
Hostuviroid, 231
Hotta, Susumu, 291
Houghton, Michael, 295
HPV vaccine, 288*t*
Human, Mary, 45, 272
Human adenovirus A-G (HAdV A-G), 306
Human B19V. *See* Erythrovirus B19
Human herpegvirus 1 (HHpgV-1), 295
Human herpes virus 4 (HHV4) (EBV), 306
Human herpes virus 5 (HHV5) (CMV), 306
Human immunodeficiency virus (HIV), 177–192,
  183*f*, 185*f*–188*f*
  classification and structure of, 179–181,
    180*f*, 181*f*
  current status of, 192
  diagnosis, treatment, and prevention of,
    190–192
  dissemination in host body, 188–191, 189*f*
  historical background of, 177–179
  host cell infection, viral replication, and tissue
    damage and, 182–188, 183*f*, 185*f*–188*f*
  host range, tropism, and susceptibility to,
    181–182
  pandemic and, 299
  pathogenesis and clinical manifestations of, 190
  scientific significance and discoveries and, 192
  transmission of, 295
Human immunodeficiency virus 1 (HIV-1),
  296, 298*t*
Human immunodeficiency virus 2 (HIV-2), 296
Human metapneumovirus, 306
Human papillomavirus (HPV), 307, 308*t*, 309
  HPV vaccine and, 288*t*
Human parainfluenza viruses, 306
Human polyomavirus, 162–173
  classification and structure of, 163–165, 164*f*
  current status of, 172–173
  diagnosis, treatment, and prevention of,
    171–172

Human polyomavirus (*Cont.*):
  dissemination in host body, 169
  historical background of, 162–163
  host infection, viral replication, and tissue
    damage and, 165–169, 166*f*, 167*f*
  host range, tropism, and susceptible host cell
    and, 165
  pathogenesis and clinical manifestations of,
    169–171, 170*f*
  scientific significance and discoveries and, 173
Human T-cell leukemia virus (HTLV), 307
Human T-cell lymphotrophic virus (HTLV-1), 308*t*
Hydrops fetalis, B19V and, 129
Hydroxyurea, for B19V, 131
Hyperhydration, for human polyomavirus, 172
Hyperthermia, 29

Ibalizumab-uiyk (Trogarzo), for HIV, 191
Ichthamaparvovirus, 122
Idnoreoivirus, 104
Idoxuridine, for herpes simplex virus, 160
Iltovirus, 140
Immune response, 42–43
  activation of, 16
  definition of, 16
Immune systems. *See* Adaptive immune system;
    Innate immune system; Mucosal immune
    system
Immunization. *See also* Vaccines; *specific vaccines*
  against polio, global, 51–52. *See also* Polio
    vaccines
Immunodeficiency treatments, lentiviral vectors
    in, 278
Immunoglobulin A (IgA), 17
  isotope switching and, 41
Immunoglobulin D (IgD), isotope switching
    and, 41
Immunoglobulin G (IgG), isotope switching and,
    41–42
Inactivated polio vaccine (IPV), 51, 59, 287*t*
  recommendations for, 285*t*
Inactivated vaccines, 286, 287*t*
Indinavir (Crixivan), for HIV, 179
Indirect transmission, 8
Infection. *See also under specific viruses*
  factors affecting, 11–12
  host immune responses to. *See* Host immune
    responses
Infectious dose, likelihood of infection and,
    11–12
"Infectious rash," 128

Inflammatory response, 25–28
  activation of, 25–28, 27*f*
  definition of, 25
Influenza A flu vaccines, 90
Influenza A virus, 75–91, 306
  Asian flu and, 76
  classification and structure of, 76, 77*t*–78*t*,
    78–79, 79*f*
  current status of, 90–91
  dissemination in host body, 85, 86, 87*t*–89*t*,
    89–90
  historical background of, 75
  Hong Kong flu and, 76
  host cell infection, viral replication, and tissue
    damage and, 81–84, 81*f*–84*f*
  host range, transmission, tropism, and
    susceptible host cell and, 80–81
  pathogenesis and clinical manifestation of,
    85–86
  scientific significance and discoveries and, 91
  Spanish flu and, 75–76
Influenza B virus, 306
Influenza C virus, 306
Influenza vaccines, 90, 285*t*, 287*t*
Informed consent, 284
Innate immune system, 16, 19–25, 20*f*
  acute-phase, 31–33, 32*f*
  inflammatory response and, 25–28
    activation of, 25–28, 27*f*
    definition of, 25
  phagocytosis and, 21–22, 22*f*
  pyrexia and, 29–31, 30*f*
  respiratory burst and, 22–23, 23*f*
  signal transduction pathways leading to
    regulation of gene expression and,
    23–25, 24*f*
  steps in, 25–26
Innate lymphoid cells, 18
Inoviridae, 122, 256
Interleukins, release of, bone marrow stromal
    cells and, 34
Intraepithelial lymphocytes (IELs), 17, 18–19
Intravenous immunoglobulin (IVIG), for human
    polyomavirus, 172
Iresine viroid, 230
Iridoviridae, 245
Iron lung, 58
Isavirus, 76
Ishino, Yoshizumi, 47–48
Isotope switching, 41–42
Iteradensovirus, 122
Ivanowski, Dmitrii, 1, 223

J

Jakob, Alfons, 223
Jamestown canyon virus, 293*t*
Jansen, Ruud, 48
Japanese encephalitis virus, 293*t*
Jenner, Edward, 1, 15

K

Kaposi sarcoma herpes virus (KSHV) (HHV-8), 307, 308*t*
Kemerovo virus, 294*t*
Killed vaccines, 286, 287*t*
Kimura, Ren, 291
Koch, Robert, 15
Kolmioviridae, 216
Koop, C. Everett, 299
Kuru, 223, 224

L

La Crosse virus, 293*t*
Lagevrio, for SARS-CoV-2, 71, 73
Lambda (λ) phage, 257–261
    lysogenic function and, 260–261
    lysogeny to lytic cycle switch and, 261
    lytic cycle and, 258–260
Lamina propia, 19
Lamivudine (Epivir), for HIV, 191
Landsteiner, Karl, 51
Lanzhou lamb rotavirus vaccine, 116
Lassa virus, 298*t*
Lavidaviridae, 250
Lederberg, Joshua, 300
Leflunomide, for human polyomavirus, 172
Lentiviral vectors
    in immunodeficiency treatments, 278–279
    SIN, applications of, 279
Lentivirus, 179
Leukocytes, targeted, extravasation of, inflammatory response and, 28
Levofloxacin, for human polyomavirus, 172
Linn, Stuart, 45
Lipothrixviridae, 140, 265
Live-attenuated vaccines, 286, 287*t*
Loening, Ulrich, 272
Long COVID, 63, 72
Lopinavir-ritonavir (Kaletra), for HIV, 191
Loriparvovirus, 122
Luria, Salvador, 45, 272
Lymphocryptovirus, 140

M

Macavirus, 140
Macnamara, Jean, 51
Macrophages, 19
Mad cow disease, 224
Malacoherpesviridae, 140
Mannan-binding lectin (MBL), hepatocytes and, 31, 32
Maraviroc (Selzentry), for HIV, 191
Marburgvirus, 92
Mardivirus, 140
Marseilleviridae, 245
Marshall, George, 299
Martin, W. H., 229
Matshushitaviridae, 256
Mavirus, 253
Mayer, Adolph, 1
McReynolds, Katherine D., 192
Measles vaccine, 285*t*, 287*t*
Measles virus, 306
Megavirus mammoth, 249
Meningitis, aseptic, poliovirus and, 57–58
Merkel cell carcinoma, 171
Merkel cell polyomavirus (MCPyV), 308, 308*t*
Metaviridae, 179
Metchnikoff, Elias, 15
Mexican papita viroid, 230
Microfold (M) cells, 17, 18
Microviridae, 122, 256
Middle East respiratory syndrome (MERS), 63
Millman, Irving, 200
Mimiviridae, 140, 245
Mimivirus. *See Acanthamoeba polyphaga* Mimivirus *(APMV)*
Mimivirus-dependent virus Sputnik, 250
Mimivirus-dependent virus Zamilon, 250
Mimivirus virophage resistance element (MIMIVIRE), 247
Mimoreovirus, 104
Miniambidensovirus, 122
Minus-strand single-stranded RNA. *See* Ebola virus; Influenza A virus
miRNA inhibitors, for herpes simplex virus, 162
Mode of transmission, likelihood of infection and, 11
Modes of transmission, 7–8
Mojica, Francisco, 48
Molliviridae, 245
Mollivirus sibericum, 249

Monoclonal antibodies (mAbs)
    for Ebola virus, 99
    for SARS-CoV-2, 73
"Morbus quintus," 128
Morris, J, A., 223
Morrow, John, 272
Mucosa-associated lymphoid tissues (MALTs), 18
Mucosal immune system, 16–19
    epithelium and, 17–19
    lamina propria and, 19
    muscularis mucosa and, 19
Multiplicity of infection (MOI), 10
Mumps vaccine, recommendations for, 285t
Mumps virus, 306
Murakami, Seishi, 307
Muromegalovirus, 140
Murray Valley encephalitis virus, 293t
Muscodensovirus, 122
Muscularis mucosa, 19
MVA-BN-Filo (Mvabea), 99
Mycoreovirus, 104
Myoviridae, 140

N

Nanotechnology, 73
Nanoviridae, 122
Natalizumab, for human polyomavirus, 172
Nathans, Daniel, 46, 272
Natural reservoirs, 7
Neonatal herpes, 159
Nevirapine (Viramune), for HIV, 179
Nicotinamide adenine dinucleotide phosphate
    (NADPH) oxidase (NOX2), 22–23, 23f
Nimaviridae, 140
Nitric oxide (NO), phagocytosis and, 22
NL63, 306
Nonreservoir hosts, 7
Norovirus (Norwalk virus), 303, 304, 304t
Nucleic acid test, for HIV, 190
Nucleocytoplasmic large DNA viruses
    (NCLDVs), 244
Nucleocytoviricota, 244
Nucleoside (nucleotide) reverse transcriptase
    inhibitors, for HIV, 191

O

OC43, 306
Okochi, Kazuo, 307
Oksanen, Hanna, 263
ONYX-015, 274–275
Open reading frames (ORFs), 4

Oral polio vaccine (OPV), 51, 59–60, 285t, 287t
Orbivirus, 104
Organic Lake virophage, 254
Orthomyxoviridae, 76, 301t, 306
Orthopoxviridae, 301t
Orthoreovirus, 104
Orthoretrovirinae, 179, 193
Oryzavirus, 104
Oseltamivir (Tamiflu), for influenza A virus, 86
Ovaliviridae, 265

P

Pacmanvirus lupus, 249
Pandemics
    HIV, 299
    influenza, 75–76, 305
    "Russian flu" as, 75
    SARS-CoV-2 declared as, 62
    Spanish flu as, 75–76
Pandoraviridae, 245
Pandoravirus mammoth, 249
Pandoravirus yedona, 249
Paneth cells, 18
Papillomaviridae, 140
Paralytic poliomyelitis, 58
Paramyxoviridae, 76, 302t, 306
Partitiviridae, 104, 268
Parvoviridae, 122, 306
Parvovirus B19. See Erythrovirus B19
Pasteur, Louis, 1, 15
Pathogen-associated molecular pattern
    (PAMP), 20
Pathogenicity, 11
Pattern recognition receptors (PRRs)
    PAMP recognition by, 20–21, 20f
    regulation of gene expression and, 23–25, 24f
Pattison, I. H., 223
Pauliniviridae, 256
Paxlovid, for SARS-CoV-2, 71, 73
pBABE-puro, 280
PCV2 vaccination, 135–136
pdCas9-humanized, 280
Peach latent mosaic viroid, 231
Pear blister canker viroid, 230
Pefuambidensovirus, 122
Pelamoviroid, 231
Penciclovir, for herpes simplex virus, 160
Penstylhamaparvovirus, 122
Percavirus, 140
Pertussis vaccine, 285t, 287t
Phaeocystis globosa virus virophage (PgVV), 254

Phagocytosis, 21–22, 22f
Phagolysosome, 21
Phenuiviridae, 302t
Phycodnaviridae, 245
Phyconaviridae, 140
Phytoreovirus, 104
Picobinrnaviridae, 104, 116, 117f, 257
Picobirnavirus, 116–119
    classification and structure of, 116–117, 117f
    diagnosis, treatment, and prevention of, 119
    historical background of, 116
    infection cycle of, 117–118, 118f
    pathogenesis and clinical manifestations of, 118–119
    scientific significance and discoveries and, 119–120
Picornaviridae, 52, 306
Piezophilic proteins, 264
Pimozide, for B19V, 131
Pithoviridae, 245
Pithovirus mammoth, 249
Pithovirus sibericum, 249
Plantovirus Saccamoebae "comedo," 254
Plant viruses, 266–270
    tobacco mosaic virus as, 268–270, 268f, 269f
    transmission to plant host, 266–268
Plaque-forming units (PFUs), 10
Plasmairidae, 257
Plasmaviridae, 140
Platform-based vaccines, 289
Plectroviridae, 257
Pleolipoviridae, 257, 265
Plummer, I. H., 226
Plus-strand single-stranded RNA. *See* Poliovirus; SARS-CoV-2
pMKO.1 puro, 280
Pneumococcal vaccine, recommendations for, 285t
Podoviridae, 140
Polio vaccines, 51–52, 58–60
    inactivated (IPV), 51, 59, 285t, 287t
    oral (OPV), 51, 59–60, 285t, 287t
    Sabin, 51
    Salk, 51
Poliovirus, 51–60. *See also* Polio vaccines
    classification and structure of, 52, 53f
    current status of, 60
    diagnosis, treatment, and prevention of, 58–60
    dissemination in host body, 56–57, 57f
    historical background of, 51–52
    host cell infection, viral replication, and tissue damage and, 54–56, 55f, 56f

Poliovirus (*Cont.*):
    host range, transmission, tropism, and susceptibility to, 53–54, 54f
    pathogenesis and clinical manifestation of, 57–58
    scientific significance and discoveries and, 60
Polyarthropathy syndrome, 128
Polydnaviridae, 140
Polymerases, viral genome replication and, 5–6, 6f
Polyomaviridae, 140, 163
Porcilis, 281
Porcine circovirus 2, 131–136
    classification and structure of, 132, 133f
    current status of, 136
    diagnosis, treatment, and prevention of, 135–136
    historical background of, 131–132
    host infection, viral replication, and tissue damage and, 132–134, 134f
    pathogenesis and clinical manifestations of, 134–135, 135f
    scientific significance and discoveries and, 136
Porcine circovirus-associated disease (PCVAD), 132
Portogloboviridae, 257
Pospiviroid, 231
Pospiviroidae, 230, 231
Post-COVID-19 condition, 63, 72
Post-polio syndrome, 58
Posttranscriptional gene silencing, 44–45, 45f
Postweaning multisystemic wasting syndrome (PMWS), 131–132
Potato spindle tuber viroid (PSTVd), 229
Potyviridae, 268
Powassan virus, 293t
Poxviridae, 140, 245
Prince, Alfred, 307
Prions, 223–229
    current status of, 229
    diagnosis, treatment, and prevention of, 228–229
    historical background of, 223–224
    host range, transmission, tropism, replication, and dissemination of, 226–228, 227f
    pathogenesis and clinical manifestations of, 228
    prion disease and, 224
    structure and function of, 224–226, 225f
Proboscivirus, 140
Prosimiispumavirus, 179
Prostaglandin E_2 (PGE$_2$), pyrexia and, 30
Protease inhibitors, for HIV, 191

Protoambidensovirus, 122
Protoparvovirus, 122
Provenge (sipuleucel-T), 281
Prusiner, Stanley, 224
Pseudoviridae, 179
PSTVd, 230
Psychrophilic proteins, 264
Public Health Emergency of International
    Concern (PHEIC), SARS-CoV-2 as, 61
Pyrexia, innate immune system and, 29–31, 30f

Q

Qinghai Lake virophage, 254
Quadriviridae, 104
Quaranjavirus, 76

R

Rabies vaccine, 287t
Race, Richard, 224
Raltegravir (Isentress), for HIV, 191
Rate of viral reproduction (R_0), likelihood of
    infection and, 12
Reactive oxygen species (ROS), phagocytosis and,
    21–22
Receptor-binding protein (RBP), 272
Recombinant vector (recombinant virus)
    vaccines, 289
Recombinant vesicular stomatitis virus-Zaire
    ebolavirus (rVSV-ZEBOV [Ervebo]), 99
Reed, Walter, 291
REGN-EB3 (Inmazeb), for Ebola virus, 99
Reoviridae, 104, 106, 268, 292, 294t
Replication competent (RC) vectors, 274–275
Replication defective (RD) vectors, 274, 275
Reservoir hosts, 7
Reservoirs of infection, 7
Respiratory burst, 22–23, 23f
Respiratory syncytial virus (RSV), 306
Respiratory viruses, 305–306
    common features of, 305–306
    historical background of, 305
Restriction digestion, bacterial defense against
    bacteriophages and, 45–47, 46f
Retrovidiridae, 179
Retroviral vectors, biotechnology applications of,
    279–280
Retroviridae, 193, 302t
Retroviruses, 308
Rhabdoviridae, 268, 302t
Rhadinovirus, 140
Rhinoviruses, 52, 306

Rice, Charles, 295
Rilpivirine (Edurant), for HIV, 191
Rimantadine, for influenza A virus, 86
Ritonavir (Norvir), for HIV, 179
Rizzetto, Mario, 216
RNA interference (RNAi), 44–45, 45f
RNA vaccines, 289
RNA with reverse transcriptase. *See* Human
    immunodeficiency virus (HIV); Rous sar-
    coma virus (RSV)
Rolling adhesion, 27–28, 27f
Roseolovirus (HSV-6), 140, 306
Ross River virus, 294t
Rotarix, 115
Rotasiil, 116
RotaTeq, 115
Rotavac, 116
Rotavin-M1, 116
Rotavirus, 103–116, 304, 304t
    classification and structure of, 104–105, 105f
    current status of, 115
    diagnosis, treatment, and prevention of,
        114–115
    dissemination in host body, 112
    historical background of, 103
    host cell infection, viral replication, and tissue
        damage and, 106, 107f–111f, 108–110
    host range, transmission, tropism, and
        susceptible host cell and, 106
    pathogenesis and clinical manifestations of,
        113–114, 113f
    scientific significance and discoveries and,
        115–116
Rotavirus vaccines, 115–116, 285t, 287t
Rountreeviridae, 257
Rous, Francis Peyton, 193, 307
Rous sarcoma virus (RSV), 193–197, 307
    classification and structure of, 193–194, 194f
    infection cycle of, 194
    scientific significance and discoveries and,
        195–197, 195f, 196f
rQNestin34.5v, 277
Rubella vaccine, 285t, 287t
Rubella virus, 298t
Rubeola virus, 298t
Rudiviridae, 140, 265
Rumen virophage, 254

S

Sabin, Albert, 51, 273, 291
Sabin polio vaccine, 51
Salasmaviridae, 257

Salk, Jonas, 51
Salk polio vaccine, 51
Salmon River virus, 294*t*
Sänger, Heinz, 230
Saquinavir (Invirase), for HIV, 178
SARS-CoV-1, 61, 62, 67
SARS-CoV-2, 61–73, 306
  classification and structure of, 63–64, 64*f*
  current status of, 72
  dissemination in host body, 69–70
  host cell infection, viral replication,
      and tissue damage and, 66–69,
      67*f*, 68*f*
  host range, transmission, tropism, and
      susceptibility to, 64–66, 65*f*
  pathogenesis and clinical manifestations of, 70
  scientific significance and discoveries and, 73
Satellites, 238–241
  historical background of, 238
  pathogenesis and clinical manifestations of,
      240–241
  satellite nucleic acid disease, 238
  satellite nucleic acid structure and function,
      238–239
  satellite virus structure and function,
      239–240
Sawyer, Patrick Oliver, 91–92
Schitoviridae, 257
Scindoambidensovirus, 122
Scrapie, 223
Scutavirus, 140
Seadornavirus, 104
Secreted glycoprotein (sGP), 96
Semancik, Joseph, 230
Sencer, David, 300
Sequiviridae, 268
Severe acute respiratory syndrome (SARS), 63
Shakespeare, William, 140
Shingles vaccine, 288*t*
Simiispumavirus, 179
Simplexvirus, 140
Simuloviridae, 257
Single-stranded DNA (ssDNA). *See* Erythrovirus
    B19; Porcine circovirus 2
Siphoviridae, 140, 257
Siptah (pharaoh), 51
Sitimagene ceradenovec, 275
Slapped cheek rash, B19V and, 128, 129
Smacoviridae, 122
Smallpox vaccine, recommendations for, 285*t*
Smith, Hamilton O., 45–46, 272
Solvpiviridae, 257

Soper, Fred, 291
Sorivudine, for herpes simplex virus, 160, 161
Spacer acquisition, 48
Spanish flu, 75–76
Sphaerolipoviridae, 256
Spinal paralytic poliomyelitis, 58
Spiraviridae, 122, 265
Sputnik, 243, 250, 251, 252
Stanley, Wendell M., 1
Steitzviridae, 257
Steward, William, 299
Stravoviridae, 257
Striavirus, 92
Subunit vaccines, 288*t*, 289
Sulfoglycodendrimers (SGDs), 192
Summer, Max, 280
Supumaretroviriae, 179
Susceptibility. *See under specific viruses*

T

Talimogene laherparepvec (IMLYGIC), 277
TCDVd, 230
Tectiviridae, 140, 257
Temin, Howard, 193
Tenofovir disoproxil fumarate (Viread), for
    HIV, 191
Tetanus vaccine, 285*t*, 288*t*
Tetraparvovirus, 122
Tetuambidensovirus, 122
Th17 cells, 19, 39
Thamnovirus, 92
Thaspiviridae, 265
Thermophilic proteins, 264
Thogotovirus, 76
Thurisazvirus, 216
Tick-borne encephalitis virus (TBEV), 291, 293*t*
Tissue damage. *See under specific viruses*
T-lymphocytes
  maturation of, 36–37
  naïve, mature T-lymphocyte activation and,
      37–39, 37*f*
Tobacco mosaic virus (TMV)
  discovery of, 1
  infection cycle of, 268–270, 268*f*, 269*f*
Togaviridae, 52, 292, 294*t*, 302*t*
Toll-like receptors (TLRs), regulation of gene
    expression and, 23–25, 24*f*
Tomato apical stunt viroid, 230
Tomato planta macho viroid (TPMVd), 230
Tombusviridae, 268
Totiviridae, 104
Townley, Rudge, 103

Toxoid vaccines, 288, 288*t*
TPMVd, 230
Transcriptional cascade, 150
Translation, 9
Transmissible spongiform encephalopathies (TSEs), 224
Transmission. *See also under specific viruses*
  between hosts, 10–11
Trentin, John, 307
Tribec virus, 294*t*
Trifluridine, for herpes simplex virus, 160, 161
Tristromaviridae, 265
Tropism. *See under specific viruses*
Tumor viruses, 307–310
  common features of, 307, 308*t*–309*t*, 309–310
  historical background of, 307
Turriviridae, 257
Twort, William, 255
229E, 306

U

Uncoating, 9
Underwood, Michael, 51

V

Vaccination policies, 284, 285*t*
Vaccine development, 282–289
  considerations in, 284, 285*t*, 286–289
  function of vaccination and, 283–284
  herpes simplex virus vector in, 277
  historical background of, 282–283
Vaccines, 286, 287*t*
  adenovirus-based, 275
  cholera, 288*t*
  conjugate, 289
  COVID-19, 71–72, 288*t*
    Moderna, 62, 63
    Pfizer-BioNTech, 62, 63
    recommendations for, 285*t*
  diphtheria, 285*t*, 288*t*
  DNA, 289
  ebola, 99, 100
  hepatitis A, 285*t*, 287*t*
  hepatitis B, 200, 211*t*–213*t*, 285*t*, 288*t*, 295
  Hib, recommendations for, 285*t*
  HIV, 192
  HPV, 288*t*
  inactivated, 286, 287*t*
  influenza, 285*t*, 287*t*
  influenza A, 90
  killed, 286, 287*t*

Vaccines (*Cont.*):
  Lanzhou lamb rotavirus, 116
  live-attenuated, 286, 287*t*
  measles, 285*t*, 287*t*
  mumps, recommendations for, 285*t*
  pertussis, 285*t*, 287*t*
  platform-based, 289
  pneumococcal, recommendations for, 285*t*
  polio. *See* Polio vaccines
  rabies, 287*t*
  recombinant vector (recombinant virus), 289
  RNA, 289
  rotavirus, 115–116, 285*t*, 287*t*
  rubella, 285*t*, 287*t*
  Sabin, 51, 59–60
  shingles, 288*t*
  smallpox, recommendations for, 285*t*
  subunit, 288*t*, 289, 288288*t*
  tetanus, 285*t*, 288*t*
  toxoid, 288, 288*t*, 288288*t*
  varicella, 285*t*, 287*t*
  yellow fever, 287*t*
Valacyclovir, for herpes simplex virus, 160
Valganciclovir, for herpes simplex virus, 160
Variable protease-sensitive prionopathy (VPSPr), 224
Variant CJD (vCJD), 224
Varicella vaccine, 285*t*, 287*t*
Varicella zoster virus (varicella virus), 306
Varicellovirus, 140
Varmus, Harold, 193
Vascular permeability, inflammatory response and, 26–27
Vasoconstriction, inflammatory response and, 26
Vasodilation, inflammatory response and, 26, 27
Vectors, 7
Veklury, for SARS-CoV-2, 71, 73
Venezuelan equine encephalitis virus, 294*t*
Vidal, Jean Baptiste Émile, 140
Vidarabine, for herpes simplex virus, 160
Viral classification, based on epidemiological criteria, 291–311
Viral dissemination. *See under specific viruses*
Viral genome, 3, 3*f*, 4*f*
  replication strategies for, 5–6, 6*f*
Viral load, 10
  likelihood of infection and, 12
Viral particles
  release of, 9
  replication of. *See under specific viruses*
  size of, 4
  transmission of, 7

Viral replication. *See under specific viruses*
Viral structure, 4–5, 5f
Virchow, Rudolf, 15
Virgaviridae, 268
Virion assembly or packaging, 9
Viroids, 229–237. *See also specific viroids*
 current status of, 237
 diagnosis, treatment, and prevention
  of, 237
 historical background of, 229–230
 host range, transmission, tropism, susceptibility,
  replication, and dissemination of, 232–235,
  234f, 235f
 intracellular host defense against, 236
 pathogenesis and clinical manifestations of,
  235–236
 structure and function of, 231–232, 232f
 viroid disease and, 230–231
Virophages, 250–254
 classification of, 250–251
 current status of, 253–254
 historical background of, 250
 infection cycle of, 252–253, 252f
 structure of, 251–252, 251f
Viruses. *See also specific viruses*
 classification of, 2–4
 definition of, 2
 naming conventions for, 2

Virus-specific defense systems, 43–45
 interferons as, 43–44
 RNA interference as, 44–45, 45f

W

Weathers, Lewis, 230
Weigle, Jean, 45, 272
Western equine virus, 294t
West Nile virus (WNV), 291, 293t, 298t
Wilcox, K. W., 272

Y

Yellow fever vaccine, 287t
Yellow fever virus, 293t, 298t
Yellowstone Lake virophages 1-4, 254

Z

Zamilon, 250, 253
Zanamivir (Relenza), for influenza A virus, 86
Zetapolyomavirus, 163
Zidovudine (AZT; Retrovir), 178, 191
Zika virus, 291, 293t, 299t
Zilber, Lev, 291
Zobelloviridae, 257
"Zombie viruses," 249
Zoonoses, 7
Zoonotic viruses, 7
zur Hausen, Harald, 307